FORMULAE, CHARTS AND TABLES IN THE AREA OF SOIL MECHANICS AND FOUNDATION ENGINEERING

FORMELN, TAFELN UND TABELLEN AUS DEM GEBIET GRUNDBAU UND BODENMECHANIK

Formulae, Charts and Tables in the Area of Soil Mechanics and Foundation Engineering

Stresses in Soils

Formeln, Tafeln und Tabellen aus dem Gebiet Grundbau und Bodenmechanik

Bodenspannungen

By / Von

ALEMAYEHU TEFERRA & EDGAR SCHULTZE

CRC Press
Taylor & Francis Group
Boca Raton London New York

CRC Press is an imprint of the
Taylor & Francis Group, an **informa** business
A BALKEMA BOOK

Published by:
CRC Press/Balkema
P.O. Box 447, 2300 AK Leiden, The Netherlands
e-mail: Pub.NL@taylorandfrancis.com
www.crcpress.com – www.taylorandfrancis.com

ISBN-13: 978-90-6191-804-2 (hbk)

Visit the Taylor & Francis Web site at
http://www.taylorandfrancis.com

and the CRC Press Web site at
http://www.crcpress.com

CIP-DATA KONINKLIJKE BIBLIOTHEEK, DEN HAAG

Teferra, Alemayehu

Formulae, charts and tables in the area of soil mechanics and foundation engineering: stresses in soils = Formeln, Tafeln und Tabellen aus dem Gebiet Grundbau und Bodenmechanik: Bodenspannungen / by/von Alemayehu Teferra & Edgar Schultze. – Rotterdam [etc.]: Balkema. – Ill.
ISBN 90 6191 804 9 bound
SISO 692.4 UDC 631.431+624.13
Subject headings: soil mechanics / foundation engineering.

Contents / Inhalt

Preface

Formulae, Charts and Tables in the Area of Soil Mechanics and Foundation Engineering is the result of many years of arduous work of systematically collecting the scattered scientific literature that is highly valuable to students and practicing engineers in the fields of soil mechanics and foundation engineering.

In each chapter all important formulae, charts and tables have been presented in a comprehensive manner. In order to give the user the additional possibility of referring to the original works, references are given at the end of each chapter.

The merit of the handbook lies not only in the compilation of important information that is normally not easily accessible to students and practicing engineers, but also in the provision of adequate illustrative examples that will facilitate the application of the formulae, charts and tables. The fact that the handbook is presented in two languages makes it possible to cover a wider circle of readers.

Due to practical reasons, it has been found necessary to bring out the handbook in more than one volume. The present volume deals with stresses in soil. Future volumes, which will deal with settlement, contact pressure distribution and soil plasticity, are planned.

The German readers will not fail to notice the improvised modifications used in writing the text. It is hoped that these modifications will in no way hinder the full usage of the book.

Suggestions regarding modifications, omissions, errors, and presentation are sincerely welcome. These suggestions would undoubtedly help us in the preparation of the other volumes which are forthcoming in the near future.

The authors would like to express their warm appreciation to the German Academic Exchange Service (DAAD) and Addis Ababa University for granting them the necessary financial support for the preparation of the manuscript. Acknowledgement is also extended to the Institut für Verkehrswasserbau, Grundbau und Bodenmechanik der Technischen Hochschule Aachen for allowing the use of the institute's library.

Dr-Ing. Alemayehu Teferra
Prof. (em) Dr-Ing., Dr-Ing. E.h. Edgar Schultze

Vorwort

Die *Formeln, Tafeln und Tabellen aus dem Gebiet des Grundbaus und der Bodenmechanik* sind die Ergebnisse mühsamer Arbeit von mehreren Jahren in dem man recht verstreute aber zahlreiche wissenschaftliche Literaturen aus dem Gebiet des Grundbaus und der Bodenmechanik, die sehr wertvoll für Studenten und Bauingenieure sind, zusammengestellt hat.

Alle wichtige Formeln, Tafeln und Tabellen für jeden Abschnitt wurden umfassend präsentiert. Um den Lesern die Möglichkeit zu geben, sich über die originellen Werke einen Überblick zu verschaffen, sind Literaturquellen am Ende jedes Abschnittes angegeben worden.

Die Besonderheit des vorliegenden Werkes liegt nicht nur darin, dass man für den Studenten und den Ingenieur schwer auffindbare Materialien zusammengestellt hat, sondern auch zahlreiche Anwendungsbeispiele zur Erläuterung der betreffenden Formeln, Tafeln und Tabellen ausgearbeitet hat. Die Tatsache, dass das Werk in zwei Sprachen herausgegeben ist, verschafft die Möglichkeit einen breiteren Leserkreis umzufassen.

Aus praktischen Gründen, wird das Handbuch in mehreren Bändern erscheinen. Das vorliegende Band beschäftigt sich mit Bodenspannungen. In zukünftigen Bändern werden Setzungen, Sohldruckverteilungen und Plastizität des Bodens zusammengestellt.

Der deutsche Leser wird sofort zur Kenntnis nehmen, dass die Schreibweise des deutschen Textes improvisierter Abänderung untergelaufen ist. Wir hoffen, dass solche Modifizierung den vollen Gebrauch des Handbuches nicht verhindert.

Wertvolle Anregungen hinsichtlich Veränderungen, Weglassungen, Fehler und Eingabe sind herzlich willkommen. Sicherlich werden diese Ratschläge von grosseren Nützen sein, während der Vorbereitung der anderen zu erscheinenden Bändern. Wegen ihrer finanziellen Unterstützung für die Vorbereitung des Manuskripts möchten die Autoren ihren herzlichen Dank dem Deutschen Akademischen Austausch Dienst (DAAD) und Addis Abeba University zum Ausdruck bringen. Dem Institut für Verkehrswasserbau, Grundbau und Bodenmechanik der Technischen Hochschule Aachen gilt auch unser Dank für die Benutzung der Institutsbibliothek.

Dr-Ing. Alemayehu Teferra
Prof. (em) Dr-Ing., Dr-Ing. E.h. Edgar Schultze

* 1 *
Elastic-isotropic half-space
Elastisch-isotroper Halbraum

1.1 LASTEN AUF DER OBERFLAECHE

1.1.1 Lotrechte Last

1.1.1.1 Einzellast

a) Halbraum nach Boussinesq [1]

(i) Formeln /Tafeln/ Tabellen

(i)a Gleichungen nach Gray [3]

1.1 SURFACE LOADS

1.1.1 Vertical Load

1.1.1.1 Point Load

a) Half-space according to Boussinesq [1]

(i) Formulae /Charts/ Tables

(i)a Equations of Gray [3]

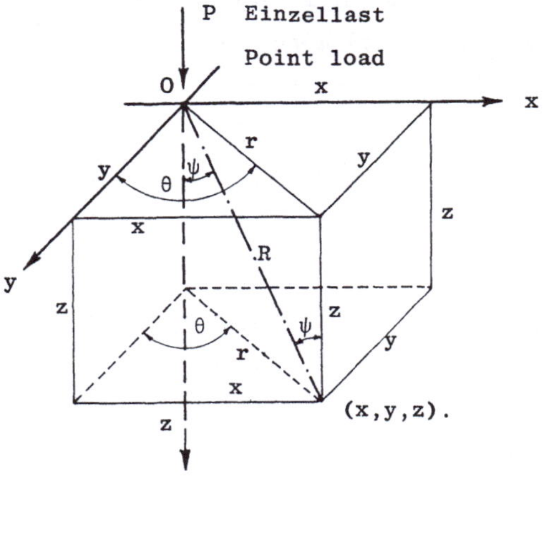

$$x^2 + y^2 = r^2$$

$$x^2 + y^2 + z^2 = z^2 + r^2 = R^2$$

$$\theta = \text{arc tan}\,\frac{x}{y}$$

$$\psi = \text{arc tan}\,\frac{r}{z}$$

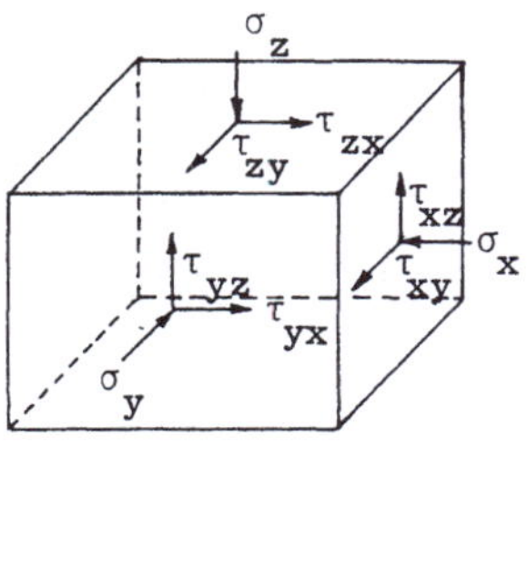

$$\sigma_z = \frac{3P}{2\pi} \cdot \frac{z^3}{R^5} = \frac{3P}{2\pi R^2} \cdot \cos^3\psi$$

$$\sigma_x = \frac{P}{2\pi}\left[3\frac{x^2 z}{R^5} - \frac{m-2}{m}\left\{\frac{x^2-y^2}{Rr^2\ (R+z)} + \frac{y^2 z}{R^3 r^2}\right\}\right]$$

$$\sigma_y = \frac{P}{2\pi}\left[3\frac{y^2 z}{R^5} - \frac{m-2}{m}\left\{\frac{y^2-x^2}{Rr^2\ (R+z)} + \frac{x^2 z}{R^3 r^2}\right\}\right]$$

$$\sigma_r = \frac{P}{2\pi}\left[3\frac{r^2 z}{R^5} - \frac{m-2}{m}\left\{\frac{R-z}{Rr^2}\right\}\right]$$

$$\sigma_\theta = \frac{P}{2\pi}\left[\frac{m-2}{m}\left\{\frac{1}{r^2} - \frac{z}{R^3} - \frac{z}{Rr^2}\right\}\right]$$

$$\tau_{rz} = \frac{3P}{2\pi} \cdot \frac{rz^2}{R^5} \qquad \tau_{zx} = \frac{3P}{2\pi} \cdot \frac{z^2 x}{R^5} \qquad \tau_{zy} = \frac{3P}{2\pi} \cdot \frac{z^2 y}{R^5}$$

$$\tau_{r\theta} = \tau_{z\theta} = 0$$

m = Poissonzahl

= Poisson's number

Fig. 1.1 Gleichungen fuer Bodenspannungen infolge einer Einzellast [3]

Fig. 1.1 Equations for stresses in soil due to a point load [3]

(i)b Einflussbeiwert nach Gilboy [2]

(i)b Enfluence value of Gilboy [2]

$$\sigma_z = \frac{3P}{2\pi} \frac{z^3}{(r^2+z^2)^{5/2}} = i_B \frac{P}{z^2}$$

$$i_B = \frac{z^2\sigma_z}{P} = \frac{3}{2\pi} \frac{1}{\left[(\frac{r}{z})^2+1\right]^{5/2}}$$

r/z	i_B	r/z	i_B	r/z	i_B	r/z	i_B	r/z	i_B	r/z	i_B
0.00	0.4775	0.70	0.1762	1.40	0.0317	2.10	0.0070	2.80	0.0021	3.50	
1	0.4773	1	0.1721	1	0.0309	1	0.0069	1	0.0020	↓	0.0007
2	0.4770	2	0.1681	2	0.0302	2	0.0068	2	0.0020	3.61	
3	0.4764	3	0.1641	3	0.0295	3	0.0066	3	0.0020		
4	0.4756	4	0.1603	4	0.0288	4	0.0065	4	0.0019	3.62	
5	0.4745	5	0.1565	5	0.0282	5	0.0064	5	0.0019	↓	0.0006
6	0.4732	6	0.1527	6	0.0275	6	0.0063	6	0.0019	3.74	
7	0.4717	7	0.1491	7	0.0269	7	0.0062	7	0.0019		
8	0.4899	8	0.1455	8	0.0263	8	0.0060	8	0.0018	3.75	
9	0.4679	9	0.1420	9	0.0257	9	0.0059	9	0.0018	↓	0.0005
0.10	0.4657	0.80	0.1386	1.50	0.0251	2.20	0.0058	2.90	0.0018	3.90	
1	0.4633	1	0.1353	1	0.0245	1	0.0057	1	0.0017		
2	0.4607	2	0.1320	2	0.0240	2	0.0056	2	0.0017	3.91	
3	0.4579	3	0.1288	3	0.0234	3	0.0055	3	0.0017	↓	0.0004
4	0.4548	4	0.1257	4	0.0229	4	0.0054	4	0.0017	4.12	
5	0.4576	5	0.1226	5	0.0224	5	0.0053	5	0.0016		
6	0.4482	6	0.1196	6	0.0219	6	0.0052	6	0.0016	4.13	
7	0.4446	7	0.1166	7	0.0214	7	0.0051	7	0.0016	↓	0.0003
8	0.4409	8	0.1138	8	0.0209	8	0.0050	8	0.0016	4.43	
9	0.4370	9	0.1110	9	0.0204	9	0.0049	9	0.0015		
0.20	0.4329	0.90	0.1088	1.60	0.0200	2.30	0.0048	3.00	0.0015	4.44	
1	0.4286	1	0.1057	1	0.0195	1	0.0047	1	0.0015	↓	0.0002
2	0.4242	2	0.1031	2	0.0191	2	0.0047	2	0.0015	4.90	
3	0.4197	3	0.1005	3	0.0187	3	0.0046	3	0.0014		
4	0.4151	4	0.0981	4	0.0183	4	0.0045	4	0.0014	4.91	
5	0.4103	5	0.0956	5	0.0179	5	0.0044	5	0.0014	↓	0.0001
6	0.4054	6	0.0933	6	0.0175	6	0.0043	6	0.0014	6.15	
7	0.4004	7	0.0910	7	0.0171	7	0.0043	7	0.0014		
8	0.3954	8	0.0887	8	0.0167	8	0.0042	8	0.0013		
9	0.3902	9	0.0865	9	0.0163	9	0.0041	9	0.0013		
0.30	0.3849	1.00	0.0844	1.70	0.0160	2.40	0.0040	3.10	0.0013		
1	0.3796	1	0.0823	1	0.0157	1	0.0040	1	0.0013		
2	0.3742	2	0.0803	2	0.0153	2	0.0039	2	0.0013		
3	0.3687	3	0.0783	3	0.0150	3	0.0038	3	0.0012		
4	0.3632	4	0.0764	4	0.0147	4	0.0038	4	0.0012		
5	0.3577	5	0.0744	5	0.0144	5	0.0037	5	0.0012		
6	0.3521	6	0.0727	6	0.0141	6	0.0036	6	0.0012		
7	0.3465	7	0.0709	7	0.0138	7	0.0036	7	0.0012		
8	0.3408	8	0.0691	8	0.0135	8	0.0035	8	0.0012		
9	0.3351	9	0.0674	9	0.0132	9	0.0034	9	0.0011		
0.40	0.3294	1.10	0.0658	1.80	0.0129	2.50	0.0034	3.20	0.0011		
1	0.3238	1	0.0641	1	0.0126	1	0.0033	1	0.0011		
2	0.3181	2	0.0626	2	0.0124	2	0.0033	2	0.0011		
3	0.3124	3	0.0610	3	0.0121	3	0.0032	0	0.0011		

r/z	i_B	r/z	i_B	r/z	i_B	r/z	i_B	r/z	i_B
4	0.3068	4	0.0595	4	0.0119	4	0.0032	4	0.0011
5	0.3011	5	0.0581	5	0.0116	5	0.0031	4	0.0011
6	0.2955	6	0.0567	6	0.0114	6	0.0031	6	0.0010
7	0.2899	7	0.0553	7	0.0112	7	0.0030	7	0.0010
8	0.2843	8	0.0539	8	0.0109	8	0.0030	8	0.0010
9	0.2788	9	0.0526	9	0.0107	9	0.0029	9	0.0010
0.50	0.2733	1.20	0.0513	1.90	0.0105	2.60	0.0029	3.30	0.0010
1	0.2679	1	0.0501	1	0.0103	1	0.0028	1	0.0009
2	0.2625	2	0.0489	2	0.0101	2	0.0028	2	0.0009
3	0.2571	3	0.0477	3	0.0099	3	0.0027	3	0.0009
4	0.2518	4	0.0466	4	0.0097	4	0.0027	4	0.0009
5	0.2466	5	0.0454	5	0.0095	5	0.0026	5	0.0009
6	0.2414	6	0.0443	6	0.0093	6	0.0026	6	0.0009
7	0.2363	7	0.0433	7	0.0091	7	0.0025	7	0.0009
8	0.2313	8	0.0422	8	0.0089	8	0.0025	8	0.0009
9	0.2263	9	0.0412	9	0.0087	9	0.0025	9	0.0009
0.60	0.2214	1.30	0.0402	2.00	0.0085	2.70	0.0024	3.40	0.0009
1	0.2165	1	0.0393	1	0.0084	1	0.0024	1	0.0008
2	0.2117	2	0.0384	2	0.0082	2	0.0023	2	0.0008
3	0.2070	3	0.0374	3	0.0081	3	0.0023	3	0.0008
4	0.2024	4	0.0365	4	0.0079	4	0.0023	4	0.0008
5	0.1978	5	0.0357	5	0.0078	5	0.0022	5	0.0008
6	0.1934	6	0.0348	6	0.0076	6	0.0022	6	0.0008
7	0.1889	7	0.0340	7	0.0075	7	0.0022	7	0.0008
8	0.1846	8	0.0332	8	0.0073	8	0.0021	8	0.0008
9	0.1804	9	0.0324	9	0.0072	9	0.0021	9	0.0008

Tabelle 1.1 Einflussbeiwert, i_B, zur Berechnung der lotrechten Spannung im Boden infloge einer Einzellast [2]

Influence value, i_B, for determining the vertical stress in soil due to a point load [2]

(ii) Erlaeuterung zur Tabelle 1.1 Bzw. Fig. 1.2

Aus Tabelle 1.1 bzw Fig. 1.2 laesst sich das Verhaeltnis r/z den Wert i_B ablesen. Mit der gewaehlten Tiefe z und der last P kann σ_z berechnet werden.

(ii) Explannatory Note to Table 1.1 and Fig. 1.2

From Table 1.1 or Fig. 1.2 one obtains the value of i_B for a given r/z ratio. For a given depth z and load P, one determines σ_z.

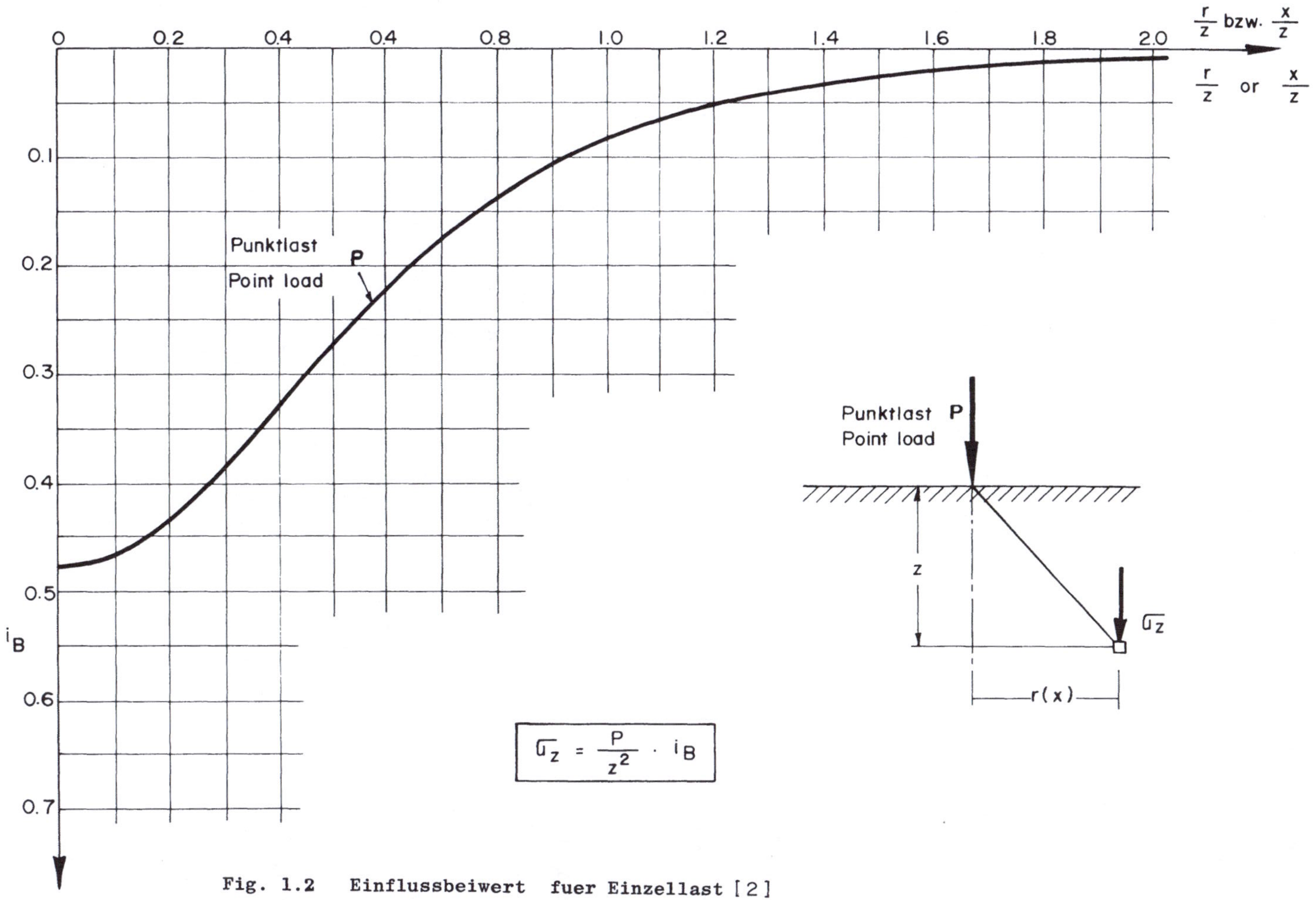

Fig. 1.2 **Einflussbeiwert fuer Einzellast** [2]
Influence value for a point load [2]

(iii) Anwendungsbeispiel

Gegeben

P = 100 kN

Gesucht

σ_z = im Abstand r = 4.0 in der Tiefe

z = 6.0 m

Loesung

Fuer $r/_z = \frac{4.0}{6.0} = 0.67$ ergibt sich

$i_B = 0.190$

$\tau_z = \frac{P}{z^2} \cdot i_B = \frac{100}{6^2}(0.190) = 0.53\ kN/m^2$

(iii) Example

Given

P = 100 kN

Required

σ_z at a distance of r = 4.0 in a depth of z = 6.0 m

Solution

For $r/_z = \frac{4.0}{6.0} = 0.67$ the value of

$i_B = 0.190$

$\tau_z = \frac{P}{z^2} \cdot i_B = \frac{100}{6^2}(0.190)$

$= 0.53\ kN/m^2$

(iv) Quellen

References

[1] Boussinesq (1885) - Application des potentiels à L'étude de L'équilibre et due mouvement des solides élastique, Paris, Gauthier - Villars

[2] Gilboy (1933) - Influence Tables for Solution of Boussinesq Equation. Proc.Am. Soc. Civ. Eng. 59, PP 781,

s.a Grundbautaschenbuch (1966), Bd. 1, Abschn. 2.1 Berlin, W. Ernst &Sohn. S. 94

s.a. DIN 4019, Blatt 1 (Juni 1958) - Baugraund. Setzungsberechnungen bei lotrechter, mittiger Belastung. Richtlinien. S. 3.

[3] Gray (1936) - Stresss distribution in elastic solids Proc. 1 ICSMFE Cambridge, Vol. 2 PP 157

b) Halbraum nach Froehlich [1]

(i) Formeln/Tabellen

b) Half-space according to Froehlich [1]

(i) Formulae/Tables

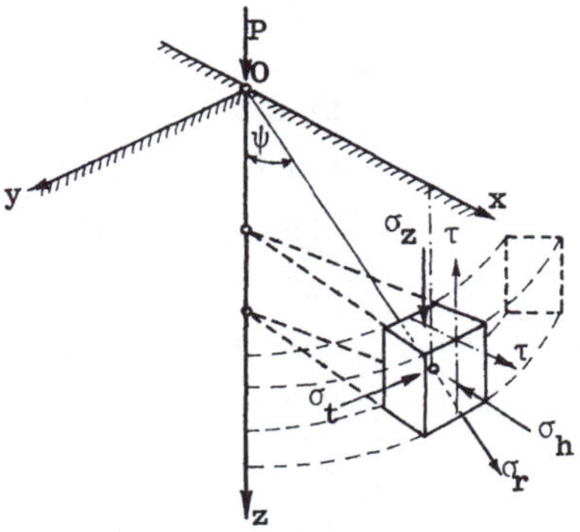

$$\sigma_z = \frac{\nu . P}{2\pi R^2} \cdot \cos^{\nu}\psi \qquad = \frac{\nu . P}{2\pi z^2} \cdot \cos^{\nu+2}\psi$$

$$\sigma_h = \frac{\nu . P}{2\pi R^2} \cdot \cos^{\nu-2}\psi . \sin^2\psi \quad = \frac{\nu P}{2\pi z^2} . \cos^{\nu}\psi \, . \, \sin^2\psi$$

$$\tau = \frac{\nu . P}{2\pi R^2} \cdot \cos^{\nu-1}\psi . \sin\psi \quad = \frac{\nu P}{2\pi z^2} \cdot \cos^{\nu+1}\psi \cdot \sin \psi$$

$$\sigma_r = \frac{\nu P}{2\pi R^2} . \cos^{\nu-2}\psi \qquad = \frac{\nu P}{2\pi P^2} . \cos^{\nu}\psi$$

ν = Konzentrationsfaktor

= concentration factor

Fig. 1.3 Gleichungen fuer Bodenspannungen infolge einer Einzellast [1]

Equations for stresses in soils due to a point load [1]

$$\sigma_z = \frac{P}{z^2} \cdot i_F$$

	i_F							
r/z	ν=3.5	ν=4	ν=4.5	ν=5	ν=5.5	ν=6	ν=6.5	ν=7
0.0	.557042	.636620	.716197	.795775	.875352	.954930	1.034507	1.114085
.1	.542007	.617897	.693407	.768538	.843292	.917669	.991671	1.065301
.2	.500088	.565953	.630485	.693703	.755628	.816278	.875673	.933831
.3	.439507	.491587	.541248	.588569	.633627	.676496	.717251	.755960
.4	.370364	.407855	.442124	.473355	.501725	.527399	.550538	.571291
.5	.301568	.325949	.346797	.364423	.379115	.391139	.400743	.408153
.6	.239142	.253083	.263652	.271271	.276320	.279136	.280023	.279250
.7	.186048	.192452	.195965	.197078	.196216	.193743	.189973	.185174
.8	.142912	.144327	.143480	.140876	.136936	.132007	.126371	.120260
.9	.108961	.107361	.104130	.099751	.094599	.088973	.083100	.077155
1.0	.082805	.079578	.075281	.070337	.065061	.059683	.054370	.049236
1.2	.047925	.043824	.039447	.035069	.030865	.026941	.023352	.020122
1.4	.028173	.024547	.021054	.017835	.014957	.012440	.010274	.008435
1.6	.016959	.014110	.011556	.009348	.007486	.005945	.004689	.003676
1.8	.010487	.008352	.006548	.005070	.003887	.002955	.002231	.001674
2.0	.006664	.005093	.003832	.002847	.002094	.001528	.001107	.000797
2.2	.004348	.003196	.002313	.001653	.001170	.000821	.000572	.000396
2.4	.002908	.002061	.001438	.000991	.000676	.000457	.000307	.000205
2.6	.001990	.001362	.000918	.000611	.000403	.000263	.000171	.000110
2.8	.001390	.000922	.000601	.000387	.000247	.000156	.000098	.000061
3.0	.000991	.000637	.000403	.000252	.000156	.000095	.000058	.000035
3.5	.000457	.000274	.000161	.000094	.000054	.000031	.000018	.000010
4.0	.000230	.000130	.000072	.000039	.000021	.000011	.000006	.000003
4.5	.000125	.000066	.000035	.000018	.000009	.000005	.000002	.000001
5.0	.000072	.000036	.000018	.000009	.000004	.000002	.000001	.000000
5.5	.000043	.000021	.000010	.000005	.000002	.000001	.000000	.000000
6.0	.000027	.000013	.000006	.000003	.000001	.000001	.000000	.000000
6.5	.000018	.000008	.000003	.000001	.000001	.000000	.000000	.000000
7.0	.000012	.000005	.000002	.000001	.000000	.000000	.000000	.000000

Tabelle 1.2 Einflussbeiwert i_F, zur Berechnung der lotrechten Spannung im Boden infolge Einzellast [1]

Table 1.2 Influence value i_F, for determining the vertical stress in soil due to point load [1]

(ii) Erlauterung

Durch das Prinzip der geradlinigen Spannungsausbreitung, leitet Froehlich seine Gleichungen fuer die Spannungsberechnung ab. Dabei ist die Konzentrationszahl (Ordnungszahl) ν eine unbekannte Groesse, die von der Bodenart abhaengig ist. Die Groesse von ν bestimmt den Grad der Spannungskonzentration im Halbraum. Bei $\nu = 3$ entspricht der Halbraum von Froehlich dem Halbraum von Boussinesq (elastisch, isotrop, homogen), wenn die Querdehnungszahl m = 2 ist.
- volumen bestaendiger Halbraum.

(ii) Explanatory note

Froehlich derives his equations for determining stresses in soil by applying the principle of straight line propagation of stress. He introduces a factor known as concentration factor, ν, which is dependent on the soil type. The magnitude of ν determines the degree of stress concentration in the half-space. For $\nu = 3$, the half-space of Froehlich corresponds to that of Boussinesq (elastic, isotropic, homogeneous), where the Poisson's number m = 2 - volume constant half-space.

(iii) Anwendungsbeispiel

Gegeben

P = 1000 kN

Gesucht

σ_z im Abstand r = 5.0 m in der Tiefe z = 15.0 m fuer $\nu = 3.5$

Loesung

Fuer $r/_z = 5/_{15} = 0.33$ ergibt sich

$i_F = 0.416$ (Tabelle 1.2)

$$\sigma_z = \frac{P}{z^2} \cdot i_F = \frac{1000}{15^2} (0.416)$$

$$= 1.85 \text{ kN/m}^2$$

(iii) Example

Given

P = 1000 kN

Required

σ_z at a distance of r = 5.0 m and a depth of z = 15.0 m for $\nu = 3.5$

Solution

For $r/_z = 5/_{15} = 0.33$, the value of

$i_F = 0.416$ (Table 1.2)

$$\sigma_z = \frac{P}{z^2} \cdot i_F = \frac{1000}{15^2} (0.416)$$

$$= 1.85 \text{ kN/m}^2$$

(iv) Quelle

Reference

[1] Froehlich (1934) - Druckverteilung im Baugrunde. Wien, Verlag von Julius Springer

c) **Halbraum nach Westergaard [2]**

(i) Formeln/Tabellen

c) Half-space according to Westergaard [2]

(i) Formulae/Tables

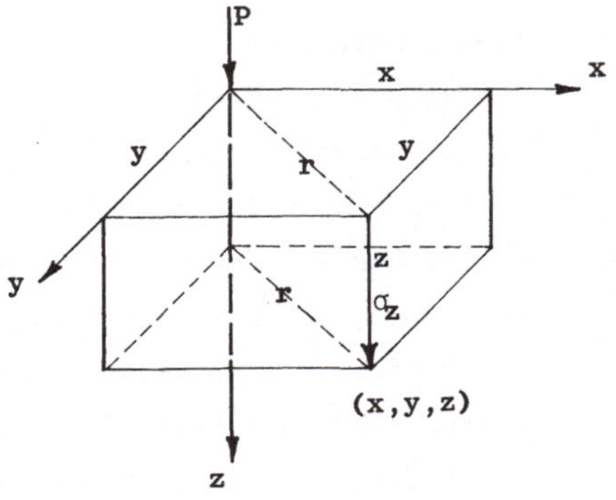

$$\sigma_z = \frac{P.K}{2\pi.z^2} \cdot \frac{1}{\left[(\frac{r}{z})^2 + K^2\right]^{3/2}}$$

$$K = \sqrt{\frac{1-2\mu}{2(1-\mu)}}, \quad \mu = \frac{1}{m}$$

m = Querdehnungszahl (Poissonzahl)

= Poisson's number

Fig. 1.4 Gleichung fuer Bodenspaunung infolge einer Einzellast [1]
Equation for stress in soil due to a point load [1]

$$\sigma_z = \frac{P.\sqrt{2}}{4\pi z^2} \cdot \frac{1}{\left[(\frac{r}{z})^2 + \frac{1}{2}\right]^{3/2}} = \frac{P}{z^2} \cdot i_w$$

$$i_w = \frac{\sqrt{2}}{4\pi} \frac{1}{\left[(\frac{r}{z})^2 + \frac{1}{2}\right]^{3/2}}$$

$\mu = 0$

r/z	i_w	r/z	i_w	r/z	i_w	r/z	i_w	r/z	i_w
0.00	0.3183	0.10	0.3090	0.20	0.2836	0.30	0.2483	0.40	0.2099
1	0.3182	1	0.3071	1	0.2804	1	0.2445	1	0.2061
2	0.3179	2	0.3050	2	0.2771	2	0.2407	2	0.2023
3	0 3175	3	0.3028	3	0.2737	3	0.2369	3	0.1986
4	0.3168	4	0.3005	4	0.2703	4	0.2330	4	0.1948
5	0.3159	5	0.2980	5	0.2668	5	0.2291	5	0.1911
6	0.3149	6	0.2953	6	0.2632	6	0.2253	6	0.1875
7	0.3137	7	0.2926	7	0.2595	7	0.2214	7	0.1839
8	0.3123	8	0.2897	8	0.2558	8	0.2176	8	0.1803
9	0.3107	9	0.2867	9	0.2521	9	0.2137	9	0.1768

r/z	i_w	r/z	i_w	r/z	i_w	r/z	i_w	r/z	i_w
0.50	0.1733	0.60	0.1411	0.70	0.1143	0.80	0.0925	0.90	0.0751
1	0.1698	1	0.1382	1	0.1119	1	0.0905	1	0.0735
2	0.1664	2	0.1353	2	0.1095	2	0.0887	2	0.0720
3	0.1631	3	0.1325	3	0.1072	3	0.0868	3	0.0706
4	0.1598	4	0.1297	4	0.1050	4	0.0850	4	0.0692
5	0.1566	5	0.1270	5	0.1028	5	0.0833	5	0.0678
6	0.1534	6	0.1244	6	0.1006	6	0.0815	6	0.0664
7	0.1502	7	0.1218	7	0.0985	7	0.0799	7	0.0651
8	0.1471	8	0.1192	8	0.0964	8	0.0782	8	0.0638
9	0.1441	9	0.1167	9	0.0944	9	0.0766	9	0.0625
1.00	0.0613	1.10	0.0503	1.20	0.0417	1.30	0.0347	1.40	0.0292
1	0.0601	1	0.0494	1	0.0409	1	0.0341	1	0.0287
2	0.0589	2	0.0484	2	0.0401	2	0.0335	2	0.0282
3	0.0577	3	0.0475	3	0.0394	3	0.0329	3	0.0277
4	0.0566	4	0.0466	4	0.0387	4	0.0324	4	0.0273
5	0.0555	5	0.0458	5	0.0380	5	0.0318	5	0.0268
6	0.0544	6	0.0449	6	0.0373	6	0.0313	6	0.0264
7	0.0534	7	0.0441	7	0.0366	7	0.0307	7	0.0259
8	0.0523	8	0.0432	8	0.0360	8	0.0302	8	0.0255
9	0.0513	9	0.0424	9	0.0354	9	0.0297	9	0.0251
1.50	0.0247	1.60	0.0210	1.70	0.0180	1.80	0.0156	1.90	0.0135
1	0.0243	1	0.0207	1	0.0178	1	0.0153	1	0.0133
2	0.0239	2	0.0204	2	0.0175	2	0.0151	2	0.0131
3	0.0235	3	0.0201	3	0.0172	3	0.0149	3	0.0130
4	0.0231	4	0.0198	4	0.0170	4	0.0147	4	0.0128
5	0.0228	5	0.0195	5	0.0167	5	0.0145	5	0.0126
6	0.0224	6	0.0192	6	0.0165	6	0.0143	6	0.0124
7	0.0220	7	0.0189	7	0.0163	7	0.0141	7	0.0123
8	0.0217	8	0.0186	8	0.0160	8	0.0139	8	0.0121
9	0.0214	9	0.0183	9	0.0158	9	0.0137	9	0.0120

r/z	i_w	r/z	i_w	r/z	i_w	r/z	i_w	r/z	i_w
2.00	0.0118	2.10	0.0103	2.20	0.0091	2.30	0.0081	2.40	0.0072
1	0.0116	1	0.0102	1	0.0090	1	0.0080	1	0.0071
2	0.0115	2	0.0101	2	0.0089	2	0.0079	2	0.0070
3	0.0113	3	0.0100	3	0.0088	3	0.0078	3	0.0069
4	0.0112	4	0.0098	4	0.0087	4	0.0077	4	0.0069
5	0.0110	5	0.0097	5	0.0086	5	0.0076	5	0.0068
6	0.0109	6	0.0096	6	0.0085	6	0.0075	6	0.0067
7	0.0108	7	0.0095	7	0.0084	7	0.0074	7	0.0066
8	0.0106	8	0.0094	8	0.0083	8	0.0074	8	0.0066
9	0.0105	9	0.0092	9	0.0082	9	0.0073	9	0.0065
2.50	0.0064	2.60	0.0058	2.70	0.0052	2.80	0.0047	2.90	0.0042
1	0.0064	1	0.0057	1	0.0051	1	0.0046	1	0.0042
2	0.0063	2	0.0056	2	0.0051	2	0.0046	2	0.0042
3	0.0062	3	0.0056	3	0.0050	3	0.0045	3	0.0041
4	0.0061	4	0.0055	4	0.0050	4	0.0045	4	0.0041
5	0.0061	5	0.0054	5	0.0049	5	0.0045	5	0.0040
6	0.0060	6	0.0054	6	0.0049	6	0.0044	6	0.0040
7	0.0059	7	0.0053	7	0.0048	7	0.0044	7	0.0040
8	0.0059	8	0.0053	8	0.0048	8	0.0043	8	0.0039
9	0.0058	9	0.0052	9	0.0047	9	0.0043	9	0.0039
3.00	0.0038	3.10	0.0035	3.20	0.0032	3.30	0.0029	3.40	0.0027
2	0.0038	2	0.0034	2	0.0031	2	0.0029	.2	0.0026
4	0.0037	4	0.0034	4	0.0031	4	0.0028	.4	0.0026
6	0.0036	6	0.0033	6	0.0030	6	0.0028	.6	0.0026
8	0.0036	8	0.0033	8	0.0030	8	0.0027	8	0.0025
3.50	0.0025	4.00	0.0017	4.50	0.0012	5.00	0.0009	6.00	0.0005
.6	0.0023	.1	0.0016	.6	0.0011	.2	0.0008	7.00	0.0003
.7	0.0021	.2	0.0015	.7	0.0011	.4	0.0007	8.00	0.0002
.8	0.0020	.3	0.0014	.8	0.0010	.5	0.0007	9.00	0.0002
.9	0.0018	.4	0.0013	.9	0.0009	.6	0.0006	10.00	0.0001
						.8	0.0006	∞	0.0000

Tabelle 1.3 Einflussbeiwert, i_w, zur Berechnung der lotrechten Spannung im Boden infolge Einzellast [1]

Table 1.3 Influence value i_w, for determining the vertical stress in soil due to point load [1]

(ii) Voraussetzung

Westergaard hat die Gleichung mit der Annahme eines Halbraums, der in der waagerechten Richtung mit unendlich starren, duennen plaettchen streifenartig durchwebt ist, die keine waagerechten Verschiebungen ermoeglichen, abgeleitet.

(ii) Assumption

Westergaard derived the equation by assuming a half-space that is reinforced by infinitely rigid horizontal sheets which do not allow horizontal desplacements to occur.

(iii) Quellen

References

[1] Fadum (1948) - Influence values for estimating stresses in Elastic foundations. Proc. 2 ICSMFE, Rotterdam Vol 3 PP 77.

[2] Westergaard (1939) - A problem of elasticity suggested by a problem in soil mechanics: soft material reinforced by numerous strong horizontal sheets. Contributions to the mechanics of solids, on the occasion of the 60th anniversary of S. Timoshenko, Macmillan Co., New York, 1939.

1.1.1.2 Linienlast

a) Halbraum nach Boussinesq

(i) Formeln/Tabellen

(i)a Gleichungen nach Gray [1]

1.1.1.2 Line Load

a) Half-space according to Boussinesq

(i) Formulae/Tables

(i)a Equations of Gray [1]

Unendlich lange Linienlast

Infinite line load

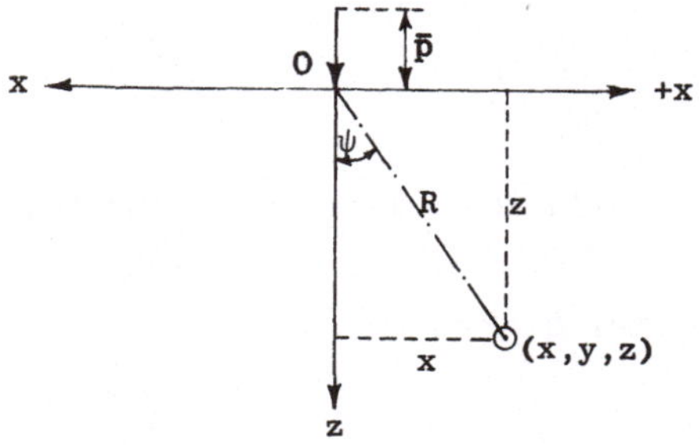

$$\sigma_x = \frac{2\bar{p}}{\pi} \cdot \frac{x^2 z}{R^4} \qquad \sigma_y = \frac{2\bar{p}}{\pi m} \cdot \frac{z}{R^2}$$

$$\sigma_z = \frac{2\bar{p}}{\pi} \cdot \frac{z^3}{R^4} \qquad \tau_{xz} = \frac{2\bar{p}}{\pi} \cdot \frac{xz^2}{R^4}$$

Hauptspannungen:

Principal Stresses:

$$\sigma_1 = \sigma_R = \frac{2\bar{p}}{\pi} \cdot \frac{z}{R^2} \qquad \sigma_2 = \sigma_\psi = 0$$

$$\tau_{max} = \frac{\bar{p}}{\pi} \cdot \frac{z}{R^2}$$

Orte von konstantem σ_1, σ_2 und τ_{max}	Kreise die x-Achse in O tangierend
Trajektorien von σ_1	Radien durch O
Trajektorien von σ_2	Schar von Halbkreisen, Mittelpunkt in O
Trajektorien von τ_{max}	Zwei orthogonale scharen von gleichwinkligen spiralen, die die Radien unter 45° schneiden

Loci of constant σ_1, σ_2 and τ_{max}	Circles tangent to x-axis at O.
Trajectories of σ_1	Radial lines through O.
Trajectories of σ_2	Family of semi-circles, centers at O.
Trajectories of τ_{max}	Two orthogonal families of equiangular spirals intersecting the radial lines under 45°

Fig. 1.5 Gleichungen fuer Bodenspannungen infolge einer Linienlast [1]
Equations for stresses in soils due to a line load [1]

(i)b Spannungsbeiwert nach Szechy [2]

(i)b Influence coefficient according to Szechy [2]

$$\sigma_z = \frac{2\bar{p}}{\pi} \frac{z^3}{(x^2 + z^2)^2} = i_s \cdot \frac{\bar{p}}{z}$$

$$i_s = \frac{2}{\pi} \frac{1}{\left[(\frac{x}{z})^2 + 1\right]^2}$$

$\frac{x}{z}$	i_s	$\frac{x}{z}$	i_s	$\frac{x}{z}$	i_s	$\frac{x}{z}$	i_s	$\frac{x}{z}$	i_s
0.00	0.6366	0.50	0.4074	1.00	0.1591	1.50	0.0602	2.00	0.0254
0.02	0.6361	0.52	0.3944	1.02	0.1528	1.52	0.0580	2.02	0.0246
0.04	0.6346	0.54	0.3815	1.04	0.1468	1.54	0.0559	0.04	0.0238
0.06	0.6320	0.56	0.3689	1.06	0.1410	1.56	0.0539	0.06	0.0231
0.08	0.6284	0.58	0.3564	1.08	0.1355	1.58	0.0520	2.08	0.0224
0.10	0.6241	0.60	0.3441	1.10	0.1302	1.60	0.0501	2.10	0.0217
0.12	0.6187	0.62	0.3321	1.12	0.1251	1.62	0.0484	2.12	0.0210
0.14	0.6124	0.64	0.3203	1.14	0.1203	1.64	0.0467	2.14	0.0204
0.16	0.6052	0.66	0.3088	1.16	0.1156	1.66	0.0450	2.16	0.0197
0.18	0.5973	0.68	0.2976	1.18	0.1111	1.68	0.0435	2.18	0.0192
0.20	0.5886	0.70	0.2867	1.20	0.1068	1.70	0.0420	2.20	0.0186
0.22	0.5791	0.72	0.2760	1.22	0.1027	1.72	0.0406	2.22	0.0180
0.24	0.5691	0.74	0.2657	1.24	0.0988	1.74	0.0392	2.24	0.0175
0.26	0.5586	0.76	0.2557	1.26	0.0950	1.76	0.0378	2.26	0.0169
0.28	0.5474	0.78	0.2460	1.28	0.0914	1.78	0.0366	2.28	0.0165
0.30	0.5358	0.80	0.2366	1.30	0.0879	1.80	0.0353	2.30	0.0160
0.32	0.5238	0.82	0.2276	1.32	0.0846	1.82	0.0341	2.32	0.0155
0.34	0.5115	0.84	0.2188	1.34	0.0814	1.84	0.0330	2.34	0.0151
0.36	0.4989	0.86	0.2102	1.36	0.0783	1.86	0.0319	2.36	0.0147
0.38	0.4861	0.88	0.2021	1.38	0.0754	1.88	0.0309	2.38	0.0143
0.40	0.4731	0.90	0.1942	1.40	0.0726	1.90	0.0299	2.40	0.0138
0.42	0.4600	0.92	0.1866	1.42	0.0699	1.92	0.0289	2.42	0.0134
0.44	0.4468	0.94	0.1793	1.44	0.0673	1.94	0.0280	2.44	0.0131
0.46	0.4337	0.96	0.1723	1.46	0.0648	1.96	0.0271	2.46	0.0127
0.48	0.4204	0.98	0.1655	1.48	0.0625	1.98	0.0262	2.48	0.0124

$\frac{x}{z}$	i_s	$\frac{x}{z}$	i_s	$\frac{x}{z}$	i_s	$\frac{x}{z}$	i_s
2.50	0.0120	2.70	0.0092	2.90	0.0071	3.50	0.0036
2.52	0.0117	2.72	0.0089	2.92	0.0070	3.60	0.0032
2.54	0.0113	2.74	0.0087	2.94	0.0068	3.70	0.0029
2.56	0.0111	2.76	0.0085	2.96	0.0066	3.80	0.0026
2.58	0.0108	2.78	0.0083	2.98	0.0064	3.90	0.0024
2.60	0.0105	2.80	0.0080	3.00	0.0063	4.00	0.0021
2.62	0.0102	2.82	0.0078	3.10	0.0056	5.00	0.0009
2.64	0.0099	2.84	0.0077	3.20	0.0050	6.00	0.0004
2.66	0.0097	2.86	0.0075	3.30	0.0045	7.00	0.0002
2.68	0.0094	2.88	0.0073	3.40	0.0040	8.00	0.0001

Tabelle 1.4 Einflussbeiwert i_s, zur Berechnung der lotrechten Spannung im Boden infolge einer Linienlast [2]

Table 1.4 Influence coefficient i_s, for determining the vertical stress in soil due to line load

Quellen:

References:

[1] Gray (1936) - Stress distribution in elastic solids
Proc. 1. ICSMFE Cambridge, Vol 2. PP 157

[2] Szechy (1963) - Der Grundbau
Springer Verlag
Bd I. S. 257

b) Halbraum nach Froehlich [1]

(i) Formeln

b) Half-space according to Froehlich [1]

(i) Formulae

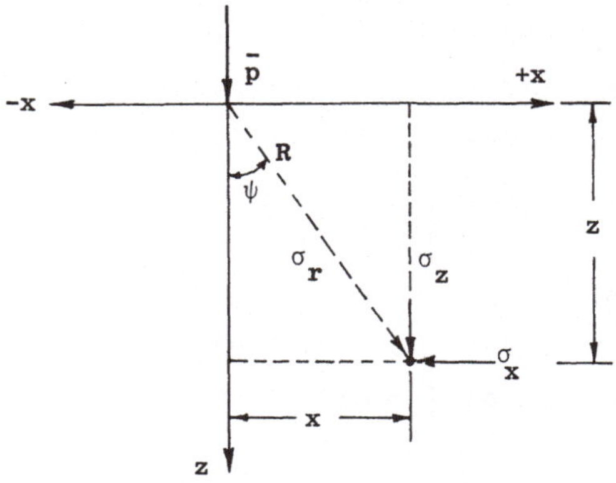

$$\sigma_z = f \cdot \frac{\bar{p}}{R} \cdot \cos^{\nu}\psi$$

$$\sigma_x = f \cdot \frac{\bar{p}}{R} \cdot \cos^{\nu-2}\psi \cdot \sin^2\psi$$

$$\tau = f \cdot \frac{\bar{p}}{R} \cdot \cos^{\nu-1}\psi \cdot \sin\psi$$

$$\sigma_r = f \cdot \frac{\bar{p}}{R} \cdot \cos^{\nu-2}\psi$$

ν = Konzentrationsfaktor
Concentration factor

Fuer ν = 1 bis 6 ergeben sich dann folgende spannungen.

for ν = 1 to 6, the following stresses are given:

ν	σ_r	σ_z	σ_x	τ	Faktor fuer alle Spannungen Factor for all Stresses
1	$\frac{1}{\cos\psi}$	$\cos\psi$	$\sin\psi \tan\psi$	$\sin\psi$	$\frac{1}{\pi} \cdot \frac{\bar{p}}{R}$
2	1	$\cos^2\psi$	$\sin^2\psi$	$\sin\psi \cos\psi$	$\frac{1}{2} \cdot \bar{p}/R$
3	$\cos\psi$	$\cos^3\psi$	$\sin^2\psi \cos\psi$	$\sin\psi \cos^2\psi$	$\frac{2}{\pi} \cdot \frac{\bar{p}}{R}$
4	$\cos^2\psi$	$\cos^4\psi$	$\sin^2\psi \cos^2\psi$	$\sin\psi \cos^3\psi$	$\frac{3}{4} \cdot \bar{p}/R$
5	$\cos^3\psi$	$\cos^5\psi$	$\sin^2\psi \cos^3\psi$	$\sin\psi \cos^4\psi$	$8/3\pi \cdot \bar{p}/R$
6	$\cos^4\psi$	$\cos^6\psi$	$\sin^2\psi \cos^4\psi$	$\sin\psi \cos\psi$	$\frac{15}{16} \cdot \frac{\bar{p}}{R}$

Fig. 1.6 Gleichungen fuer Bodenspannungen infolge einer Linienlast [1]
Equations for stresses in soils due to a line load [1]

(ii) Quelle

Reference

[1] Froehlich (1934): Druckverteilung im Baugraunde Wien, Verlag von Julius Springer

c) Halbraum nach Westergaard [1]

(i) Formeln/Tafeln

c) Half-space according to Westergaard [1]

(i) Formulae/Charts

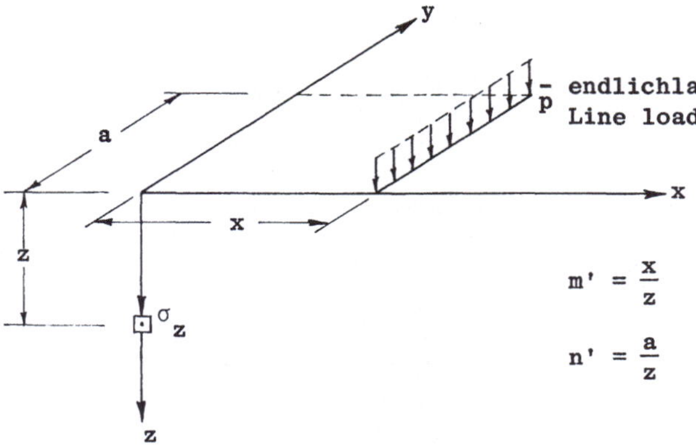

$$m' = \frac{x}{z} \qquad \mu = 0$$

$$n' = \frac{a}{z}$$

$$\sigma_z = \frac{\bar{p}}{2\pi} \cdot \frac{az^3}{(x^2 + z^2)} \cdot \frac{1}{\sqrt{x^2 + a^2 + z^2}} \left[\frac{1}{(x^2 + a^2 + z^2)} + \frac{2}{(x^2 + z^2)} \right]$$

Fig. 1.7 Gleichung fuer Bodenspannung infolge einer endlichlangen Linienlast [1]

Fig. 1.7 Equation for stress in soil due to a line load of finite length [1]

(ii) Erlaeuterung

Die Tafel (Fig. 1.8) is auch fuer Streifengruendungen anwendbar, wenn man diese vereinfacht als Linienlast betrachten darf.

(ii) Explanatory note

The chart (Fig. 1.8) may also be used for strip foundations, if for simplicity, one considers the load as being a line load.

(iii) Anwendungsbeispiel

Gegeben

Eine Streifengruendung von a = 10 m Laenge und b = 1.5 m Breite, die mit p = 300 kN/m^2 belastet ist.

Gesucht

Die senkrechte Druckspannung σ_z in der Tiefe z = 2.0 m in einem Abstand von x_1 = 0.75 m von der Gruendungsaussenkante.

(iii) Example

Given

A strip foundation with a = 10 m and b = 1.5 m acted upon by a uniform load p = 300 kN/m^2.

Required

The vertical stress σ_z at a depth of z = 2.0 m and at a distance of x_1 = 0.75 m from the edge of the foundation.

$\sigma_z = \frac{\bar{p}}{z} \cdot i_{WF}$: Fuer $\mu = 0$

For $\mu = 0$

$$i_{WF} = \frac{1}{2\pi} \cdot \frac{n'}{(m'^2+1)\sqrt{m'^2+n'^2+1}} \left[\frac{1}{(m'^2+n'^2+1)} + \frac{2}{(m'^2+1)}\right]$$

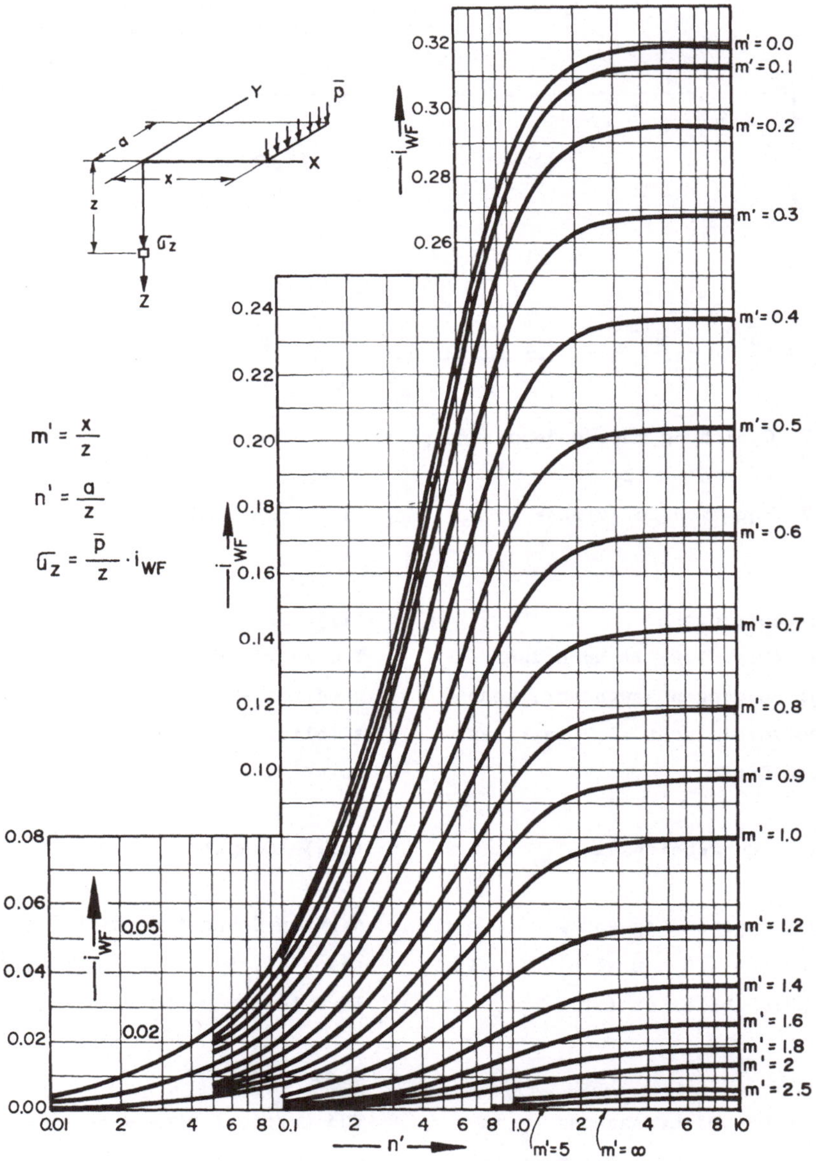

Fig. 1.8 Einflussbeiwert, i_{WF}, fuer endlichlange Linienlast [1]

Influence value, i_{WF}, for a line load of finite length [1]

Loesung

Die Linienlast wird $\bar{p} = p.b$

$= 300\ (1.5) = 450$ kN/m

$x = x_1 + \frac{1}{2} \cdot b = 0.75 + 0.5(1.50) = 1.50$m

$a = 10.0$ m

$m' = \frac{x}{z} = \frac{1.50}{2.00} = 0.75$

$n' = \frac{a}{z} = \frac{10.0}{2.00} = 5.00$

Fuer diese Werte von m' und n' liest man aus der Tafel (Fig. 1.8)

$i_{wF} = 0.13$

$\sigma_z = \frac{\bar{p}}{z} \cdot (i_{WF}) = \frac{450}{2.0}\ (0.13) = 29.2\ \text{kN/m}^2$

Solution

The line load will be $\bar{p} = p.b$

$= (300)(1.5) = 450$ kN/m

$x = x_1 + \frac{1}{2} \cdot b = 0.75 + 0.5(1.50) = 1.50$m

$a = 10.0$ m

$m' = \frac{x}{z} = \frac{1.50}{2.00} = 0.75$

$n' = \frac{a}{z} = \frac{10.0}{2.00} = 5.00$

For these values of m' and n' one reads from the chart (Fig. 1.8)

$i_{wF} = 0.13.$

$\sigma_z = \frac{\bar{p}}{z}\ (i_{WF}) = \frac{450}{2.0}\ (0.13) = 29.2\ \text{kN/m}^2$

(iv) Quelle

Reference

[1] Fadum (1948): Influence value for estimating stresses in elastic foundations.
Proc. 2. ICSMFE, Rotterdam, Vol. 3, PP 77
Bd 3, S.77

1.1.1.3 Streifenlast

1.1.1.3.1 Gleichmaessige Last

a) Halbraum nach Boussinesq

(i) Formeln/Tafeln/Tabellen

(i)a Gleichungen nach Gray [1]

1.1.1.3 Strip Load

1.1.1.3.1 Uniform Load

a) Half-space according to Boussinesq

(i) Formulae/Charts/Tables

(i)a Equations of Gray [1]

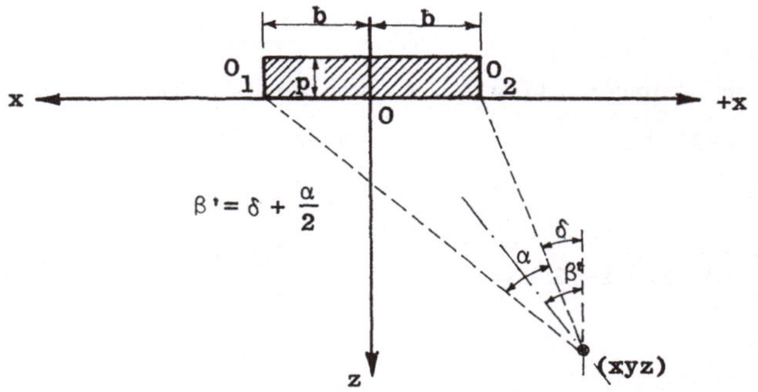

$$\sigma_x = \frac{p}{\pi}\left[\alpha - \sin\alpha \cdot \cos 2\beta'\right]$$

$$\sigma_z = \frac{p}{\pi}\left[\alpha + \sin\alpha \cdot \cos 2\beta'\right]$$

$$\tau_{xz} = \frac{p}{\pi} \sin\alpha \cdot \sin 2\beta'$$

$$\sigma_y = \frac{2p}{\pi m} \cdot \alpha$$

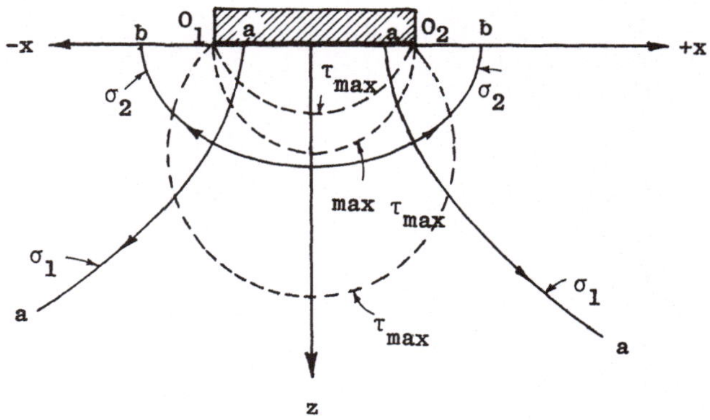

Hauptspannungen

$$\sigma_1 = \frac{p}{\pi}(\alpha + \sin\alpha) \qquad \sigma_2 = \frac{p}{\pi}(\alpha - \sin\alpha)$$

$$\tau_{max} = \frac{p}{\pi}\sin\alpha$$

Ort Konstanten σ_1 und σ_2	Kreise durch O_1 und O_2
Ort Konstanten τ_{max}	Kreise durch O_1 und O_2
Trajektorien von σ_1	Schar konfokaler Hyperbeln (aa)*). Brennpunkte in O_1 und O_2.
Trajektorien von σ_2	Schar konfokaler Ellipsen (bb). Brennpunkte in O_1 und O_2.
Trajektorien von τ_{max}	Zwei orthogonale Scharen: gleichwinklige Spiralen, die die Ellipsen unter 45° schneiden.
Maximum $\tau_{max} = p/\pi$, vorhanden an allen Punkten des Halbkreises durch O_1 und O_2.	
Maximum $\sigma_1 = p$, vorhanden an den Punkten $(x,0)$, $-b < x < b$. Minimum $\sigma_2 = 0$, vorhanden an den Punkten $(x,0)$, $-b < x < b$.	

*) Diese Kurven halbieren den Winkel α.

Principal stresses

$$\sigma_1 = \frac{p}{\pi}(\alpha + \sin\alpha) \qquad \sigma_2 = \frac{p}{\pi}(\alpha - \sin\alpha)$$

$$\tau_{max} = \frac{p}{\pi}\sin\alpha$$

Loci of constant σ_1 and σ_2	Circles passing through O_1 and O_2
Loci of constant τ_{max}	Circles passing through O_1 and O_2
Trajectories of σ_1	Family of confocal hyperbolas, (aa)*. Foci at O_1 and O_2
Trajectories of σ_2	Family of confocal ellipses, (bb). Foci at O_1 and O_2
Trajectories of τ_{max}	Two orthogonal families: equiangular spirals intersecting the ellipses under 45°
Maximum $\tau_{max} = p/\pi$, occurs at all points of the semi-circle through O_1 and O_2	
Maximum $\sigma_1 = p$, occurs at points $(x,0)$, $-b < x < b$. Maximum $\sigma_2 = 0$, occurs at points $(x,0)$, $-b < x < b$.	

* These curves bisect the angle, α, at all points.

Fig. 1.9 Gleichungen fuer Bodenspannungen infolge gleichmaessig verteilter Streifenlast [1]

Equations for stresses in soils due to uniformly distributed strip loading [1]

(i)b Gleichungen nach Szechy [2]

(i)b Equations of Szechy [2]

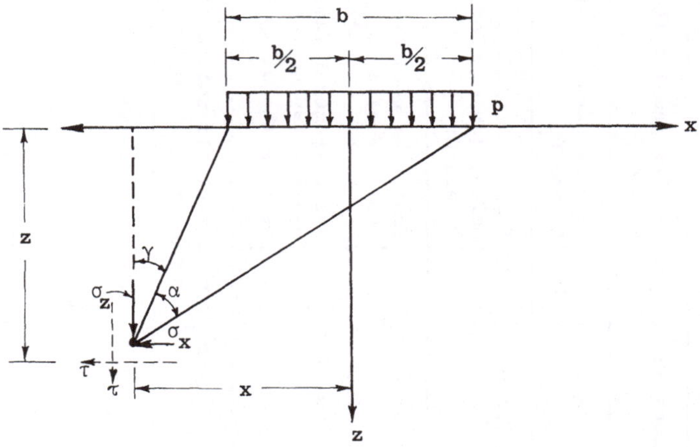

$$\sigma_z = \frac{p}{\pi}\left[\alpha + \sin\alpha \cdot \cos(\alpha + 2\gamma)\right]$$

$$\sigma_x = \frac{p}{\pi}\left[\alpha - \sin\alpha \cdot \cos(\alpha + 2\gamma)\right]$$

$$\tau = \frac{p}{\pi} \cdot \sin\alpha \cdot \sin(\alpha + 2\gamma)$$

Fig. 1.10 Gleichungen fuer Bodenspannungen infolge gleichmaessig verteilter Streifenlast nach Szechy [2]

Equations for stresses in soil due to uniformly distributed strip loading according to Szechy [2]

(i)c Spannungsbeiwerte nach Szechy [2]

(i)c Influence coefficients according to Szechy [2]

$$\sigma_z = p \cdot \frac{1}{\pi}\left[\alpha + \sin\alpha \cdot \cos(\alpha + 2\gamma)\right]$$

$$= p \cdot i_{s1}$$

$$\sigma_x = p \cdot \frac{1}{\pi}\left[\alpha - \sin\alpha \cdot \cos(\alpha + 2\gamma)\right]$$

$$= p \cdot i_{s2}$$

$$\tau = p \cdot \frac{1}{\pi} \cdot \sin\alpha \cdot \sin(\alpha + 2\gamma)$$

$$= p . i_{s3}$$

$\frac{z}{b}$ \ $\frac{x}{b}$	0			0.25			0.50			1.0			1.5			2.0		
	i_{s1}	i_{s2}	i_{s3}	i_{s1}	i_{s2}	i_{s3}	i_{s1}	i_{s2}	i_{s3}	i_{s1}	i_{s2}	i_{s3}	i_{s1}	i_{s2}	i_{s3}	i_{s1}	i_{s2}	i_{s3}
0	1.00	1.00	0	1.00	1.00	0	0.50	0.50	0.32	0	0	0	0	0	0	0	0	0
0.25	0.96	0.45	0	0.90	0.39	0.13	0.50	0.35	0.30	0.02	0.17	0.05	0	0.07	0.01	0.0	0.04	0.0
0.50	0.82	0.18	0	0.74	0.19	0.16	0.48	0.23	0.26	0.08	0.21	0.13	0.02	0.12	0.04	0.0	0.07	0.02
0.75	0.67	0.08	0	0.61	0.10	0.13	0.45	0.14	0.20	0.15	0.22	0.16	0.04	0.14	0.07	0.02	0.10	0.04
1.00	0.55	0.04	0	0.51	0.05	0.10	0.41	0.09	0.16	0.19	0.15	0.16	0.07	0.14	0.10	0.03	0.13	0.05
1.25	0.46	0.02	0	0.44	0.03	0.07	0.37	0.06	0.12	0.20	0.11	0.14	0.10	0.12	0.10	0.04	0.11	0.07
1.50	0.40	0.01	0	0.38	0.02	0.06	0.33	0.04	0.10	0.21	0.08	0.13	0.11	0.10	0.10	0.06	0.10	0.07
1.75	0.35	--	0	0.34	0.01	0.04	0.30	0.03	0.08	0.21	0.06	0.11	0.13	0.09	0.10	0.07	0.09	0.08
2.00	0.31	--	0	0.31	--	0.03	0.28	0.02	0.06	0.20	0.05	0.10	0.13	0.07	0.10	0.08	0.08	0.08
3.00	0.21	--	0	0.21	--	0.02	0.20	0.01	0.03	0.17	0.02	0.06	0.13	0.03	0.07	0.10	0.04	0.07
4.00	0.16	--	0	0.16	--	0.01	0.15	--	0.02	0.14	0.01	0.03	0.12	0.02	0.05	0.10	0.03	0.05
5.00	0.13	--	0	0.13	--	--	0.12	--	--	0.12	--	--	0.11	--	--	0.09	--	--
6.00	0.11	--	0	0.10	--	--	0.10	--	--	0.10	--	--	0.10	--	--	0.10	--	--

Tabelle 1.5 Spannungsbeiwerte fuer gleichmaessig verteilte Streifenbelastung nach Szechy [2]

Table 1.5 Influence coefficients for uniformly distributed strip loading according to Szechy [2]

(ii) Quellen

References

[1]	Szechy (1963)	Der Grundbau, Springer Verlag Bd I, S. 257
[1]	Gray (1936)	Stress distribution in elastic solids Proc. 1. ICSMFE Cambridge, Vol. 2 PP 157

b) Halbraum nach Froehlich

i) Formeln/Tafel/Tabelle

(i)a Gleichungen nach Froehlich [1]

b) Half-space according to Froehlich

(i) Formulae/Chart/Table

(i)a Equations of Froehlich [1]

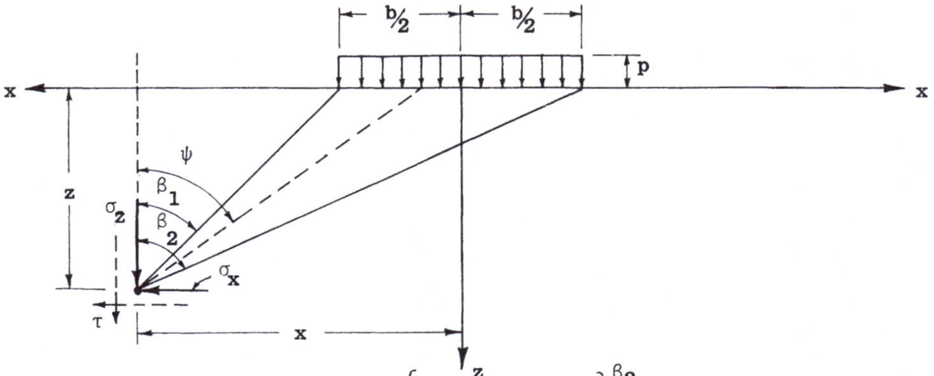

$$\sigma_z = \frac{3}{4}\,p\left\{\sin\psi - \frac{1}{3}\sin^3\psi\right\}_{\beta_1}^{\beta_2}$$

$$\sigma_x = \frac{p}{4}\left\{\sin^3\psi\right\}_{\beta_1}^{\beta_2}$$

$$\tau = \frac{p}{4}\left\{\cos^3\psi\right\}_{\beta_1}^{\beta_2}$$

$\nu = 4$

$$\sigma_z = \frac{2}{3}\,\frac{p}{\pi}\left\{\sin\psi\cdot\cos^3\psi + \frac{3}{2}(\sin\psi\ \cos\psi + \psi)\right\}_{\beta_1}^{\beta_2}$$

$$\sigma_x = -\,\frac{2}{3}\,\frac{p}{\pi}\left\{\sin\psi\cdot\cos^3\psi + 2(\sin\psi\cdot\cos\psi + \psi)\right\}_{\beta_1}^{\beta_2}$$

$$\tau = -\,\frac{2p}{3\pi}\left\{\cos^4\psi\right\}_{\beta_1}^{\beta_2}$$

$\nu = 5$

$$\sigma_z = \frac{15}{16}\,p\left\{\sin\psi - \frac{2}{3}\sin^3\psi + \frac{1}{5}\sin^5\psi\right\}_{\beta_1}^{\beta_2}$$

$$\sigma_x = \frac{15}{16}\,p\left\{\frac{1}{3}\sin^3\psi - \frac{1}{5}\sin^5\psi\right\}_{\beta_1}^{\beta_2}$$

$$\tau = -\,\frac{3}{16}\,p\left\{\cos^5\psi\right\}_{\beta_1}^{\beta_2}$$

$\nu = 6$

Fig. 1.11 Gleichungen fuer Bodenspannungen infolge gleichmaessig verteilter Streifenlast nach Froehlich [1]

Equations for stresses in soil due to uniformly distributed strip loading according to Froehlich [1]

(i)b Spannungsbeiwerte nach Ladanij [2]

(i)b Influence coefficients according to Ladanij [2]

$$\sigma_z = p\left\{i_z(\beta_2) - i_z(\beta_1)\right\}$$

$$\sigma_x = p\left\{i_x(\beta_2) - i_x(\beta_1)\right\}$$

$\nu = 4$

tan β	i_z	i_x
0.0	0.0000	0.0000
0.1	0.0744	0.0002
0.2	0.1432	0.0019
0.3	0.2096	0.0059
0.4	0.2657	0.0128
0.5	0.3131	0.0224
0.6	0.3518	0.0340
0.7	0.3829	0.0471
0.8	0.4076	0.0609
0.9	0.4269	0.0748
1.0	0.4419	0.0884
1.1	0.4537	0.1013
1.2	0.4628	0.1133
1.3	0.4700	0.1245
1.4	0.4756	0.1347
1.5	0.4800	0.1440
1.6	0.4836	0.1525
1.7	0.4864	0.1601
1.8	0.4886	0.1670
1.9	0.4905	0.1732
2.0	0.4919	0.1789
2.2	0.4942	0.1886
2.4	0.4957	0.1966
2.6	0.4967	0.2033
2.8	0.4975	0.2088
3.0	0.4981	0.2135
3.5	0.4989	0.2222
4.0	0.4993	0.2283
4.5	0.4996	0.2326
5.0	0.4997	0.2357
5.5	0.4998	0.2381
6.0	0.4999	0.2399
7.0	0.4999	0.2425
8.0	0.4999	0.2443
9.0	0.5000	0.2454
10.0	0.5000	0.2463
+∞	0.5000	0.2500

Tabelle 1.6 Spannungsbeiwerte i_z und i_x bei $\nu = 4$ fuer Halbraum nach Ladanij [2]
Influence coefficients i_z and i_x by $\nu = 4$ for a half-space according to Ladanij [2]

(i) c Kurven nach Soos [3] (i) c Curves according to Soos [3]

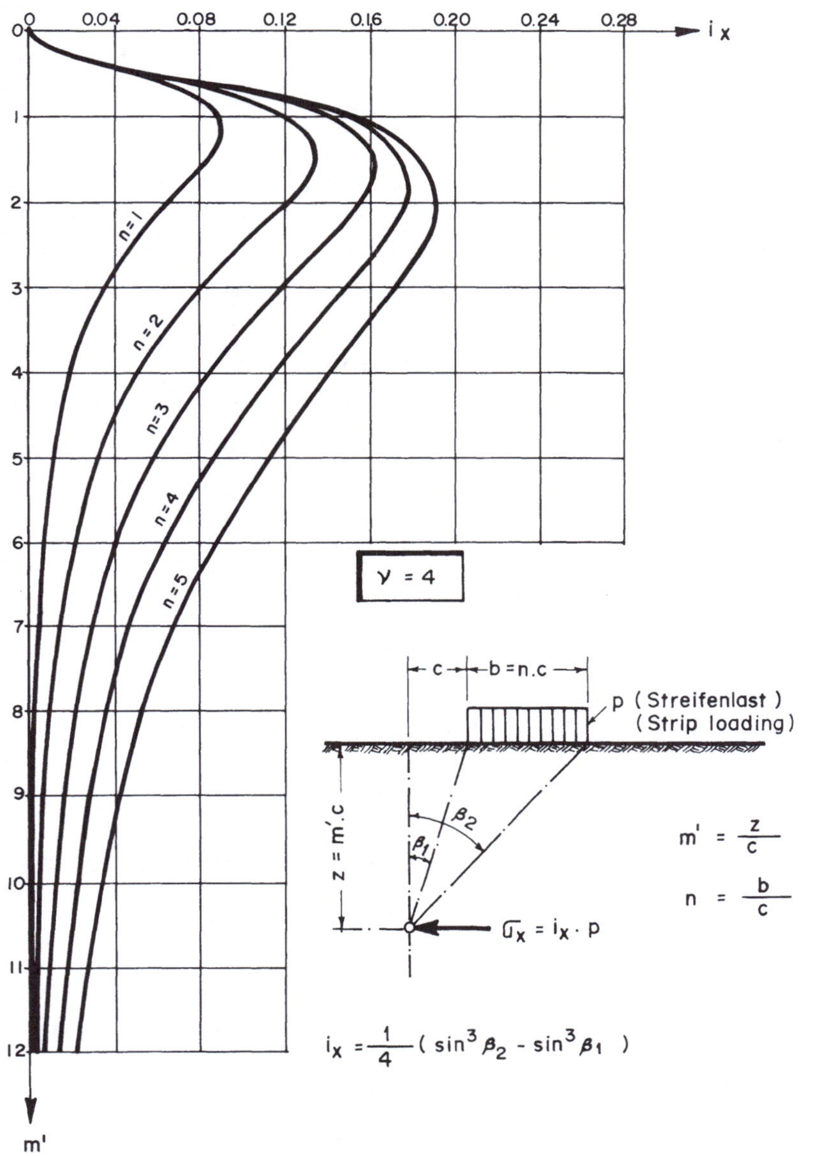

Fig. 1.12 Kurven zur Ermittlung der waagerechten Spannungen, σ_x, bei $\nu = 4$ nach Soos [3]

Curves for determining the horizontal stresses σ_x by $\nu = 4$ according to Soos [3]

(ii) Anwendungsbeispiel

Gegeben

Streifenlast mit b = 4.0 m,
p = 200 kN/m^2

Gesucht

Die waagerechte spannung σ_x, im punkt x = 4,0 m; z = 4.0 m bei ν = 4

Loesung

(ii)a Nach den Formeln von Froehlich (Fig. 1.11)

(ii) Example

Given

Strip loading with b = 4.0 m,
p = 200 kN/m^2

Required

The horizontal stress, σ_x, for a point located at x = 4.0 m, z = 4.0m for the case ν = 4

Solution

(ii)a According to the equations of Froehlich (Fig. 1.11)

$$\tan\beta_1 = \frac{2}{4} = 0.5 \Rightarrow \beta_1 = 26.6^o$$

$$\tan\beta_2 = 6/4 = 1.5 \Rightarrow \beta_2 = 56.3^o$$

$$\sigma_z = \frac{3}{4}p\left\{ \sin\psi - \frac{1}{3}\sin^3\psi \right\}_{\beta_1}^{\beta_2}$$

$$= \frac{3}{4}p\left[(\sin\beta_2 - \frac{1}{3}\sin^3\beta_2) - (\sin\beta_1 - \frac{1}{3}\sin^3\beta_1)\right]$$

$$= \frac{3}{4}(200)\left[(\sin 56.3 - \frac{1}{3}\sin^3 56.3) - (\sin 26.6 - \frac{1}{3}\sin^3 26.6)\right]$$

$$= 24.3 \text{ kN/m}^2$$

(ii)b Nach der Tabelle von Ladanij (Tabelle 1.6)

(ii)b According to the Table of Ladanij (Table 1.6)

$$\tan\beta_1 = 0.5 \quad : \quad i_x = 0.0224$$

$$\tan\beta_2 = 1.5 \quad : \quad i_x = 0.1440$$

$$\sigma_x = p\left[i_x(\beta_2) - i_x(\beta_1) \right]$$

$$= 200\left[0.1440 - 0.0224 \right] = 24.3 \text{ kN/m}^2$$

(ii)c Nach der Kurven von Soos (Fig.1.12)

$c = x - b/2 = 4 - 2 = 2.0$ m

$n = b/c = \frac{4}{2} = 2.0$ m

$m' = z/c = 4/2 = 2.0$ m

Daraus ergibt sich $i_x = 0.121$

$\sigma_x = i_x \cdot p = 0.121(200) = 24.2$ kN/m^2

(ii)c According to the curves of Soos (Fig. 1.12)

$c = x - b/2 = 4 - 2 = 2.0$ m

$n = b/c = \frac{4}{2} = 2.0$ m

$m' = z/c = 4/2 = 2.0$ m

From which one gets $i_x = 0.121$

$\sigma_x = i_x \cdot p = 0.121(200) = 24.2$ kN/m^2

(iii) Quellen

References

[1] Froehlich (1934) Druckverteilung im Baugrunde
Wien, Verlag von Julius Springer

[2] Ladanij (1955) Numerical calculation of stresses in soil under a uniformly loaded infinite strip
Proc. Yuog. Soc. Soil Mech. Found. Eng.
No. 11, PP 13.

[3] Soos (1967) Abstuetzen von Baugruben spundwaenden.
Baugrundkursus Muenchen, 1967.
Unveroeffentlicht

1.1.1.3.2 Parabolisch verteilt symmetrische Last von unendlicher Laenge

a) Halbraum nach Boussinesq

(i) Formeln/Tabellen

(i)a Gleichungen nach Fischer [1]

1.1.1.3.2 Parabolic distributed symmetrical load with infinite length

a) Half-space according to Boussinesq

(i) Formulae/Tables

(i)a Equations of Fischer [1]

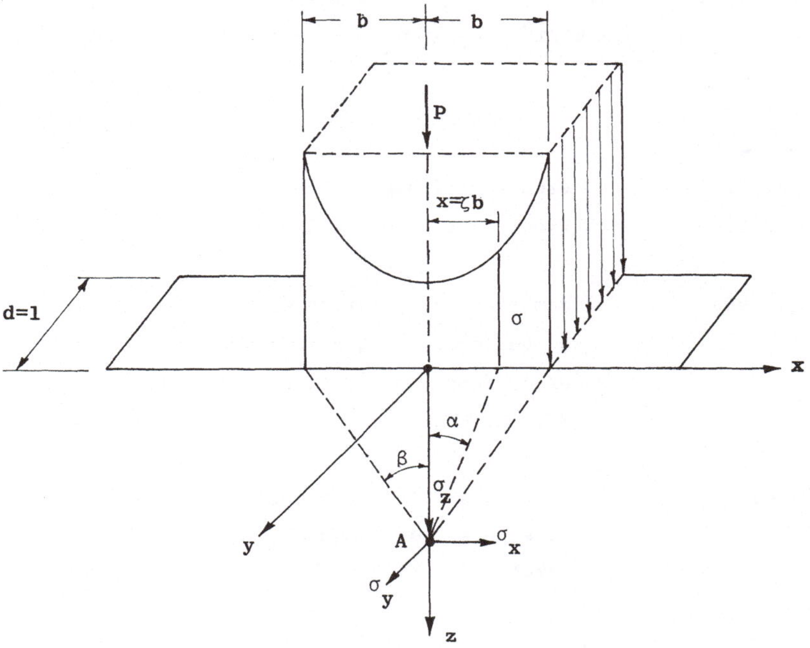

$$p = \frac{P}{2b}\ ; \quad \sigma = \frac{2}{\pi}\frac{p \cdot b}{\sqrt{b^2 - x^2}} = \frac{2}{\pi}\frac{p}{\sqrt{1-\zeta^2}}$$

$$\sigma_z = \frac{2}{\pi}\, p \cdot \sin\beta(1 + \cos^2\beta)$$

$$\sigma_x = \frac{2}{\pi} \cdot p \cdot \sin^3\beta$$

$$\sigma_y = \frac{4}{\pi \cdot m} \cdot p \cdot \sin\beta$$

$$\tau_{xy} = 0$$

m = Poissonzahl
= Poisson's number

Fig. 1.13 Gleichungen fuer Bodenspannungen im Punkt A eines starren mittig belasteten Streifens von unendlicher Laenge nach Fischer [1]

Equation for stresses in soil at point A of a rigid centrically loaded strip with infinite length according to Fischer[1]

(i)b Spannungsbeiwerte nach Fischer [1]

(i)b Influence coefficients according to Fischer [1]

$$\sigma_z = p\cdot(1+\mu)\frac{2}{\pi}.\sin\beta\,[1 - 2\mu + \cos^2\beta]$$
$$= p\cdot i_F$$

$\frac{z}{b}$	i_F			
	μ = 0.50	μ = 0.33	μ = 0.25	μ = 0.00
0	0.000	0.283	0.398	0.637
1	0.338	0.500	0.563	0.675
2	0.342	0.430	0.463	0.512
3	0.272	0.331	0.352	0.383
4	0.218	0.262	0.278	0.300
5	0.180	0.216	0.228	0.245
6	0.153	0.182	0.193	0.206
7	0.132	0.158	0.167	0.178
8	0.117	0.139	0.147	0.157
9	0.104	0.124	0.131	0.140
10	0.094	0.112	0.118	0.126
$\mu = \frac{1}{m}$				

Tabelle 1.7 Spannungsbeiwerte nach Fischer [1]

Table 1.7 Influence coefficients according to Fischer [1]

(iii) Quelle

Reference

[1] Fischer (1957) - Zur Berechnung der Setzung eines starren, mittig belasteten Plattenstreifens auf geschichteter Unterlage
Beton und Stahlbetonbau 52, H. 10

1.1.1.3.3 Unendlich breite einseitig unendlich lange Gleichlast

a) Halbraum nach Boussinesq

i) Formeln/Tabellen

ia) Gleichungen nach Gray [2]

1.1.1.3.3 Uniformly distributed load covering a half infinite plane

a) Half-space according to Boussinesq

i) Formulae/Tables

ia) Equations of Gray [2]

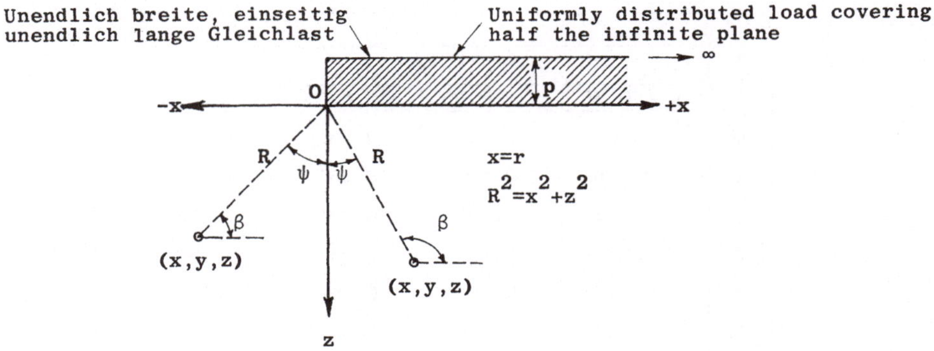

$$\sigma_x = \frac{p}{\pi}\left(\beta - \frac{zx}{R^2}\right)$$

$$\sigma_z = \frac{p}{\pi}\left(\beta + \frac{xz}{R^2}\right)$$

$$\tau_{xz} = -\frac{p}{\pi}\cdot\sin^2\beta$$

$$\sigma_y = \frac{2p}{\pi m}\beta$$

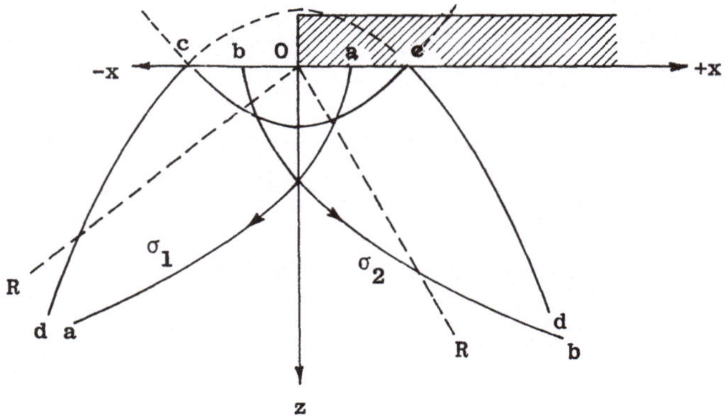

Hauptspannungen

$$\sigma_1 = \frac{p}{\pi}(\beta + \sin\beta)$$

$$\sigma_2 = \frac{p}{\pi}(\beta - \sin\beta)$$

$$\tau_{max} = \frac{p}{\pi}\sin\beta$$

Ort konstanten σ_1 und σ_2	Radien, OR.
Ort konstanten τ_{max}	Radien, OR.
Trajektorien von σ_1	Schar konfokaler Parabeln (aa)+. Brennpunkt in O, horizontale Achse.
Trajektorien von σ_2	Schar konfokaler Parabeln (bb). Brennpunkt in O, horizontale Achse.
Die zwei scharen sind orthogonal.	
Trajektorien von τ_{max}	Zwei orthogonale Scharen von Parabeln (cc) und (dd). Brennpunkt in O, senkrechte Achse.
Maximum $\tau_{max} = p/\pi$, vorhanden an den Punkten $(0,z)\, z>0$	
Maximum $\sigma_1 = p$, " " " " $(x,0)\, x>0$	
Minimum $\sigma_2 = 0$, " " " " $(x,0)\, x>0$	

+ Diese Kurven halbieran den Winkel β.

Principal stresses

$$\sigma_1 = \frac{p}{\pi}(\beta + \sin\beta)$$

$$\sigma_2 = \frac{p}{\pi}(\beta - \sin\beta)$$

$$\tau_{max} = \frac{p}{\pi}\sin\beta$$

Loci of constant σ_1 and σ_2	Radial lines, OR.
Loci of constant τ_{max}	Radial lines, OR.
Trajectories of σ_1	Family of confocal parabolas (aa)*. Focus at O. Horizontal axis.
Trajectories of σ_2	Family of confocal parabolas (bb). Focus at O. Horizontal axis.
The two	families are orthogonal.
Trajectories of τ_{max}	Two orthogonal families of parabolas, (cc) and (cd). Focus at O. Vertical axis.
Maximum $\tau_{max} = p/\pi$, occurs at point $(0,z)$ $z > 0$	
Maximum $\sigma_1 = p$, " " " $(x,0)$ $x > 0$	
Minimum $\sigma_2 = 0$, " " " $(x,0)$ $x < 0$	

* These curves bisect the angle β, at all points.

Fig. 1.14 Gleichungen fuer Bodenspannungen infolge unendlich breiter einseitig unendlich langer Gleichlast [2]

Equation for stresses in soil due to uniformly distributed load covering half infinite plane [2]

ib) Einflussbeiwerte nach Barber [1]

ib) Influence coefficients of Barber [1]

$\sigma_x = p.i_x$

$\sigma_z = p.i_z$; $\tau = p.i_\tau$

Beiwert / Factor	Koordinaten / Coordinates	-10	-9	-8	-7	-6	-5	-4	-3	-2	-1	0
i_z	0	0.000	0.000	0.000	0.000	0.000	0.000	0.000	0.000	0.000	0.000	0.5
	1	0.000	0.000	0.000	0.001	0.001	0.002	0.003	0.007	0.020	0.019	0.5
	2	0.002	0.002	0.003	0.005	0.007	0.011	0.020	0.040	0.091	0.225	0.5
	3	0.005	0.007	0.010	0.014	0.020	0.032	0.052	0.091	0.166	0.302	0.5
	4	0.011	0.015	0.020	0.028	0.040	0.059	0.091	0.142	0.225	0.347	0.5
	5	0.020	0.026	0.035	0.044	0.065	0.091	0.130	0.188	0.269	0.276	0.5
	6	0.032	0.040	0.052	0.068	0.091	0.122	0.166	0.225	0.302	0.396	0.5
	7	0.045	0.056	0.071	0.091	0.117	0.152	0.198	0.256	0.327	0.410	0.5
	8	0.060	0.073	0.091	0.114	0.142	0.179	0.225	0.281	0.347	0.421	0.5
	9	0.075	0.091	0.110	0.135	0.166	0.203	0.249	0.302	0.363	0.430	0.5
	10	I,I91	0.108	0.130	0.156	0.188	0.225	0.269	0.320	0.376	0.437	0.5

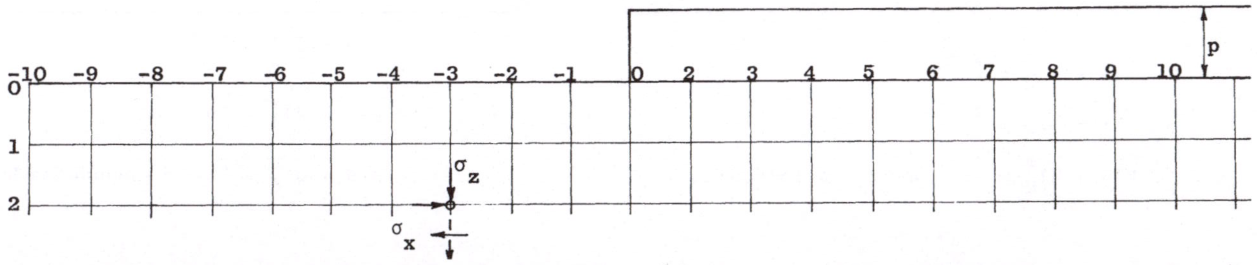

Beiwert / Factor	Koord-inaten / Coord-inates	1	2	3	4	5	6	7	8	9	10
i_z	0	1.000	1.000	1.000	1.000	1.000	1.000	1.000	1.000	1.000	1.000
	1	0.909	0.980	0.993	0.997	0.998	0.999	0.999	1.000	1.000	1.000
	2	0.775	0.909	0.960	0.980	0.989	0.993	0.995	0.997	0.998	0.998
	3	0.698	0.834	0.909	0.948	0.968	0.980	0.986	0.990	0.993	0.995
	4	0.653	0.775	0.858	0.909	0.941	0.960	0.972	0.980	0.985	0.989
	5	0.624	0.731	0.812	0.870	0.909	0.935	0.953	0.965	0.974	0.980
	6	0.604	0.698	0.775	0.834	0.878	0.919	0.932	0.948	0.960	0.968
	7	0.590	0.673	0.744	0.802	0.848	0.883	0.908	0.929	0.944	0.955
	8	0.579	0.653	0.719	0.775	0.821	0.858	0.886	0.909	0.927	0.940
	9	0.570	0.637	0.698	0.751	0.797	0.834	0.865	0.890	0.909	0.925
	10	0.563	0.624	0.680	0.731	0.775	0.812	0.844	0.870	0.892	0.909

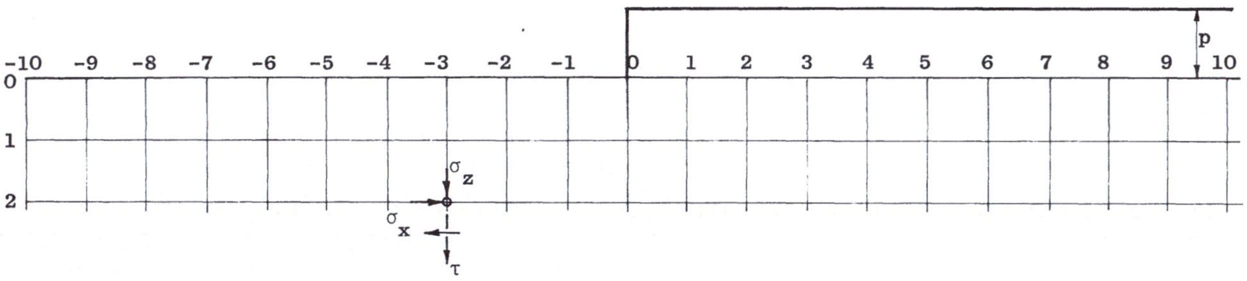

Beiwert / Factor	Koordinaten / Coordinates	-10	-9	-8	-7	-6	-5	-4	-3	-2	-1	0
i_x	0	0.000	0.000	0.000	0.000	0.000	0.000	0.000	0.000	0.000	0.000	0.5
	1	0.063	0.070	0.079	0.090	0.104	0.124	0.153	0.198	0.275	0.409	0.5
	2	0.124	0.137	0.153	0.173	0.198	0.231	0.275	0.334	0.409	0.480	0.5
	3	0.180	0.198	0.219	0.244	0.275	0.313	0.358	0.409	0.460	0.493	0.5
	4	0.231	0.251	0.275	0.302	0.334	0.370	0.409	0.448	0.480	0.497	0.5
	5	0.275	0.297	0.321	0.348	0.378	0.409	0.440	0.468	0.489	0.498	0.5
	6	0.313	0.334	0.358	0.383	0.409	0.435	0.460	0.480	0.493	0.499	0.5
	7	0.343	0.364	0.386	0.409	0.432	0.453	0.472	0.486	0.496	0.499	0.5
	8	0.370	0.389	0.409	0.429	0.448	0.465	0.480	0.490	0.497	0.500	0.5
	9	0.391	0.409	0.427	0.444	0.460	0.474	0.485	0.493	0.498	0.500	0.5
	10	0.409	0.425	0.440	0.455	0.468	0.480	0.489	0.495	0.498	0.500	0.5

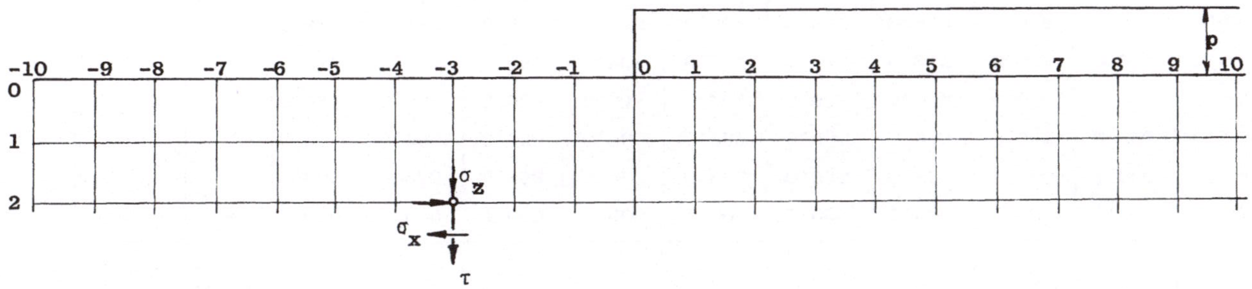

Beiwert / Factor	Koordinaten / Coordinates	1	2	3	4	5	6	7	8	9	10
	0	1.000	1.000	1.000	1.000	1.000	1.000	1.000	1.000	1.000	1.000
	1	0.591	0.725	0.802	0.847	0.876	0.896	0.910	.0921	0.930	0.937
	2	0.520	0.591	0.666	0.725	0.769	0.802	0.827	0.847	0.863	0.876
	3	0.507	0.540	0.591	0.642	0.687	0.725	0.756	0.781	0.802	0.820
	4	0.503	0.520	0.552	0.591	0.630	0.666	0.698	0.725	0.749	0.769
i_x	5	0.502	0.511	0.532	0.560	0.591	0.622	0.652	0.679	0.703	0.725
	6	0.501	0.507	0.520	0.540	0.565	0.591	0.617	0.642	0.666	0.687
	7	0.501	0.504	0.514	0.528	0.547	0.568	0.591	0.614	0.636	0.657
	8	0.500	0.503	0.510	0.520	0.535	0.552	0.571	0.591	0.611	0.630
	9	0.500	0.502	0.507	0.515	0.526	0.540	0.556	0.573	0.591	0.608
	10	0.500	0.502	0.505	0.511	0.520	0.532	0.545	0.560	0.575	0.591

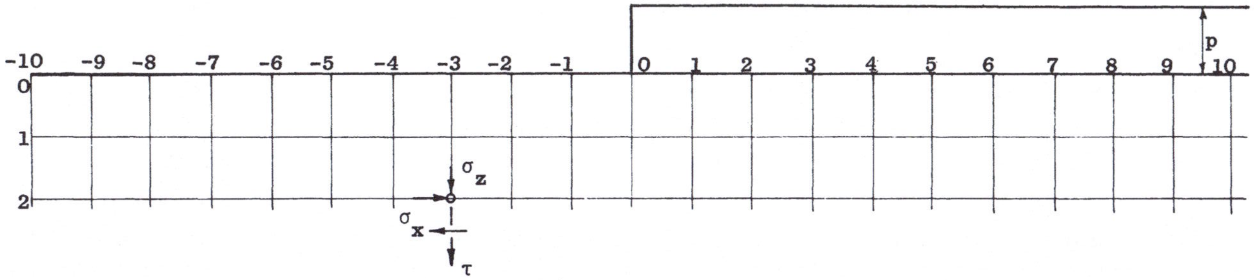

Beiwert / Factor	Koordinaten / Coordinates	-10	-9	-8	-7	-6	-5	-4	-3	-2	-1	0
i_τ	0	0.000	0.000	0.000	0.000	0.000	0.000	0.000	0.000	0.000	0.000	0.318
	1	0.003	0.004	0.005	0.006	0.008	0.012	0.019	0.032	0.064	0.159	0.318
	2	0.012	0.015	0.019	0.024	0.032	0.044	0.064	0.098	0.159	0.255	0.318
	3	0.026	0.032	0.039	0.049	0.064	0.084	0.115	0.159	0.220	0.287	0.318
	4	0.044	0.052	0.064	0.078	0.098	0.124	0.159	0.204	0.255	0.300	0.318
	5	0.064	0.075	0.089	0.108	0.130	0.159	0.194	0.234	0.275	0.306	0.318
	6	0.084	0.098	0.115	0.135	0.159	0.188	0.220	0.255	0.287	0.310	0.318
	7	0.105	0.120	0.138	0.159	0.184	0.211	0.240	0.269	0.295	0.312	0.318
	8	0.124	0.141	0.159	0.180	0.204	0.229	0.255	0.279	0.300	0.314	0.318
	9	0.142	0.159	0.178	0.198	0.220	0.243	0.266	0.287	0.304	0.315	0.318
	10	0.159	0.176	0.194	0.214	0.234	0.255	0.275	0.292	0.306	0.315	0.318

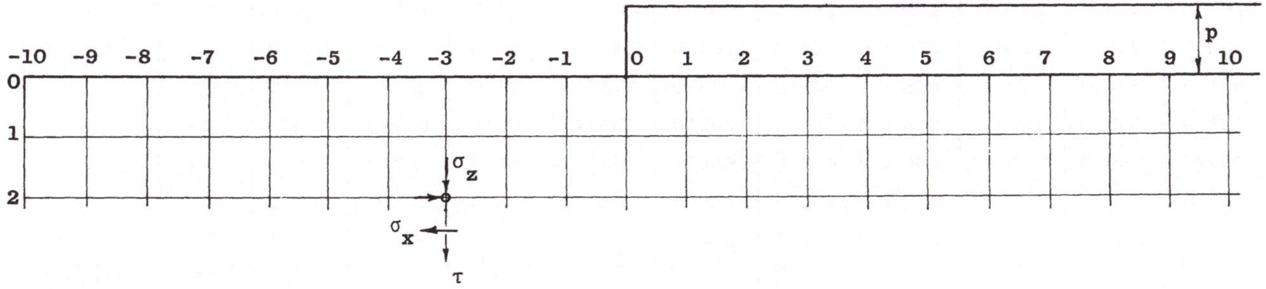

Beiwert Factor	Koord-inaten Coord-inates	1	2	3	4	5	6	7	8	9	10
	0	0.000	0.000	0.000	0.000	0.000	0.000	0.000	0.000	0.000	0.000
	1	0.159	0.064	0.032	0.019	0.012	0.009	0.006	0.005	0.004	0.003
	2	0.255	0.159	0.098	0.064	0.044	0.032	0.024	0.019	0.015	0.012
	3	0.287	0.220	0.159	0.115	0.084	0.064	0.049	0.039	0.032	0.026
	4	0.300	0.255	0.204	0.159	0.124	0.098	0.078	0.064	0.052	0.044
i_τ	5	0.306	0.275	0.234	0.194	0.159	0.130	0.108	0.089	0.075	0.064
	6	0.310	0.287	0.255	0.220	0.188	0.159	0.135	0.115	0.098	0.084
	7	0.312	0.295	0.269	0.240	0.211	0.184	0.159	0.138	0.120	0.105
	8	0.314	0.300	0.279	0.255	0.229	0.204	0.180	0.159	0.141	0.124
	9	0.315	0.304	0.287	0.266	0.243	0.220	0.198	0.178	0.159	0.142
	10	0.315	0.306	0.292	0.275	0.255	0.234	0.214	0.194	0.176	0.159

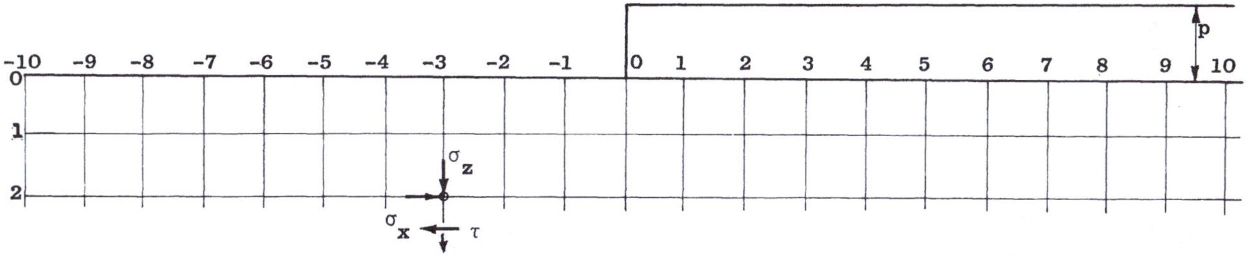

Tabelle 1.8 Einflussbeiwerte fuer Spannungen unter unendlich breiter einseitig langer Gleichlast [1]

Table 1.8 Influences coefficients for stresses under long strip with semi-infinite loaded width [1]

(iii) Quellen

References

[1] Barber (1952): Use of direct shear tests in highway design. Symposium on direct shear testing of soils ASTM No. 131, PP. 19

[2] Gray (1936): Stress distribution in elastic solids. Proc. 1. ICSMFE, Cambridge Vol. 2, PP. 157.

1.1.1.3.4 Unendlisch breite, einseitig unendlich lange Dreieckslast

a) Halbraum nach Boussinesq

i) Formeln

1.1.1.3.4 Triangular loading with infinite width and length

a) Half-space according to Boussinesq

i) Formulae

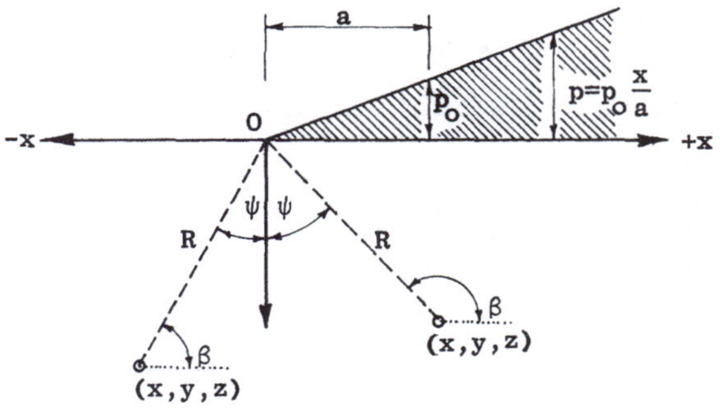

$$\sigma_x = \frac{p_o}{\pi a}\left[x\beta - z - 3z\ln R\right] \qquad \sigma_z = \frac{p_o}{\pi a}\left[x\beta + z\right]$$

$$\sigma_y = \frac{2p_o}{\pi am}\left[x\beta - z\ln R\right] \qquad \tau_{xz} = \frac{p_o}{\pi a} z\beta$$

Hauptspannungen:

Principal stresses:

$$\sigma_1 = \frac{p_o}{\pi a}(x\beta - z\ln R) + \frac{p_o z}{\pi a}\sqrt{(1 + \ln R)^2 + \beta^2}$$

$$\sigma_2 = \frac{p_o}{\pi a}(x\beta - z\ln R) - \frac{p_o z}{\pi a}\sqrt{(1 + \ln R)^2 + \beta^2}$$

$$\tau_{max} = \frac{p_o z}{\pi a}\sqrt{(1 + \ln R)^2 + \beta^2}$$

Fig. 1.15 Gleichungen fuer Bodenspannungen infolge unendlich breiter, einseitig unendlich langer Dreieckslast [1]

Equations for stresses in soil due to triangular loading with infinite width and length [1]

(ii) Quelle

Reference

[1] Gray (1936) - Stress Distribution in elastic solids
Proc. 1. ICSMFE Cambridge, Vol 2. PP 157

1.1.1.3.5 Uendlich breite, eiseitig unendlich lange Trapezlast

a) Halbraum nach Boussinesq

i) Formeln/Tabelle

ia) Gleichungen nach Gray [2]

1.1.1.3.5 Trapezoidal Loading with infinite width and length

a) Half-space according to Boussinesq

i) Formulae/Table

ia) Equations of Gray [2]

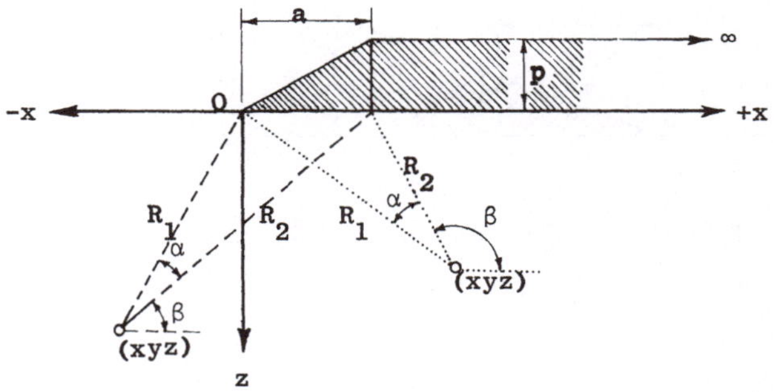

$$\sigma_x = \frac{p}{\pi a}\left(a\beta + x\alpha + 2z \ln \frac{R_2}{R_1}\right) \qquad \sigma_z = \frac{p}{\pi a}(a\beta + x\alpha)$$

$$\sigma_y = \frac{2p}{\pi am}\left(a\beta + x\alpha + z \ln \frac{R_2}{R_1}\right) \qquad \tau_{xz} = -\frac{p}{\pi a}\, z\alpha$$

Hauptspannungen:

Principal stresses:

$$\sigma_1 = \frac{p}{\pi a}\left[\left(a\beta + x\alpha + z \ln \frac{R_2}{R_1}\right) + z\sqrt{\ln^2 \frac{R_2}{R_1} + \alpha^2} \right]$$

$$\sigma_2 = \frac{p}{\pi a}\left[\left(a\beta + x\alpha + z \ln \frac{R_2}{R_1}\right) - z\sqrt{\ln^2 \frac{R_2}{R_1} + \alpha^2} \right]$$

$$\tau_{max} = \frac{pz}{\pi a}\sqrt{\ln^2 \frac{R_2}{R_1} + \alpha^2}$$

Fig. 1.16 Gleichungen fuer Bodenspannungen infolge unendlich breiter, eiseitig undlich langer Trapezlast [2]

Equations for stresses in soil due to trapezoidal loading with infinite width and length [2]

$\sigma_x = p \cdot i_x$

$\sigma_z = p \cdot i_z \qquad \tau = p \cdot i_\tau$

Beiwert / Factor	Koordinaten / Coordinates	-4	-3	-2	-1	0	1	2	3	4	5	6	7	8
i_z	0	0.000	0.000	0.000	0.000	0.000	0.250	0.500	0.750	1.000	1.000	1.000	1.000	1.000
	1	0.001	0.003	0.005	0.016	0.078	0.262	0.500	0.737	0.922	0.984	0.998	0.998	0.999
	2	0.008	0.015	0.029	0.063	0.147	0.302	0.500	0.698	0.852	0.937	0.971	0.986	0.992
	3	0.024	0.038	0.065	0.116	0.205	0.338	0.500	0.662	0.795	0.884	0.935	0.962	0.977
	4	0.045	0.068	0.104	0.163	0.250	0.366	0.500	0.634	0.750	0.837	0.896	0.932	0.955
	5	0.071	0.100	0.143	0.203	0.285	0.387	0.500	0.613	0.715	0.797	0.858	0.900	0.930
	6	0.097	0.130	0.175	0.237	0.313	0.402	0.500	0.598	0.687	0.763	0.824	0.870	0.903
	7	0.123	0.159	0.206	0.265	0.335	0.415	0.500	0.585	0.665	0.736	0.794	0.841	0.877
	8	0.148	0.186	0.231	0.288	0.352	0.424	0.500	0.576	0.647	0.712	0.768	0.815	0.852
	9	0.171	0.208	0.253	0.307	0.367	0.432	0.500	0.568	0.633	0.693	0.746	0.791	0.829
	10	0.192	0.229	0.274	0.323	0.379	0.439	0.500	0.562	0.621	0.676	0.727	0.771	0.809
i_x	0	0.000	0.000	0.000	0.000	0.000	0.250	0.500	0.750	1.000	1.000	1.000	1.000	1.000
	1	0.108	0.131	0.164	0.220	0.304	0.392	0.500	0.608	0.697	0.780	0.836	0.870	0.892
	2	0.203	0.239	0.285	0.343	0.404	0.454	0.500	0.546	0.596	0.657	0.715	0.762	0.797
	3	0.279	0.318	0.361	0.408	0.449	0.478	0.500	0.521	0.551	0.592	0.638	0.683	0.721
	4	0.337	0.372	0.409	0.443	0.470	0.489	0.500	0.511	0.529	0.556	0.591	0.628	0.664
	5	0.379	0.409	0.438	0.463	0.481	0.493	0.500	0.506	0.518	0.537	0.562	0.591	0.622
	6	0.410	0.434	0.457	0.475	0.489	0.496	0.500	0.504	0.512	0.525	0.543	0.566	0.590
	7	0.431	0.451	0.469	0.483	0.492	0.497	0.500	0.503	0.508	0.518	0.530	0.549	0.569
	8	0.447	0.463	0.477	0.487	0.495	0.497	0.500	0.503	0.506	0.513	0.523	0.537	0.553
	9	0.458	0.472	0.483	0.491	0.496	0.498	0.500	0.502	0.504	0.509	0.517	0.528	0.542
	10	0.467	0.478	0.486	0.493	0.497	0.499	0.500	0.502	0.503	0.507	0.514	0.521	0.532

Beiwert / Factor	Koordinaten / Coordinates	-4	-3	-2	-1	0	1	2	3	4	5	6	7	8
	0	0.000	0.000	0.000	0.000	0.000	0.000	0.000	0.000	0.000	0.000	0.000	0.000	0.000
	1	0.010	0.014	0.024	0.047	0.106	0.162	0.177	0.162	0.105	0.047	0.024	0.014	0.010
	2	0.035	0.049	0.074	0.116	0.177	0.230	0.250	0.231	0.176	0.116	0.074	0.049	0.035
	3	0.068	0.090	0.124	0.169	0.222	0.264	0.283	0.265	0.221	0.169	0.124	0.091	0.068
	4	0.103	0.130	0.165	0.207	0.250	0.283	0.295	0.283	0.250	0.208	0.166	0.130	0.102
i_τ	5	0.134	0.164	0.197	0.234	0.268	0.294	0.303	0.294	0.268	0.234	0.198	0.163	0.134
	6	0.162	0.191	0.221	0.253	0.280	0.301	0.308	0.301	0.280	0.253	0.222	0.191	0.162
	7	0.185	0.212	0.240	0.267	0.289	0.305	0.311	0.305	0.289	0.267	0.240	0.212	0.185
	8	0.205	0.229	0.254	0.277	0.296	0.307	0.312	0.308	0.295	0.277	0.254	0.229	0.205
	9	0.221	0.243	0.264	0.284	0.300	0.310	0.314	0.310	0.299	0.284	0.365	0.243	0.221
	10	0.234	0.255	0.273	0.290	0.303	0.311	0.314	0.311	0.302	0.289	0.273	0.254	0.235

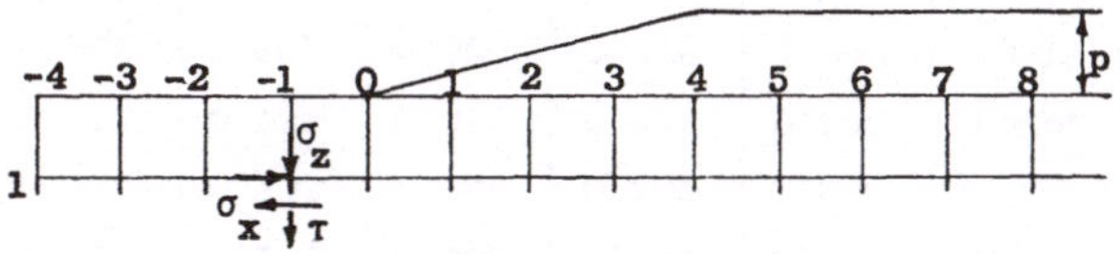

Tabelle 1.9 Einflussbeiwerte fuer unendlich breite, einseitig unendlich lange Trapezlast [1]

Table 1.9 Influence values for trapezoidal loading with infinite width and length [1]

(iii) Quellen

References

[1] Barber (1952) - Use of direct shear tests in highway design. Symposium on direct shear testing of soils, ASTM No. 131, PP 19.

[2] Gray (1936) - Stress distribution in elastic solids. Proc. 1. ICSMFE, Cambridge, Vol. 2, PP 157.

1.1.1.3.6 Unendlich breite, einsetig endlich lange Trapezlast

a) Halbraum nach Boussinesq

i) Formeln/Tafeln

ia) Gleichungen nach Gray [1]

1.1.1.3.6 Trapezoidal Loading with infinite width

a) Half-space according to Boussinesq

i) Formulae/Charts

ia) Equations of Gray [1]

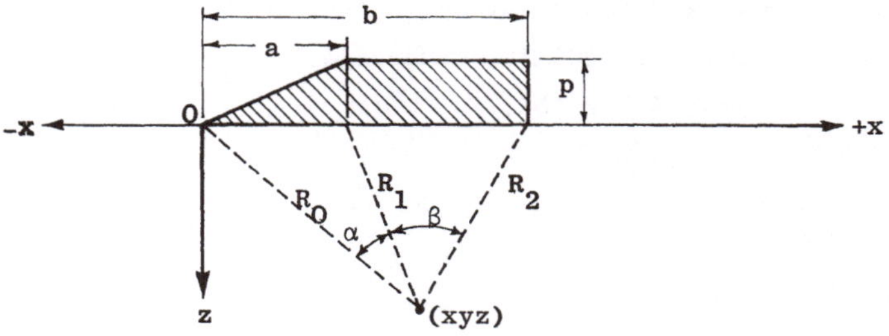

$$\sigma_x = \frac{p}{\pi a}\left[a\beta + x\alpha + \frac{az}{R_2^2}(x-b) + 2z \ \ln \frac{R_1}{R_0}\right]$$

$$\sigma_z = \frac{p}{\pi a}\left[a\beta + x\alpha - \frac{az}{R_2^2}(x-b)\right]$$

$$\sigma_y = \frac{2p}{\pi m a}\left[a\beta + x\alpha + z \ \ln \frac{R_1}{R_0}\right]$$

$$\tau_{xz} = \frac{-p}{\pi a}\left[z\alpha - a \frac{z^2}{R_2^2}\right]$$

Hauptspannungen:

Principal stresses:

$$\sigma_1 = \frac{p}{\pi a}\left[a\beta + x\alpha + z \ln \frac{R_1}{R_2} + z\sqrt{\left(\frac{a}{R_2^2}(x-b) + \ln \frac{R_1}{R_0}\right)^2 + (\alpha - \frac{az}{R_2^2})^2}\right]$$

$$\sigma_2 = \frac{p}{\pi a}\left[a\beta + x\alpha + z \ln \frac{R_1}{R_2} - z\sqrt{\left(\frac{a}{R_2^2}(x-b) + \ln \frac{R_1}{R_0}\right)^2 + (\alpha - \frac{az}{R_2^2})^2}\right]$$

$$\tau_{max} = \frac{pz}{\pi a}\sqrt{\left(\frac{a}{R_2^2}(x-b) + \ln \frac{R_1}{R_0}\right)^2 + (\alpha - \frac{az}{R_2^2})^2}$$

Fig. 1.17 Gleichungen fuer Bodenspannungen infolge unendlich breiter, einseitig endlich langer Trapezlast [1]

Equations for stresses in soil due to trapezoidal loading with infinite width [1]

ib) Tafel nach Herzog [2]

ib) Chart of Herzog [2]

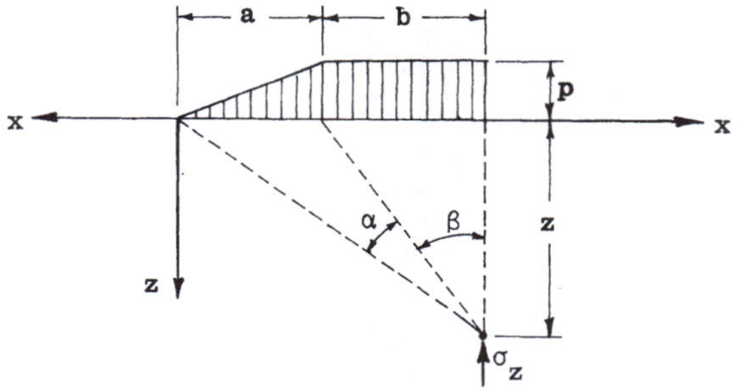

$$\sigma_z = \frac{p.1}{\pi}\left[\beta + (1 + \frac{b}{a})\alpha\right] = p.i_z$$

$$\alpha = \text{arc tan } \frac{a + b}{z} - \text{arc tan } \frac{b}{z}$$

$$\beta = \text{arc tan } \frac{b}{z}$$

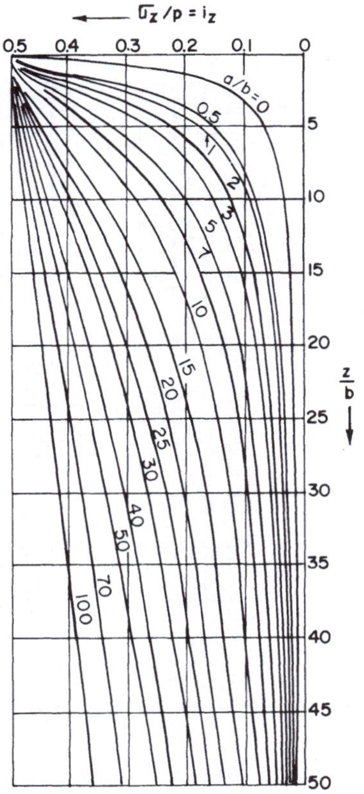

Fig. 1.18 Einflusslinien zur Bestimmung von lotrechter Spannung in der Randvertikalen infolge unendlich breiter einseitig endlich langer Trapezlast [2]

Influence chart for determining vertical stress at the vertical ends due to trapezoidal loading with infinite width [2]

ic) Tafel nach Osterberg[3] ic) Chart of Osterberg[3]

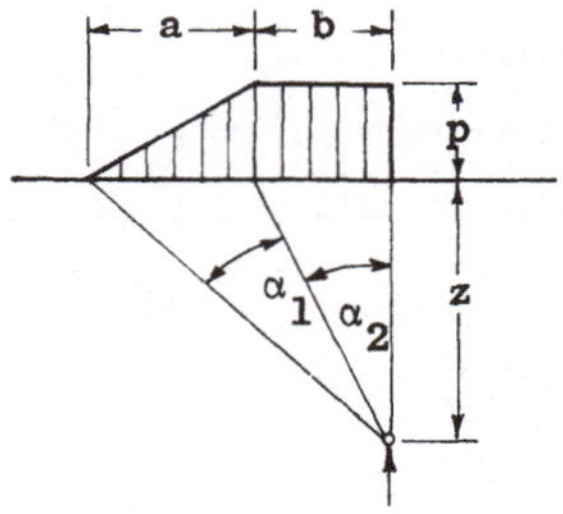

$$\sigma_z = p \cdot I_0$$

$$\sigma_z = \frac{p}{\pi}\left[\left(\frac{a+b}{a}\right)(\alpha_1 + \alpha_2) - \frac{b}{a}\alpha_2\right]$$

$$I_0 = \frac{1}{\pi}\left[\left(\frac{a+b}{a}\right)(\alpha_1 + \alpha_2) - \frac{b}{a}\alpha_2\right]$$

$$I_0 = f\left(\frac{a}{z}, \frac{b}{z}\right) \qquad \boxed{\sigma_z = I_0 \cdot p}$$

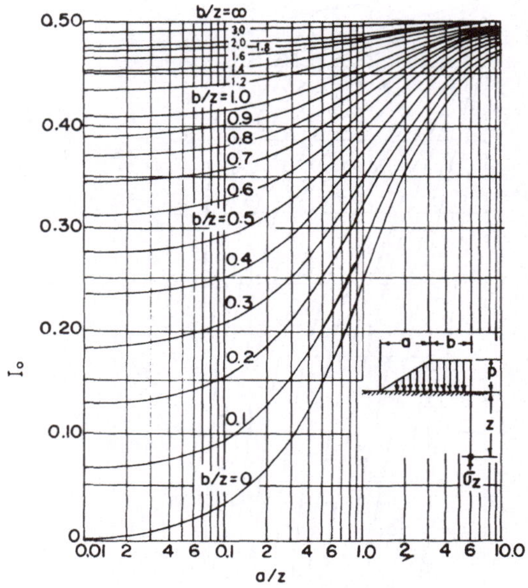

Fig. 1.19 Einflusslinien zur Bestimmung von lotrechter Spannung infolge unendlich breiter, einseitig endlich langer Trapezlast [3]

Influence chart for determining vertical stress due to trapezoidal loading with infinite width [3]

(ii) Quellen

References

[1] Gray (1936) - Stress Distribution in elastic solids
Proc. 1. ICSMFE Cambridge, Vol 2. PP 157

[2] Herzog (1961) - Ueber Setzungen und Setzungsunterschiede
Bauplan u. Baut. 15, S. 139.

[3] Osterberg (1957) - Influence values for vertical stresses in a semi-infinite mass due to an embankment loading.
Proc. 4. ICSMFE, London, Vol. 1, PP 393

1.1.1.3.7 Unendlich breite, Rechtwinklige Dreiecklast

a) Halbraum nach Boussinesq

(i) Formeln/Tabellen

(i)a Gleichungen nach Gray [2]

1.1.1.3.7 Right angle triangular loading with infinite width

a) Half-space according to Boussinesq

(i) Formulae/Tables

(i)a Equations of Gray [2]

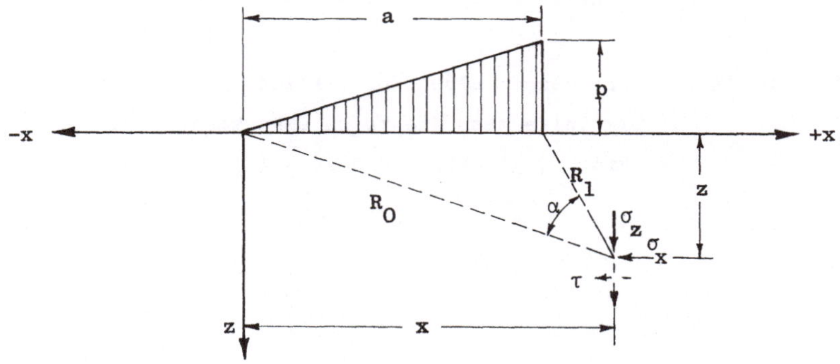

$$\sigma_z = \frac{p}{\pi}\left[\frac{x}{a}\cdot\alpha - \frac{(x-a)}{R_1^2}z\right]$$

$$\sigma_x = \frac{p}{\pi}\left[\frac{x}{a}\cdot\alpha + \frac{2(x-a)}{R_1^2} + 2\cdot\frac{z}{a}\cdot\ln\frac{R_1}{R_o}\right]$$

$$\sigma_y = \frac{2p}{\pi m}\left[\frac{x}{a}\cdot\alpha + \frac{z}{a}\cdot\ln\frac{R_1}{R_o}\right]$$

$$\tau_{xz} = -\frac{p}{\pi}\left[\frac{z}{a}\cdot\alpha - \frac{z^2}{R_1^2}\right]$$

Fig. 1.20 Gleichungen fuer Bodenspannungen infólge unendlich breiter Dreiecklast [2]

Equation for stresses in soil due to right triangular loading with infinite width [2]

(i)b Einflussbeiwerte nach Barber [1]

(i)b Influence coefficients of Barber [1]

$$\sigma_x = p \cdot i_x$$

$$\sigma_z = p.i_z; \quad \tau = p.i_\tau$$

Beiwert Factor	Koordinaten Coordinates	-4	-3	-2	-1	0	1	2	3
i_z	0	0.000	0.000	0.000	0.000	0.000	0.250	0.500	0.750
	1	0.001	0.002	0.004	0.075	0.255	0.480	0.646	0.422
	2	0.005	0.010	0.022	0.052	0.127	0.262	0.409	0.473
	3	0.014	0.024	0.045	0.084	0.153	0.247	0.334	0.360
	4	0.025	0.040	0.064	0.104	0.159	0.224	0.275	0.287
	5	0.036	0.053	0.078	0.112	0.155	0.199	0.231	0.237
	6	0.045	0.062	0.084	0.115	0.147	0.177	0.198	0.202
	7	0.052	0.068	0.089	0.113	0.137	0.159	0.173	0.175
	8	0.057	0.072	0.089	0.109	0.127	0.143	0.153	0.155
	9	0.060	0.073	0.087	0.104	0.118	0.130	0.137	0.138
	10	0.062	0.073	0.086	0.098	0.110	0.119	0.124	0.125
i_x	0	0.000	0.000	0.000	0.000	0.000	0.250	0.500	0.750
	1	0.029	0.041	0.060	0.096	0.151	0.194	0.225	0.199
	2	0.050	0.066	0.087	0.112	0.129	0.120	0.091	0.066
	3	0.060	0.074	0.086	0.095	0.091	0.069	0.040	0.028
	4	0.062	0.070	0.075	0.073	0.061	0.041	0.020	0.014
	5	0.058	0.61	0.060	0.054	0.041	0.025	0.011	0.008
	6	0.052	0.051	0.048	0.040	0.029	0.016	0.007	0.005
	7	0.045	0.042	0.037	0.030	0.020	0.011	0.004	0.004
	8	0.038	0.034	0.029	0.022	0.015	0.007	0.003	0.003
	9	0.031	0.028	0.023	0.017	0.011	0.005	0.002	0.002
	10	0.027	0.023	0.018	0.013	0.008	0.004	0.002	0.002

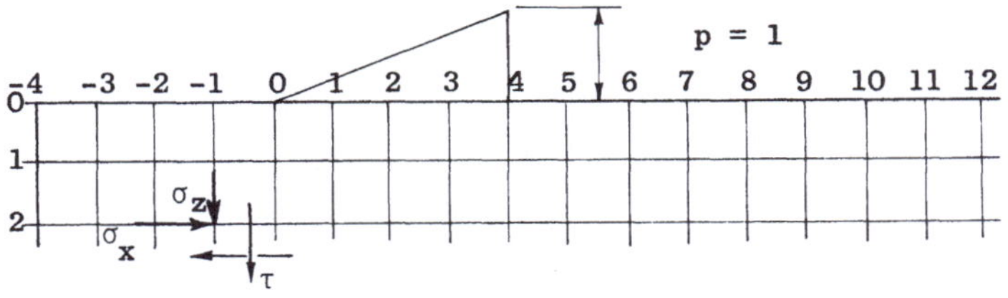

Beiwert Factor	Koordinaten Coordinates	4	5	6	7	8	9	10	11	12
	0	0.000	0.000	0.000	0.000	0.000	0.000	0.000	0.000	0.00
	1	0.422	0.075	0.015	0.005	0.002	0.001	0.001	0.000	0.00
	2	0.352	0.162	0.062	0.026	0.013	0.006	0.003	0.002	0.00
	3	0.295	0.186	0.0101	0.053	0.029	0.016	0.010	0.006	0.00
	4	0.250	0.184	0.121	0.074	0.046	0.028	0.018	0.012	0.00
i_z	5	0.215	0.173	0.127	0.088	0.060	0.041	0.027	0.019	0.01
	6	0.187	0.159	0.126	0.095	0.069	0.050	0.36	0.025	0.01
	7	0.165	0.146	0.121	0.097	0.075	0.057	0.043	0.032	0.02
	8	0.147	0.133	0.115	0.096	0.077	0.061	0.048	0.037	0.02
	9	0.133	0.123	0.109	0.093	0.078	0.064	0.052	0.042	0.03
	10	0.121	0.113	0.103	0.091	0.078	0.065	0.054	0.045	0.03
	0	0.500	0.000	0.000	0.000	0.000	0.000	0.000	0.000	0.00
	1	0.197	0.189	0.111	0.068	0.045	0.032	0.024	0.019	0.01
	2	0.096	0.137	0.124	0.096	0.072	0.055	0.042	0.034	0.02
	3	0.051	0.085	0.098	0.092	0.079	0.065	0.053	0.044	0.03
	4	0.029	0.053	0.071	0.076	0.072	0.065	0.057	0.049	0.04
i_x	5	0.018	0.035	0.051	0.059	0.062	0.060	0.055	0.050	0.04
	6	0.012	0.024	0.036	0.046	0.050	0.052	0.051	0.045	0.04
	7	0.008	0.017	0.026	0.035	0.041	0.044	0.045	0.043	0.04
	8	0.006	0.013	0.020	0.027	0.033	0.037	0.039	0.039	0.03
	9	0.004	0.009	0.015	0.021	0.027	0.031	0.033	0.035	0.03
	10	0.003	0.007	0.012	0.016	0.021	0.026	0.028	0.030	0.03

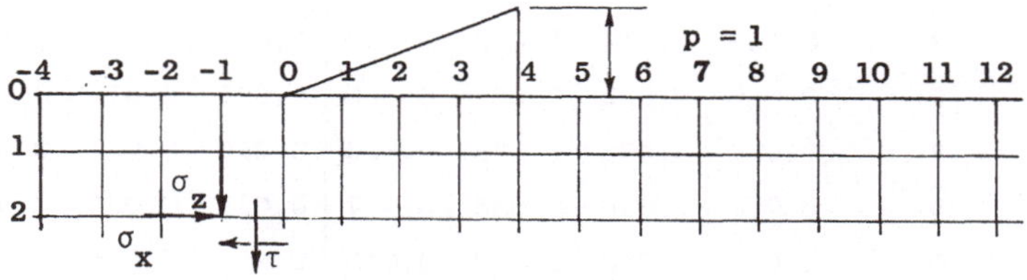

Beiwert Factor	Koordinaten Coordinates	-4	-3	-2	-1	0	1	2	3
i_τ	0	0.000	0.000	0.000	0.000	0.000	0.000	0.000	0.000
	1	0.005	0.008	0.015	0.035	0.087	0.130	0.113	0.003
	2	0.016	0.025	0.042	0.072	0.113	0.132	0.091	0.024
	3	0.029	0.041	0.060	0.085	0.107	0.105	0.063	0.022
	4	0.039	0.052	0.067	0.083	0.091	0.079	0.040	0.017
	5	0.045	0.056	0.065	0.075	0.074	0.060	0.028	0.012
	6	0.047	0.056	0.062	0.065	0.060	0.046	0.021	0.009
	7	0.047	0.053	0.056	0.056	0.049	0.036	0.016	0.007
	8	0.046	0.049	0.050	0.048	0.041	0.028	0.012	0.006
	9	0.043	0.045	0.044	0.041	0.034	0.023	0.010	0.005
	10	0.040	0.041	0.039	0.035	0.028	0.019	0.008	0.004

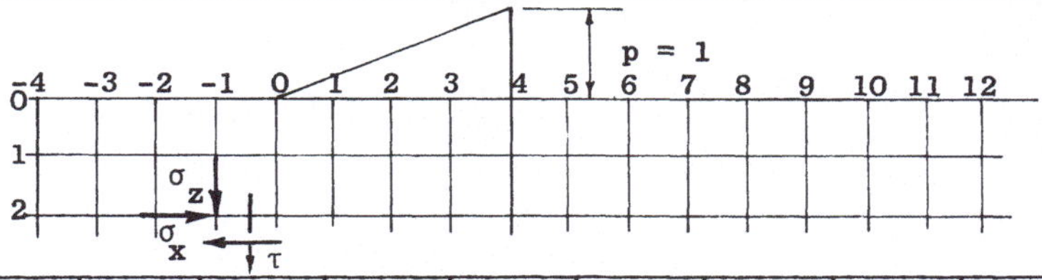

Beiwert Factor	Koordinaten Coordinates	4	5	6	7	8	9	10	11	12
i_τ	0	0.318	0.000	0.000	0.000	0.000	0.000	0.000	0.000	0.000
	1	0.213	0.112	0.040	0.018	0.009	0.005	0.004	0.002	0.002
	2	0.142	0.139	0.085	0.049	0.029	0.018	0.012	0.008	0.006
	3	0.097	0.118	0.096	0.068	0.047	0.032	0.022	0.016	0.012
	4	0.068	0.092	0.089	0.074	0.057	0.043	0.032	0.024	0.019
	5	0.050	0.072	0.077	0.071	0.060	0.048	0.039	0.031	0.024
	6	0.038	0.057	0.065	0.064	0.058	0.050	0.042	0.035	0.029
	7	0.029	0.045	0.055	0.057	0.055	0.049	0.043	0.037	0.032
	8	0.032	0.037	0.046	0.050	0.050	0.047	0.043	0.038	0.033
	9	0.019	0.031	0.039	0.044	0.045	0.044	0.041	0.038	0.034
	10	0.016	0.026	0.033	0.038	0.040	0.040	0.039	0.037	0.034

Tabelle 1.10 Einflussbeiwert fuer unendlich breite, endlich Dreiecklast [1]

Table 1.10 Influence coefficient for right triangular loading with infinite width [1]

(ii) Quellen (s. Seite 59)

References (Ref. PP. 59)

1.1.1.3.8 Unendlich breite, endlich Doppeldreickslast

a) Halbraum nach Boussinesq

(i) Formeln/Tabellen

(i)a Gleichungen nach Gray [2]

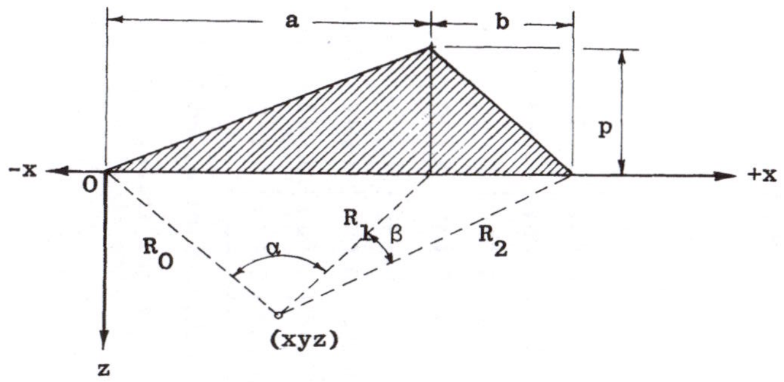

$$\sigma_x = \frac{p}{\pi}\left[\frac{x}{a}\alpha + \frac{a+b-x}{b}\beta + \frac{2z}{a}\ln\frac{R_1}{R_o} + \frac{2z}{b}\ln\frac{R_1}{R_2}\right]$$

$$\sigma_z = \frac{p}{\pi}\left[\frac{x}{a}\alpha + \frac{a+b-x}{b}\beta\right]$$

$$\sigma_y = \frac{2p}{\pi m}\left[\frac{x}{a}\alpha + \frac{a+b-x}{b}\beta + \frac{z}{a}\ln\frac{R_1}{R_o} + \frac{z}{b}\ln\frac{R_1}{R_2}\right]$$

$$\tau_{xz} = \frac{pz}{\pi}\left[\frac{\alpha}{a} - \frac{\beta}{b}\right]$$

1.1.1.3.8 Triangular loading with infinite width

a) Half-space according to Boussinesq

(i) Formulae/Tables

(i)a Equations of Gray [2]

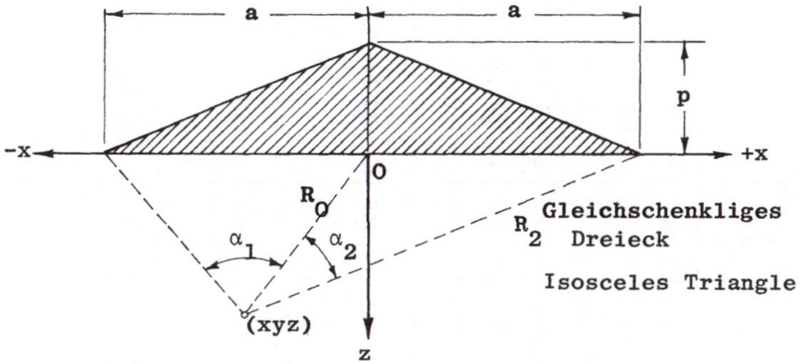

$$\sigma_x = \frac{p}{\pi a}\left[a(\alpha_1+\alpha_2) + x(\alpha_1-\alpha_2) - 2z\ln\frac{R_1R_2}{R_o^2}\right]$$

$$\sigma_z = \frac{p}{\pi a}\left[a(\alpha_1+\alpha_2) + x(\alpha_1-\alpha_2)\right]$$

$$\sigma_y = \frac{2p}{\pi am}\left[a(\alpha_1+\alpha_2) + x(\alpha_1-\alpha_2) - z\ln\frac{R_1R_2}{R_o^2}\right]$$

$$\tau_{xz} = -\frac{pz}{\pi a}\left[\alpha_1-\alpha_2\right]$$

Hauptspannungen

Prinipal Stresses

$$\sigma_1 = \frac{p}{\pi}\left[\frac{x}{a}\alpha + \frac{a+b-x}{b}\beta + \frac{z}{a}\ln\frac{R_1}{R_o} + \frac{z}{b}\ln\frac{R_1}{R_2}\right] + \frac{pz}{\pi}\sqrt{\left(\frac{1}{a}\ln\frac{R_1}{R_o} + \frac{1}{b}\ln\frac{R_1}{R_2}\right)^2 + \left(\frac{\alpha}{a} - \frac{\beta}{b}\right)^2}$$

$$\sigma_2 = \frac{p}{\pi}\left[\frac{x}{a}\alpha + \frac{a+b-x}{b}\beta + \frac{z}{a}\ln\frac{R_1}{R_o} + \frac{z}{b}\ln\frac{R_1}{R_2}\right] - \frac{pz}{\pi}\sqrt{\left(\frac{1}{a}\ln\frac{R_1}{R_o} + \frac{1}{b}\ln\frac{R_1}{R_2}\right)^2 + \left(\frac{\alpha}{a} - \frac{\beta}{b}\right)^2}$$

$$\tau_{max} = \frac{pz}{\pi}\sqrt{\left(\frac{1}{a}\ln\frac{R_1}{R_o} + \frac{1}{b}\ln\frac{R_1}{R_2}\right)^2 + \left(\frac{\alpha}{a} - \frac{\beta}{b}\right)^2}$$

Hauptspannungen

Principal Stresses

$$\sigma_1 = \frac{p}{\pi a}\left[a(\alpha_1+\alpha_2) + x(\alpha_1-\alpha_2) - z\ln\frac{R_1R_2}{R_o^2}\right] + \frac{pz}{\pi a}\sqrt{\ln^2\frac{R_1R_2}{R_o} + (\alpha_1-\alpha_2)^2}$$

$$\sigma_2 = \frac{p}{\pi a}\left[a(\alpha_1+\alpha_2) + x(\alpha_1-\alpha_2) - z\ln\frac{R_1R_2}{R_o}\right] - \frac{pz}{\pi a}\sqrt{\ln^2\frac{R_1R_2}{R_o^2} + (\alpha_1-\alpha_2)^2}$$

$$\tau_{max} = \frac{pz}{\pi a}\sqrt{\ln^2\frac{R_1R_2}{R_o^2} + (\alpha_1-\alpha_2)^2}$$

Fig. 1.21 Gleichungen fuer Bodenspannungen infolge unendlich breiter endlich Doppoldreiecklast [2]

Equations for stresses in soils due to triangular loading with infinite width [2]

(ii) Quellen (s. Seite 59)

References (Ref. PP. 59)

(i)b Einflussbeiwerte nach Barber [1]

(i)b Influence coefficients of Barber [1]

$$\tau_x = p \cdot i_x \quad ; \quad \sigma_z = p \cdot i_z$$
$$\tau = p \cdot i_\tau$$

Beiwert / Factor	Koord-inaten / Coord-inates	0	1	2	3	4	5	6	7	8
i_z	0	1.000	0.750	0.500	0.250	0.000	0.000	0.000	0.000	0.000
	1	0.844	0.721	0.495	0.260	0.077	0.015	0.005	0.002	0.001
	2	0.704	0.635	0.471	0.288	0.139	0.058	0.025	0.012	0.006
	3	0.590	0.546	0.435	0.300	0.182	0.100	0.055	0.030	0.018
	4	0.500	0.471	0.396	0.298	0.205	0.132	0.082	0.052	0.033
	5	0.430	0.410	0.358	0.287	0.215	0.153	0.105	0.072	0.049
	6	0.374	0.361	0.324	0.272	0.216	0.165	0.120	0.089	0.064
	7	0.330	0.321	0.294	0.256	0.212	0.170	0.132	0.100	0.076
	8	0.294	0.288	0.268	0.239	0.204	0.170	0.137	0.109	0.086
	9	0.266	0.261	0.246	0.223	0.196	0.168	0.139	0.115	0.093
	10	0.242	0.238	0.227	0.210	0.188	0.163	0.140	0.118	0.099
i_x	0	1.000	0.750	0.500	0.250	0.000	0.000	0.000	0.000	0.000
	1	0.394	0.388	0.336	0.262	0.196	0.128	0.084	0.060	0.044
	2	0.192	0.203	0.215	0.216	0.201	0.167	0.129	0.100	0.077
	3	0.102	0.113	0.138	0.161	0.170	0.160	0.139	0.118	0.096
	4	0.058	0.067	0.091	0.117	0.133	0.138	0.132	0.119	0.104
	5	0.036	0.043	0.062	0.084	0.103	0.114	0.115	0.111	0.102
	6	0.024	0.029	0.043	0.062	0.079	0.092	0.099	0.098	0.096
	7	0.016	0.021	0.030	0.046	0.061	0.074	0.082	0.085	0.087
	8	0.012	0.016	0.023	0.034	0.048	0.059	0.068	0.073	0.077
	9	0.008	0.011	0.017	0.026	0.038	0.048	0.056	0.063	0.066
	10	0.006	0.009	0.014	0.020	0.029	0.039	0.046	0.053	0.059

Beiwert / Factor	Koord-inaten / Coord-inates	0	1	2	3	4	5	6	7	8
	0	0.000	0.000	0.000	0.000	0.000	0.000	0.000	0.000	0.000
	1	0.000	0.115	0.153	0.148	0.096	0.040	0.019	0.010	0.007
	2	0.000	0.115	0.176	0.181	0.142	0.090	0.054	0.033	0.022
	3	0.000	0.096	0.159	0.173	0.154	0.117	0.082	0.057	0.041
	4	0.000	0.075	0.129	0.153	0.148	0.126	0.099	0.076	0.058
i_τ	5	0.000	0.060	0.105	0.131	0.134	0.123	0.106	0.087	0.069
	6	0.000	0.048	0.086	0.110	0.118	0.115	0.104	0.091	0.076
	7	0.000	0.038	0.071	0.093	0.104	0.105	0.099	0.090	0.079
	8	0.000	0.031	0.058	0.078	0.091	0.095	0.093	0.087	0.079
	9	0.000	0.026	0.049	0.067	0.079	0.085	0.085	0.083	0.077
	10	0.000	0.022	0.041	0.057	0.068	0.075	0.078	0.078	0.074

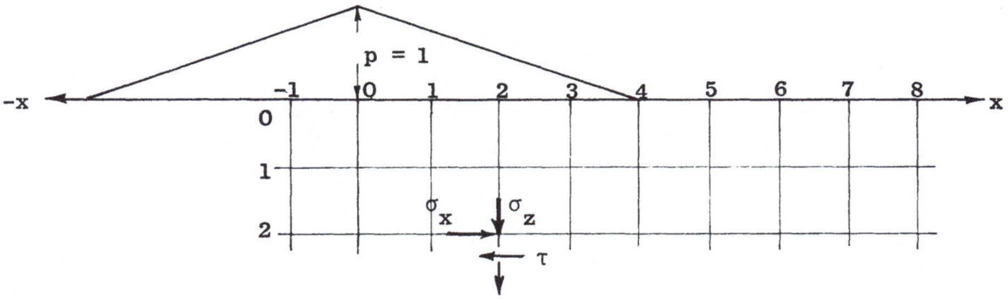

Tabelle 1.11 Einflussbeiwerte fuer unendlich breite, endlich Doppeldreiecklast [1]

Table 1.11 Influence coefficients for triangular loading with infinite width [1]

(ii) Quellen

References

[1] Barber (1952) - Use of direct shear tests in highway design. Symposium on direct shear testing of soils. ASTM No. 131, PP 19.

[2] Gray (1936) Stress distribution in elastic solids. Proc. 1, ICSMFE, Cambridge, Vol. 2, PP. 157.

1.1.1.3.9 Unendlich breite, einseitig unendlich lange Doppeltrapezlast

a) Halbraum nach Boussinesq

(i) Formeln

1.1.1.3.9 Long strip double trapezoidal loading extending infinitely in the x-direction

a) Half-space according to Boussinesq

(i) Formulae

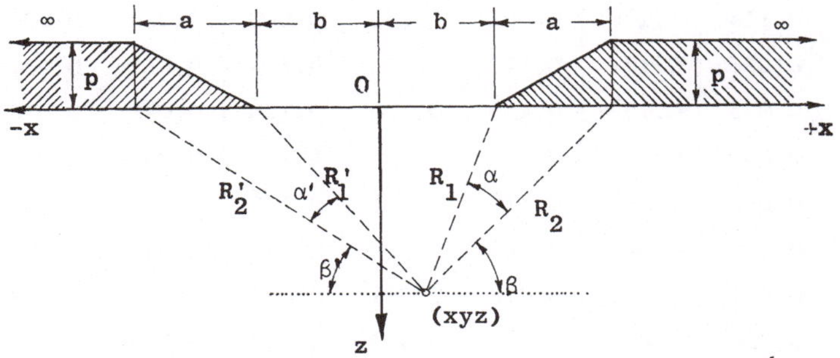

$$\sigma_x = \frac{p}{\pi a}\left[a(\beta+\beta') - b(\alpha+\alpha') + x(\alpha-\alpha') + 2z \ln\frac{R_2 R_2'}{R_1 R_1'}\right]$$

$$\sigma_z = \frac{p}{\pi a}\left[a(\beta+\beta') - b(\alpha+\alpha') + x(\alpha-\alpha')\right]$$

$$\sigma_y = -\frac{2p}{\pi a m}\left[a(\beta+\beta') - b(\alpha+\alpha') + x(\alpha-\alpha') + z \ln\frac{R_2 R_2'}{R_1 R_1'}\right]$$

$$\tau_{xz} = -\frac{pz}{\pi a}\left(\alpha-\alpha'\right)$$

Hauptspannungen

Principal stresses

$$\sigma_1 = \frac{p}{\pi a}\left[a(\beta+\beta') - b(\alpha+\alpha') + x(\alpha-\alpha') + z \ln\frac{R_2 R_2'}{R_1 R_1'}\right] + \frac{pz}{\pi a}\sqrt{\ln^2\frac{R_2 R_2'}{R_1 R_1'} + (\alpha-\alpha')^2}$$

$$\sigma_2 = \frac{p}{\pi a}\left[a(\beta+\beta') - b(\alpha+\alpha') + x(\alpha-\alpha') + z \ln\frac{R_2 R_2'}{R_1 R_1'}\right] - \frac{pz}{\pi a}\sqrt{\ln\frac{R_2 R_2'}{R_1 R_1'} + (\alpha-\alpha')^2}$$

$$\tau_{max} = \frac{pz}{\pi a}\sqrt{\ln^2\frac{R_2 R_2'}{R_1 R_1'} + (\alpha-\alpha')^2}$$

Fig. 1.22 Gleichungen fuer Bodenspannungen infolge unendlich breiter, einseitig unendlich langer Doppeltrapezlast [1]

Equations for stresses in soil due to long strip double trapezoidal loading extending infinitely in the x-direction [1]

(ii) Quelle (s. Seite 59)
Reference (Ref. PP. 59)

1.1.1.3.10 Unendlich breite, endlich lange Trapezlast

a) Halbraum nach Boussinesq

(i) Formeln

1.1.1.3.10 Long strip trapezoidal loading

a) Half-space according to Boussinesq

(i) Formulae

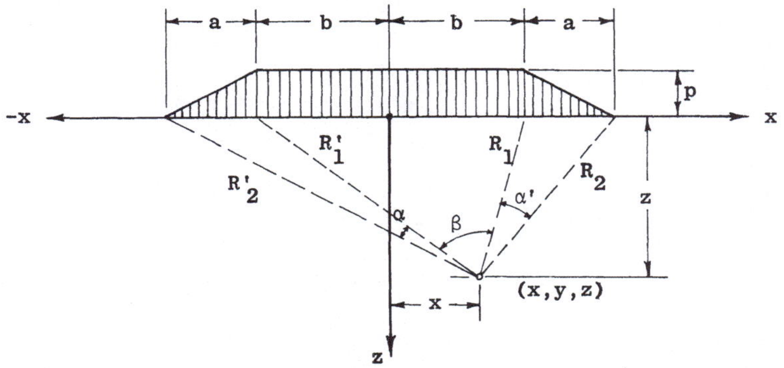

$$\sigma_x = \frac{p}{\pi}\left[(\alpha+\beta+\alpha') + \frac{b}{a}(\alpha+\alpha') + \frac{x}{a}(\alpha-\alpha') - 2\cdot\frac{z}{a}\ln\left(\frac{R'_2\cdot R_2}{R'_1\cdot R_1}\right)\right]$$

$$\sigma_z = \frac{p}{\pi}\left[(\alpha+\beta+\alpha') + \frac{b}{a}(\alpha+\alpha') + \frac{x}{a}(\alpha-\alpha')\right]$$

$$\tau_{xz} = \frac{p}{\pi}\cdot\frac{z}{a}\cdot(\alpha-\alpha')$$

Hauptspannungen

Principal stresses

$$\sigma_1 = \frac{p}{\pi}\left[(\alpha+\beta+\alpha') + \frac{b}{a}(\alpha+\alpha') + \frac{x}{a}(\alpha-\alpha') - \frac{z}{a}\cdot\ln\left(\frac{R'_2\cdot R_2}{R'_1\cdot R_1}\right)\right] + \tau_{max}$$

$$\sigma_2 = \frac{p}{\pi}\left[(\alpha+\beta+\alpha') + \frac{b}{a}(\alpha+\alpha') + \frac{x}{a}(\alpha-\alpha') - \frac{z}{a}\ln\left(\frac{R'_2\cdot R_2}{R'_1\cdot R_1}\right)\right] - \tau_{max}$$

$$\tau_{max} = \frac{p}{\pi}\cdot\frac{z}{a}\cdot\sqrt{\ln^2\left(\frac{R'_2\cdot R_2}{R'_1\cdot R_1}\right) + (\alpha-\alpha')^2}$$

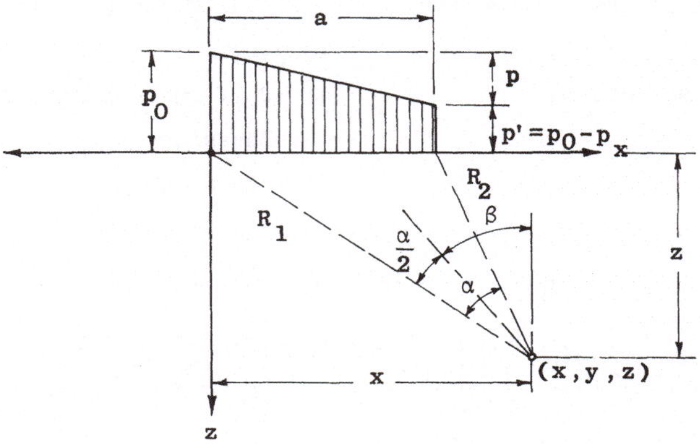

$$\sigma_x = \frac{p_o}{\pi}(\alpha - \sin\alpha \cdot \cos 2\beta) - \frac{p}{\pi}\left[\frac{x}{a}\,\alpha + \frac{z(x-a)}{R_2^2} + 2 \cdot \frac{z}{a} \cdot \ln\frac{R_2}{R_1}\right]$$

$$\sigma_z = \frac{p_o}{\pi}(\alpha + \sin\alpha \cdot \cos 2\beta) - \frac{p}{\pi}\left[\frac{x}{a} \cdot \alpha - \frac{z(x-a)}{R_2^2}\right]$$

$$\tau_{xz} = \frac{p_o}{\pi} \cdot \sin\alpha \cdot \sin 2\beta + \frac{p}{\pi}\left[\frac{z}{a}\,\alpha - \frac{z^2}{R_2^2}\right]$$

$$\tau_{max} = \frac{1}{\pi}\left\{\left[p_o \cdot \sin\alpha\ \cos 2\beta + p\,\frac{z(x-a)}{R_2^2} + p \cdot \frac{z}{a} \cdot \ln\frac{R_2}{R_1}\right]^2 + \left[p_o \cdot \sin\alpha \cdot \sin 2\beta + \frac{p}{\pi}\left(\frac{z}{a} \cdot \alpha - \frac{z^2}{R_2^2}\right)\right]^2\right\}^{\frac{1}{2}}$$

Fig. 1.23 Gleichungen fuer Bodenspannungen infolge unendlich breiter endlich langer Trapezlast [1,2]

Equations for stresses in soil due to long strip trapezoidal loading [1,2]

(ii) Quellen

References

[1] Gray (1936) - Stress sistribution in elastic solids. Proc. 1, ICSMFE, Cambridge, Vol. 2, PP. 157.

[2] Palmer (1938) - Principles of soil mechanics involved in the design of retaining walls and bridge abutments. Public Roads, December 1938, Vol. 19, No. 10.

Bearing capacity of soils. U.S. Army Corps of Engineers, Engineering Manual 1110-2-1903

1.1.1.3.11 Unendlich breite, endlich lange Parabellast

a) Halbraum nach Boussinesq

(i) Formeln

1.1.1.3.11 Long strip parabolic loading

a) Half-space according to Boussinesq

(i) Formulae

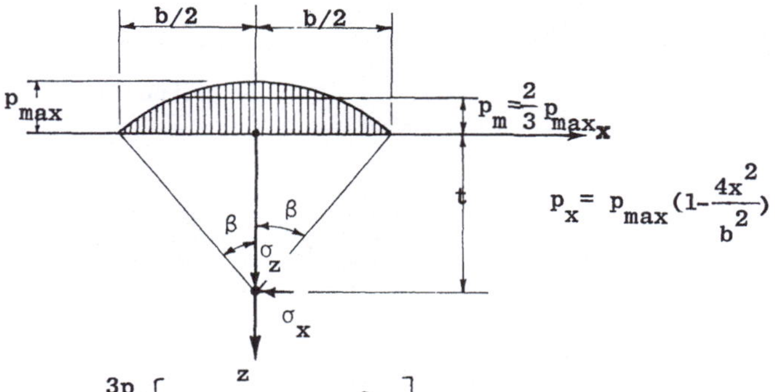

$$\sigma_z = \frac{3p_m}{\pi}\left[\cot\beta+\beta(1-\cot^2\beta)\right]$$

$$\sigma_x = \frac{3p_m}{\pi}\left[\beta-\sin\beta\cdot\cos\beta-\cot\beta(2-3\beta\cot\beta+\cos^2\beta)\right]$$

Fig. 1.24 Gleichungen fuer Bodenspannungen infolge unendlich breiter endlich langer Parabellast [1,2]

Equations for stresses in soil due to long strip parabolic loading [1,2]

(ii) Quellen

References

[1] Kollbrunner (1946) - Foundation und Konsolidation
Band 1, S. 317
Schweizer Druck und Verlagshaus

[2] Szechy (1963) - Der Grundbau
Band 1, S. 263

1.1.1.4 Rechteckiger Grundriss

1.1.1.4.1 Gleichmaessige Belastung

a) Halbraum nach Boussinesq

a.1 Lotrechte Spannung unter dem Eckpunkt

(i) Formeln/Tafeln/Tabellen

(i)a Gleichung nach Steinbrenner [5]

1.1.1.4 Rectangular Plan Area

1.1.1.4.1 Uniform Loading

a) Half-space according to Boussinesq

a.1 Vertical stress under a corner point

(i) Formulae/Charts/Tables

(i)a Equation of Steinbrenner [5]

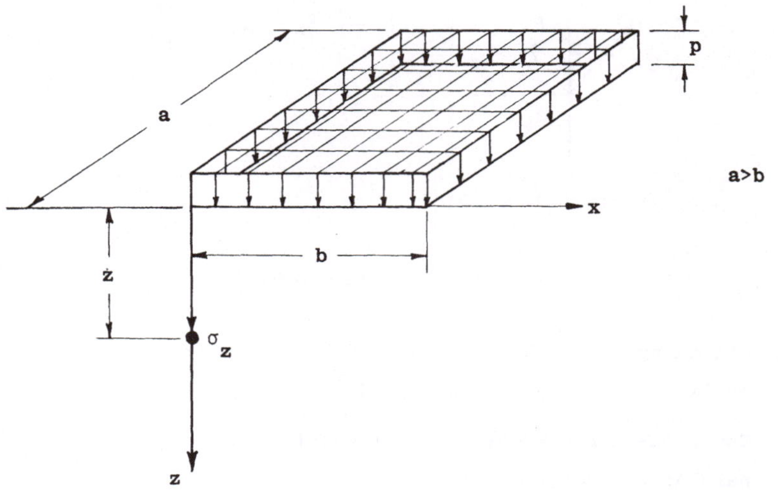

$$\sigma_z = \frac{p}{2\pi} \left\{ \arctan \frac{ab}{z\sqrt{a^2 + b^2 + z^2}} + \frac{a.b.z}{\sqrt{a^2 + b^2 + z^2}} \left(\frac{1}{a^2 + z^2} + \frac{1}{b^2 + z^2} \right) \right\}$$

$$= p \cdot i$$

$\sigma_z = p \cdot i$

$\frac{z}{b}$	$\frac{a}{b} = 1.0$	$\frac{a}{b} = 1.5$	$\frac{a}{b} = 2.0$	$\frac{a}{b} = 3.0$	$\frac{a}{b} = 5.0$	$\frac{a}{b} = 10.0$	$\frac{a}{b} = \infty$
0.25	0.2473	0.2482	0.2483	0.2484	0.2485	0.2485	0.2485
0.50	0.2325	0.2378	0.2391	0.2397	0.2398	0.2399	0.2399
0.75	0.2060	0.2182	0.2217	0.2234	0.2239	0.2240	0.2240
1.00	0.1752	0.1936	0.1999	0.2034	0.2044	0.2046	0.2046
1.50	0.1210	0.1451	0.1561	0.1638	0.1665	0.1670	0.1670
2.00	0.0840	0.1071	0.1202	0.1316	0.1363	0.1374	0.1374
3.00	0.0447	0.0612	0.0732	0.0860	0.0959	0.0987	0.0990
4.00	0.0270	0.0383	0.0475	0.0604	0.0712	0.0758	0.0764
6.00	0.0127	0.0185	0.0238	0.0323	0.0431	0.0506	0.0521
8.00	0.0073	0.0107	0.0140	0.0195	0.0283	0.0367	0.0394
10.00	0.0048	0.0070	0.0092	0.0129	0.0198	0.0279	0.0316
12.00	0.0033	0.0049	0.0065	0.0094	0.0145	0.0219	0.0264
15.00	0.0021	0.0031	0.0042	0.0061	0.0097	0.0158	0.0211
18.00	0.0015	0.0022	0.0029	0.0043	0.0069	0.0118	0.0177
20.00	0.0012	0.0018	0.0024	0.0035	0.0057	0.0099	0.0159

Tabelle 1.12a Einflussbeiwert, i, als Funktion von $\frac{z}{b}$ und $\frac{a}{b}$ [5]

Table 1.12a Influence coefficient, i as a function of $\frac{z}{b}$ and $\frac{a}{b}$ [5]

(i)b Einflussbeiwert nach Newmark [3]

(i)b Influence coefficient of Newmark [3]

$$\sigma_z = p \cdot i$$

b/z	a/z											
	0.1	0.2	0.3	0.4	0.5	0.6	0.7	0.8	0.9	1.0	1.2	1.4
0.1	0.00470	0.00917	0.01323	0.01678	0.01978	0.02223	0.02420	0.02576	0.02698	0.02794	0.02926	0.03007
0.2	0.00917	0.01792	0.02585	0.03280	0.03866	0.04348	0.04735	0.05042	0.05283	0.05471	0.05733	0.05894
0.3	0.01323	0.02585	0.03735	0.04742	0.05593	0.06294	0.06858	0.07308	0.07661	0.07938	0.08323	0.08561
0.4	0.01678	0.03280	0.04742	0.06024	0.07111	0.08009	0.08734	0.09314	0.09770	0.10129	0.10631	0.10941
0.5	0.01978	0.03866	0.05593	0.07111	0.08403	0.09473	0.10340	0.11035	0.11584	0.12018	0.12626	0.13003
0.6	0.02223	0.04348	0.06294	0.08009	0.09473	0.10688	0.11679	0.12474	0.13105	0.13605	0.14309	0.14749
0.7	0.02420	0.04735	0.06858	0.08734	0.10340	0.11679	0.12772	0.13653	0.14356	0.14914	0.15703	0.16199
0.8	0.02576	0.05042	0.07308	0.09314	0.11035	0.12474	0.13653	0.14607	0.15371	0.15978	0.16843	0.17389
0.9	0.02698	0.05283	0.07661	0.09770	0.11584	0.13105	0.14356	0.15371	0.16185	0.16835	0.17766	0.18357
1.0	0.02794	0.05471	0.07938	0.10129	0.12018	0.13605	0.14914	0.15978	0.16835	0.17522	0.18508	0.19139
1.2	0.02926	0.05733	0.08323	0.10631	0.12626	0.14309	0.15703	0.16843	0.17766	0.18508	0.19584	0.20278
1.4	0.03007	0.05894	0.08561	0.10941	0.13003	0.14749	0.16199	0.17389	0.18357	0.19139	0.20278	0.21020
1.6	0.03058	0.05994	0.08709	0.11135	0.13241	0.15028	0.16515	0.17739	0.18737	0.19546	0.20731	0.21510
1.8	0.03090	0.06058	0.08804	0.11260	0.13395	0.15207	0.16720	0.17967	0.18986	0.19814	0.21032	0.21836
2.0	0.03111	0.06100	0.08867	0.11342	0.13496	0.15326	0.16856	0.18119	0.19152	0.19994	0.21235	0.22058

	a/z											
b/z	0.1	0.2	0.3	0.4	0.5	0.6	0.7	0.8	0.9	1.0	1.2	1.4
2.5	0.03138	0.06155	0.08948	0.11450	0.13628	0.15483	0.17036	0.18321	0.19375	0.20236	0.21512	0.22364
3.0	0.03138	0.06178	0.08982	0.11495	0.13684	0.15550	0.17113	0.18407	0.19470	0.20341	0.21633	0.22499
4.0	0.03158	0.06194	0.09007	0.11527	0.13724	0.15598	0.17168	0.18469	0.19540	0.20417	0.21722	0.22600
5.0	0.03160	0.06199	0.09014	0.11537	0.13737	0.15612	0.17185	0.18488	0.19561	0.20440	0.21749	0.22632
6.0	0.03161	0.06201	0.09017	0.11541	0.13741	0.15617	0.17191	0.18496	0.19569	0.20449	0.21760	0.22644
8.0	0.03162	0.06202	0.09018	0.11543	0.13744	0.15621	0.17195	0.18500	0.19574	0.20455	0.21767	0.22652
10.0	0.03162	0.06202	0.09019	0.11544	0.13745	0.15622	0.17196	0.18502	0.19576	0.20457	0.21769	0.22654
∞	0.03162	0.06202	0.09019	0.11544	0.13745	0.15623	0.17197	0.18502	0.19577	0.20458	0.21770	0.22656

b/z	a/z 1.6	1.8	2.0	2.5	3.0	4.0	5.0	6.0	8.0	10.0	∞
0.1	0.03058	0.03090	0.03111	0.03138	0.03150	0.03158	0.03160	0.03161	0.03162	0.03162	0.03162
0.2	0.05994	0.06058	0.06100	0.06155	0.06178	0.06194	0.06199	0.06201	0.06202	0.06202	0.06202
0.3	0.08709	0.08804	0.08867	0.08948	0.08982	0.09007	0.09014	0.09017	0.09018	0.09019	0.09019
0.4	0.11135	0.11260	0.11342	0.11450	0.11495	0.11527	0.11537	0.11541	0.11543	0.11544	0.11544
0.5	0.13241	0.13395	0.13469	0.13628	0.13684	0.13724	0.13737	0.13741	0.13744	0.13745	0.13745
0.6	0.15028	0.15207	0.15326	0.15483	0.15550	0.15598	0.15612	0.15617	0.15621	0.15622	0.15623
0.7	0.16515	0.16720	0.16856	0.17035	0.17113	0.17168	0.17185	0.17191	0.17195	0.17196	0.17197
0.8	0.17739	0.17967	0.18119	0.18321	0.18407	0.18469	0.18488	0.18496	0.18500	0.18502	0.18502
0.9	0.18737	0.18986	0.19152	0.19375	0.19470	0.19540	0.19561	0.19569	0.19574	0.19576	0.19577
1.0	0.19546	0.19814	0.19994	0.20236	0.20341	0.20417	0.20440	0.20449	0.20455	0.20457	0.20458
1.2	0.20731	0.21032	0.21235	0.21512	0.21633	0.21722	0.21749	0.21760	0.21767	0.21769	0.21770
1.4	0.21510	0.21836	0.22058	0.22364	0.22499	0.22600	0.22632	0.22644	0.22652	0.22654	0.22656
1.6	0.22025	0.22372	0.22610	0.22940	0.23088	0.23200	0.23236	0.23249	0.23258	0.23261	0.23263
1.8	0.22372	0.22736	0.22986	0.23334	0.23495	0.23617	0.23656	0.23671	0.23681	0.23684	0.23686
2.0	0.22610	0.22986	0.23247	0.23614	0.23782	0.23912	0.23954	0.23970	0.23981	0.23985	0.23987
2.5	0.22940	0.23334	0.23614	0.24010	0.24196	0.24344	0.24392	0.24412	0.24425	0.24429	0.24432
3.0	0.23088	0.23495	0.23782	0.24196	0.24394	0.24554	0.24608	0.24630	0.24646	0.24650	0.24654
4.0	0.23200	0.23617	0.23912	0.24344	0.24554	0.24729	0.24791	0.24817	0.24836	0.24842	0.24846
5.0	0.23236	0.23656	0.23954	0.24392	0.24608	0.24791	0.24857	0.24885	0.24907	0.24914	0.24919
6.0	0.23249	0.23671	0.23970	0.24412	0.24630	0.24817	0.24885	0.24916	0.24939	0.24946	0.24952
8.0	0.23258	0.23681	0.23981	0.24425	0.24646	0.24836	0.24907	0.24939	0.24964	0.24973	0.24980
10.0	0.23261	0.23684	0.23985	0.24429	0.24650	0.24842	0.24914	0.24946	0.24973	0.24981	0.24989
∞	0.23263	0.23686	0.23987	0.24432	0.24654	0.24846	0.24919	0.24952	0.24980	0.24989	0.25000

	a/z										
b/z	1.6	1.8	2.0	2.5	3.0	4.00	5.0	6.0	8.0	10.0	∞
6.0	0.23249	0.23671	0.23970	0.24412	0.24630	0.24817	0.24885	0.24916	0.24939	0.24946	0.24952
8.0	0.23258	0.23681	0.23981	0.24425	0.24646	0.24836	0.24907	0.24939	0.24964	0.24973	0.24980
10.0	0.23261	0.23684	0.23985	0.24429	0.24650	0.24842	0.24914	0.24946	0.24973	0.24981	0.24989
∞	0.23263	0.23686	0.23987	0.24432	0.24654	0.24846	0.24919	0.24952	0.24980	0.24989	0.25000

Tabelle 1.12 b) Einflussbeiwert, i, als Funktion von b/z und a/z [3]

Table 1.2 b) Influence coefficient, i, as a function of b/z and a/z [3]

ic) Einflusslinien nach Fadum [2]

ic) Influence lines of Fadum [2]

$$\boxed{\sigma_z = p \cdot i}$$

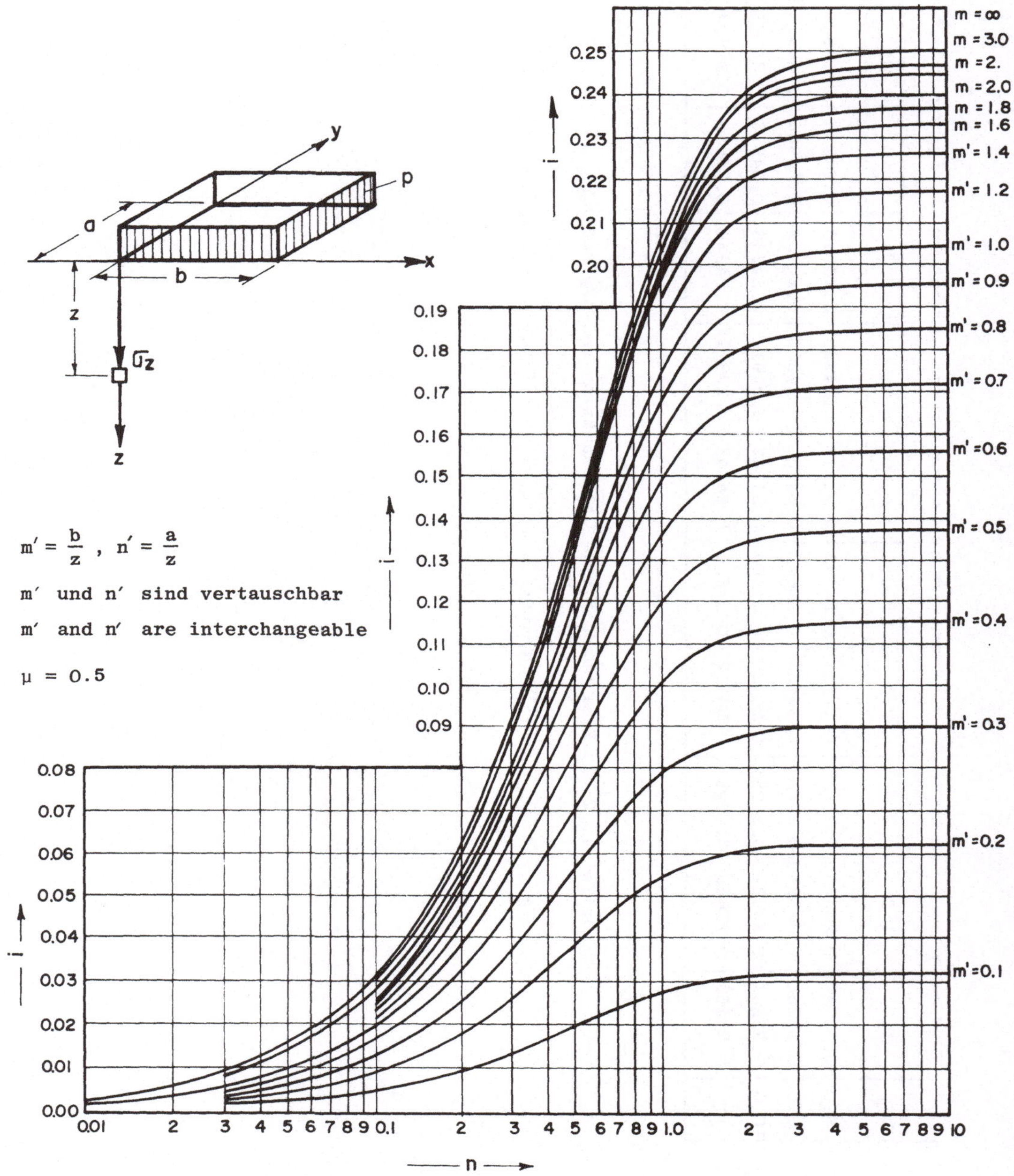

Fig. 1.25a Einflusslinien fuer i, als Funktion von b/z und a/z [2,4]

Influence lines for i, as function of b/z and z/z [2,4]

id) Einflusslinien nach DIN 4019 [1]

id) Influence Lines according DIN 4019 [1]

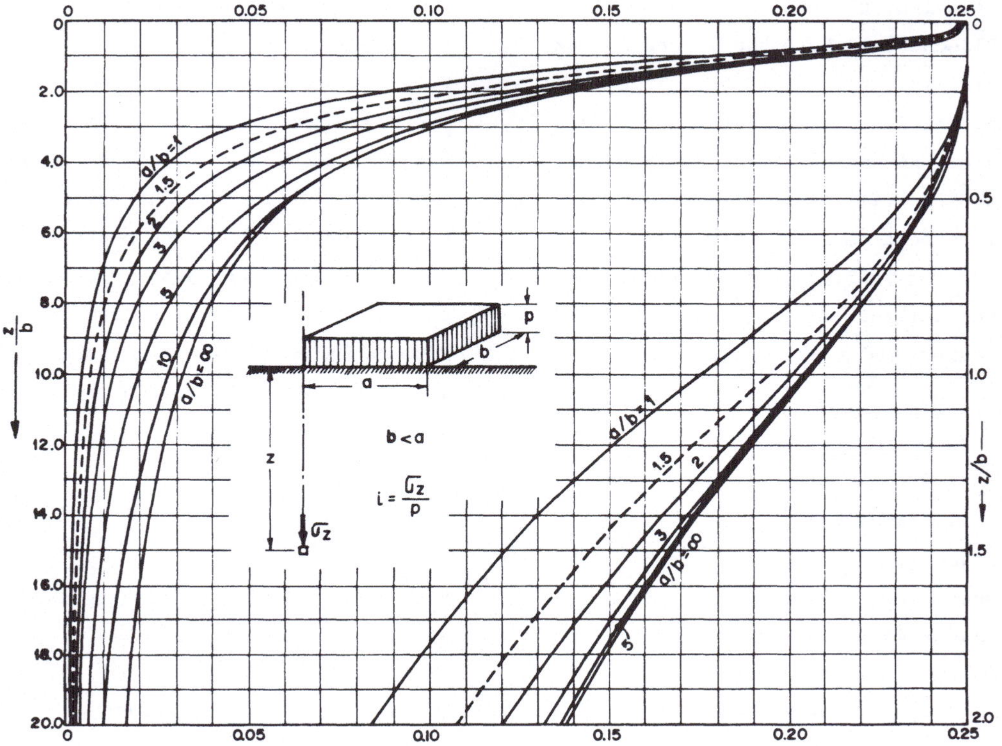

Fig. 1.25b Einflusslinien fuer i, als Funktion von z/b und a/b [1]

Influence lines for i, as a function of z/b and a/b [1]

(ii) Erlaeuterung

Nach diesen Tafeln kann man die Spannungen unter jedem Punkt einer rechteckigen, gleichmaessig verteilten Flaechenlast berechnen. Man zerlegt die Flaeche in 4 Teilrechtecke parallel zu den Aussenkanten. Fuer die 4 aneinanderstossenden Eckpunkte dieser Teilflaechen greift man entsprechend den Verhaeltnissen z/b und a/b der Teilflaechen die Einflusswerte i ab. Diese 4 i-Werte werden addiert und σ errechnet.

(ii) Explanatory note

The vertical stress at any point under the uniformly distributed rectangular load may be calculated using the above table and charts. From the projection of the point, one draws parallel lines to the edges of the loaded area thus forming 4 smaller rectangular having one common corner. For each rectangle, one calculates the ratios z/b and a/b and determines the corresponding influence value i. The four values of i are then added to obtain the required vertical stress σ.

(iii) Anwedungsbeispiel

Gegeben

Rechteckige, gleichmaessig verteilte Flaechnlast $a = 6.0$ m

$b = 4.0$ m

$p = 500$ kN/m^2

(iii) Example

Given

Uniformly distributed rectangular surface loading $a = 6.0$ m

$b = 4.0$ m

$p = 500$ kN/m^2

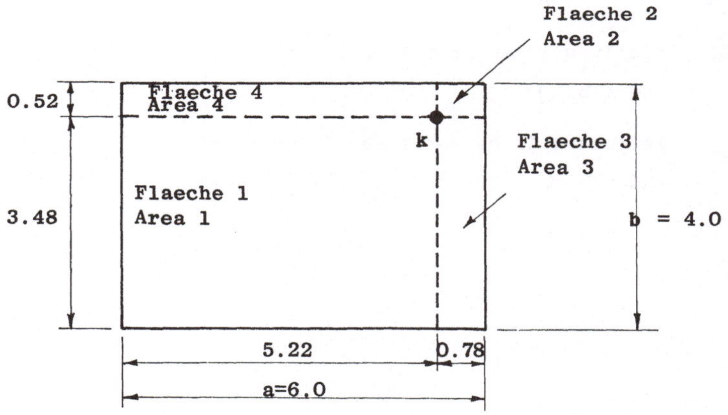

Gesucht

Lotrechte Bodenspannungen in verschiedenen Tiefen z unter einem beliebigen Punkt k, nach Steinbrenner (Tabelle 1.12)

Required

Vertical soil stress at different depths z under any given point k according to Steinbrenner (Table 1.12)

Loesung *Solution*

Tiefe Depth	Flaeche 1 Area 1 $a/b = \frac{5.22}{3.48} = 1.5$		Flaeche 2 Area 2 $a/b = \frac{0.78}{0.25} = 1.5$		Flaeche 3 Area 3 $a/b = \frac{3.48}{0.78} = 4.46$		Flaeche 4 Area 4 $a/b = \frac{5.22}{0.52} = 10.00$		Summe Sum	Lotrechte Spannung Vertical Stress
z	$z/3.48$	i_1	$z/0.52$	i_2	$z/0.78$	i_3	$z/0.52$	i_4	Σi	$\sigma_v = p \cdot \Sigma i$
m	-	-	-	-	-	-	-	-	-	kN/m^2
1.0	0.287	0.247	1.925	0.113	1.285	0.182	1.925	0.142	0.684	342
2.0	0.574	0.233	3.85	0.042	2.568	0.109	3.85	0.081	0.465	233
3.0	0.862	0.209	5.77	0.021	3.84	0.073	5.77	0.053	0.356	178
4.0	1.150	0.178	7.69	0.011	5.13	0.050	7.69	0.038	0.277	138
5.0	1.437	0.151	9.62	0.007	6.42	0.036	9.62	0.029	0.223	111
6.0	1.724	0.126	11.55	0.005	7.69	0.028	11.55	0.023	0.182	91

(iv) Quellen

References

[1] DIN 4019 Blatt 1: Setsungsberechnung bei lotrechter, mittiger Belastung. Richtlinien.

[2] Fadum (1948): Influence values for estimating stresses in elastic foundations. Proc. 2. ICSMFE Rotterdam, Vol. 3, PP 77.

[3] Newmark (1935): Simplified computation of vertical pressures in elastic foundations. University of Illinois Eng. Experiment station, Circular 24.

[4] Schultze (1966): Druckverteilung und Setzungen. Grundbau - Taschenbuch Bd. 1, 2. Auflage S. 92. Verlag von Wilhelm Ernst und Sohn, Berlin, Muenchen.

[5] Steinbrenner (1934): Tafeln zur Setzungsberechung. Schriftenreihe Strasse 1, S.121.

a.2 Lotrechte Spannung unter dem "kennzeichenden" Punkt

(i) Tabelle/Tafeln

a.2 Vertical stress under the "Characteristic" Point

(i) Table/Charts

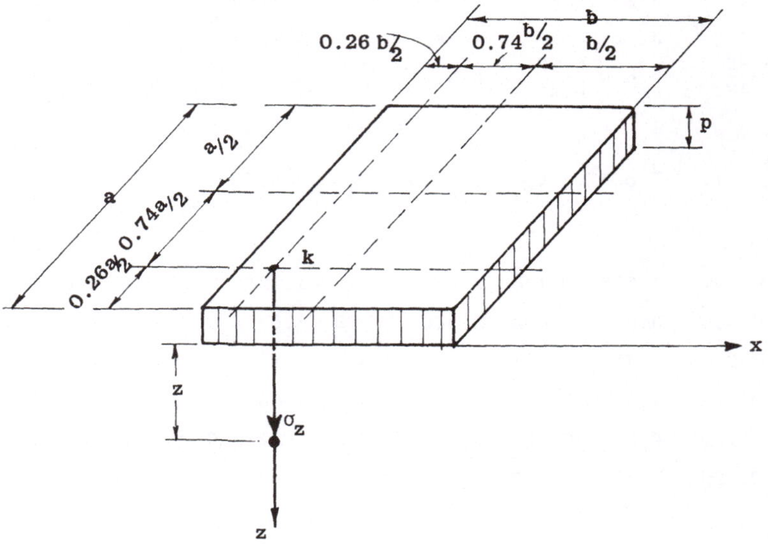

$\sigma_z = p \cdot i$

z/b	a/b=1.0	a/b=1.5	a/b=2.0	a/b=3.0	a/b=5.0	a/b=10.0	a/b=∞
0.05	0.9811	0.9819	0.9884	0.9894	0.9895	0.9897	0.9896
0.10	0.8984	0.9280	0.9372	0.9425	0.9443	0.9447	0.9447
0.15	0.7898	0.8351	0.8523	0.8755	0.8824	0.8830	0.8839
0.20	0.6947	0.7570	0.7883	0.8127	0.8335	0.8262	0.8264
0.30	0.5566	0.6213	0.6628	0.7053	0.7301	0.7376	0.7387
0.50	0.4088	0.4622	0.5032	0.5550	0.6032	0.6264	0.6299
0.70	0.3249	0.3706	0.4041	0.4527	0.5066	0.5473	0.5552
1.00	0.2342	0.2786	0.3078	0.3488	0.4008	0.4504	0.4674
1.50	0.1438	0.1830	0.2098	0.2387	0,2779	0.3303	0.3604
2.00	0.0939	0.1279	0.1475	0.1749	0.2057	0.2479	0.2883
3.00	0.0473	0.0672	0.0823	0.1043	0.1280	0.1575	0.2025
5.00	0.0183	0.0268	0.0345	0.0502	0.0646	0.0838	0.1251

z/b	a/b=1.0	a/b=1.5	a/b=2.0	a/b=3.0	a/b=5.0	a/b=10.0	a/b= ∞
7.00	0.0095	0.0141	0.0185	0.0264	0.0384	0.0541	0.0905
10.00	0.0045	0.0070	0.0093	0.0135	0.0210	0.0328	0.0633
20.00	0.0012	0.0015	0.0024	0.0035	0.0058	0.0105	0.0318

Tabelle 1.13 Einflussbeiwert, i als Funktion von z/b und a/b [1]

Table 1.13 Influence coefficient, i as a function of z/b and a/b [1]

(ii) Erlaeuterung

Die Tafel von Kany ergibt den i-Wert fuer den kennzeichnenden Punkt einer rechteckigen, gleichamaessig verteilten Flaechenlast. Dieser Punkt hat die Abstaende 0.74 $\frac{a}{2}$ bzw. 0.74 $\frac{b}{2}$ von Mittelpunkt des Rechtecks mit den Seiten a bzw. b. Mit der so ermittelten Spannung kann die Setzung unter dem kennzeichnenden Punkt bestimmt werden, die gleich der Setzung einer starren Platte gleicher Groese und Belastung ist.

(ii) Explanatory note

The Table of Kany gives the i-values for the characteristic point for a uniformly distributed rectangular load. The characteristic point is located at a distance of o.74 $\frac{a}{2}$ or 0.74 $\frac{b}{2}$ from the middle of the rectangle with side dimensions of a and b. One may calculate the settlement at the characteristic point using the vertical stress determined at this point. This settlement is equal to the settlement of a corresponding rigid plate with the same dimensions and loading.

(iii) Anwendungsbeispiel

Gegeben

Rechteckige, gleichmaessig verteilte Flaechenlast a = 6.0 m

b = 4.0 m

p = 500 kN/m^2

(iii) Example

Given

Uniformly distributed rectangular surface loading a = 6.0 m

b = 4.0 m

p = 500 kN/m^2

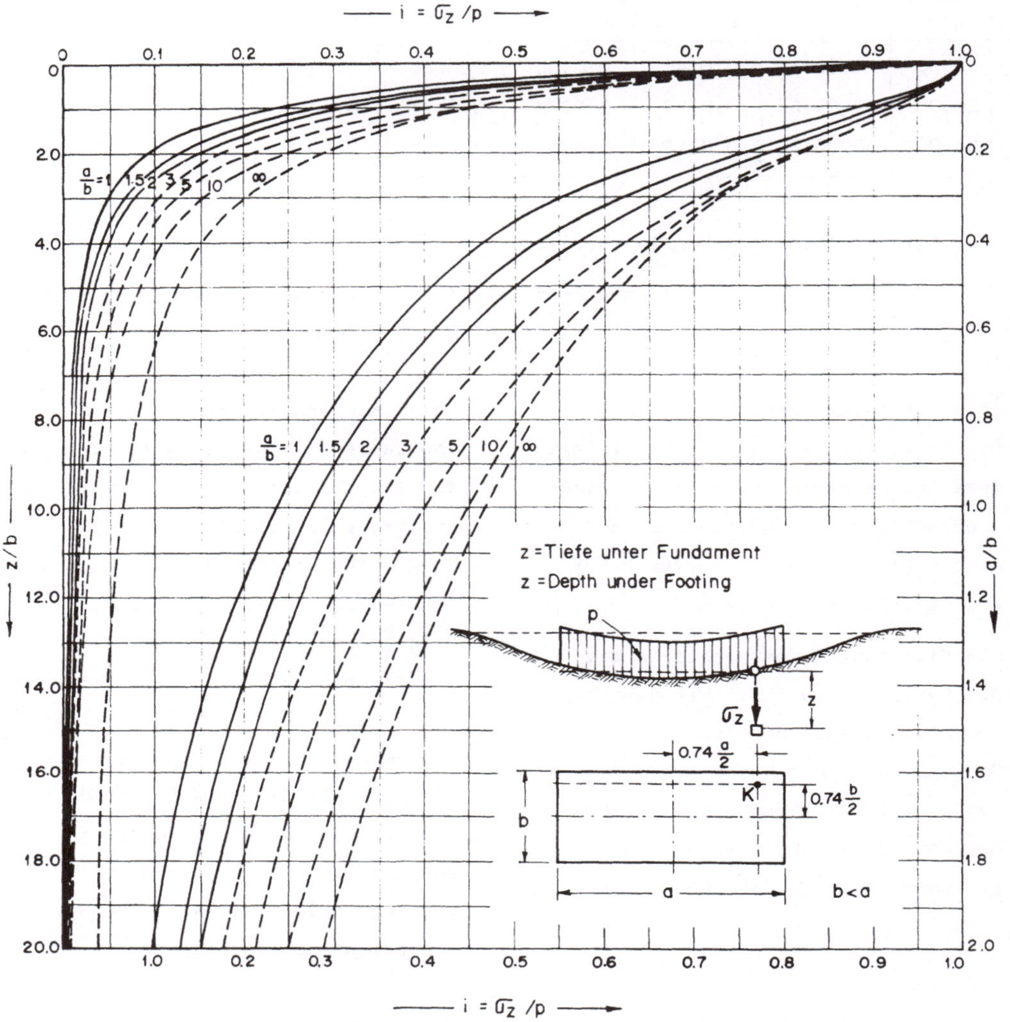

Fig. 1.26a Einflusslinien fuer i, als Funktion von z/b und a/b [1]
Influence lines for i, as function of z/b and a/b [1]

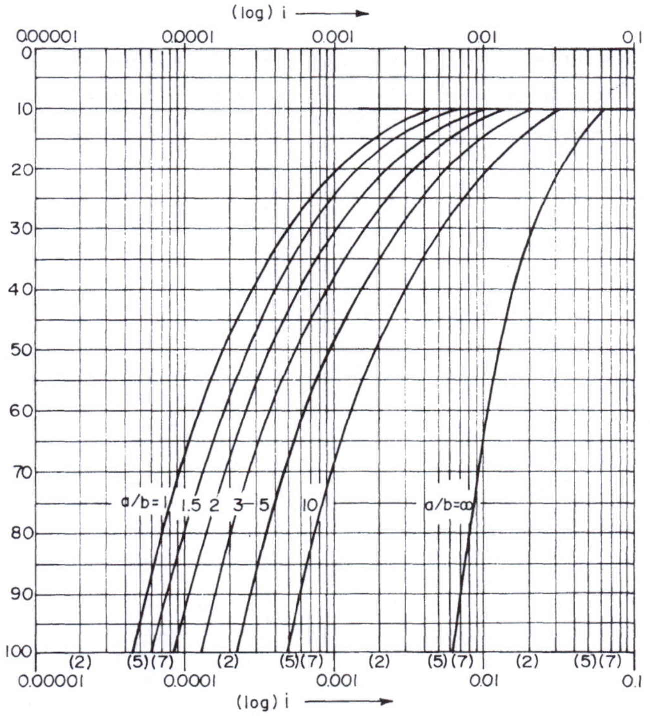

Fig. 1.26b Einflusslinien fuer i als Funktion von z/b und a/b [1]

Influence lines for i as a function of z/b and a/b [1]

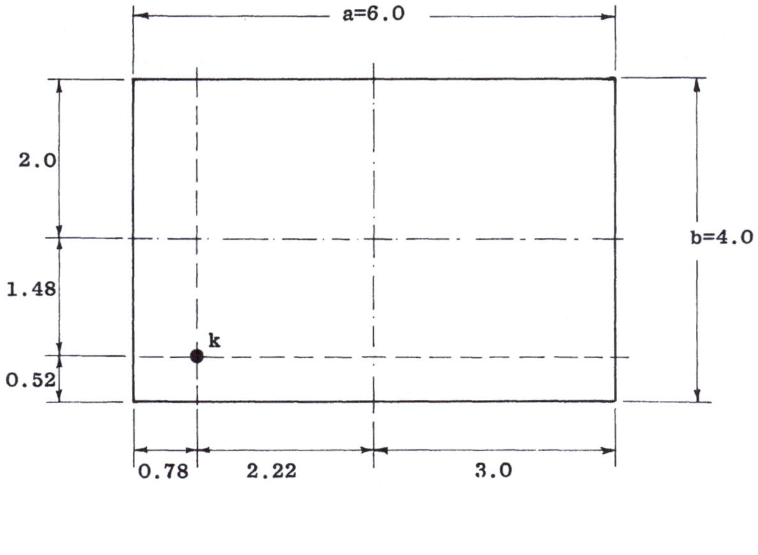

Gesucht

Lotrechte Bodenspannungen in verschiedenen Tiefen z unter dem Kennzeichenden Punkt k nach Kany.

Required

Vertical soil stress at different depths, z, under the characteristic point k according to Kany.

Loesung

Solution

a/b = 6/4 = 1.5			
Tiefe Depth z	z/b	i	$\sigma_z = p \cdot i$
m	-	-	kN/m^2
1.0	0.25	0.689	344
2.0	0.50	0.462	231
3.0	0.75	0.350	175
4.0	1.00	0.275	135
5.0	1.25	0.224	112
6.0	1.50	0.183	91

(iv) Quelle

Reference

[1] Kany (1959) Beitrag zur Berechung von Flaechengruendungen Berlin, Wilh. Ernstund Sohn.

a.3 Lotrechte Spannungen beliebiger Stelle infolge einer quadratischen Lastflaeche

(i) Tabellen/Tafeln

a.3 Vertical stress under any point due to square uniform loading

(i) Tables/Charts

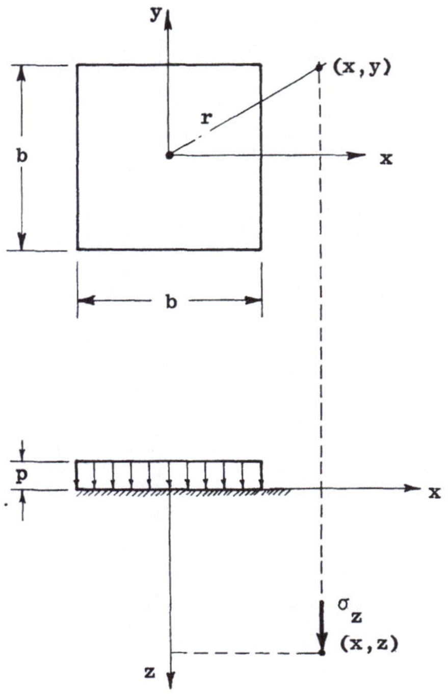

Grundriss
Plan

Schnitt
Section

x, y, z = Koordinaten des berechneter Punktes
Coordinate of the point under consideration

$$r/b = \sqrt{(\frac{x}{b})^2 + (\frac{y}{b})^2}$$

$$\boxed{\sigma_z = p \cdot i}$$

$\sigma_z = p \cdot i$

	z/b = 0.5								
5.0	0.19	0.19	0.18	0.16	0.13	0.11	0.09	0.06	0.03 / 0.26
4.0	0.60	0.57	0.51	0.43	0.34	0.26	0.19	0.10 / 0.79	0.43
3.0	2.55	2.38	1.96	1.45	1.01	0.67	0.44 / 3.19	1.44	0.68
2.5	6.42	5.82	4.43	2.96	1.84	1.12 / 7.65	4.76	1.91	0.83
2.0	19.88	17.15	11.44	6.50	3.46 / 21.73	12.22	6.97	2.47	1.00
1.5	81.48	66.27	34.79	15.11 / 77.94	38.01	18.87	9.81	3.09	1.17
1.0	559.87	376.34	122.85 / 369.20	151.64	61.92	27.12	12.92	3.66	1.31
0.5	4004.49	2325.28 / 1753.36	733.85	246.76	87.02	34.60	15.47	4.08	1.40
0	7027.15 / 3368.53	2406.80	945.77	295.39	98.39	37.72	16.47	4.23	1.44
$\frac{x}{b}$ oder $\frac{y}{b}$ / or	0	0.5	1.0	1.5	2.0	2.5	3.0	4.0	5.0

z/b = 1.0

	z/b = 1.5								
5.0	4.30	4.20	3.92	3.52	3.04	2.55	2.09	1.34	0.83 / 1.81
4.0	11.86	11.46	10.36	8.84	7.18	5.63	4.30	2.41 / 5.03	2.87
3.0	40.49	38.31	32.65	25.53	18.74	13.21	9.12 / 17.43	8.68	4.39
2.5	82.35	76.58	62.27	45.55	31.07	20.40 / 36.02	24.42	11.16	5.31
2.0	178.19	161.79	123.34	82.68	51.45 / 80.20	52.09	33.40	13.99	6.27
1.5	398.32	350.49	245.80	147.85 / 188.91	118.86	72.37	43.90	16.93	7.19
1.0	845.60	718.43	459.17 / 437.25	278.34	163.97	94.16	54.41	19.59	7.97
0.5	1465.77	1211.22 / 840.73	595.52	361.43	202.93	111.77	62.46	21.47	8.49
0	1791.81 / 1081.77	951.32	664.94	396.48	218.70	118.64	65.51	22.15	8.68
x/b oder y/b or	0	0.5	1.0	1.5	2.0	2.5	3.0	4.0	5.0

z/b = 2.0

Tabelle 1.14 a Spannungsbeiwert, $i(10^{-4})$fuer z/b = 0.5; 1.0; 1.5; 2.0 [1]

Table 1.14 a Stress coefficient, $i(10^{-4})$ for z/b = 0.5; 1.0; 1.5; 2.0 [1]

	z/b = 2.5								
5.0	13.99	13.71	12.93	11.75	10.34	8.85	7.41	4.95	3.19 / 4.89
4.0	32.85	31.94	29.42	25.80	21.69	17.64	14.00	8.42 / 12.19	7.41
3.0	84.48	81.13	72.14	59.96	47.25	35.82	26.48 / 34.74	19.50	10.81
2.5	138.27	131.69	114.37	91.84	69.47	50.47 / 61.35	45.44	24.09	12.77
2.0	224.75	212.00	179.21	138.38	100.07 / 109.88	80.76	57.93	29.04	14.75
1.5	352.42	329.00	270.25	200.32 / 193.42	143.58	102.27	71.21	33.94	16.60
1.0	511.59	473.02	378.41 / 316.78	244.75	177.00	122.79	83.41	38.16	18.12
0.5	656.30	602.63 / 447.49	374.48	284.85	202.36	137.91	92.16	41.06	19.14
0	716.56 / 507.24	476.10	396.88	300.21	211.93	143.53	95.35	42.09	19.49
x/b oder y/b or	0	0.5	1.0	1.5	2.0	2.5	3.5	4.0	5.0

z/b = 3.0

	z/b = 4								
5.0	28.71	28.28	27.04	25.13	22.75	20.14	17.48	12.60	8.72 / 12.34
4.0	53.26	52.24	49.33	44.97	39.72	34.15	28.72	19.36 / 24.50	16.99
3.0	98.18	95.80	89.12	79.35	68.01	56.50	45.80 / 49.44	33.97	22.48
2.5	130.79	127.25	117.40	103.21	87.05	71.01 / 69.43	58.26	39.11	25.32
2.0	169.74	164.68	150.69	130.80	108.59 / 95.22	80.91	67.19	44.16	28.04
1.5	212.01	205.14	186.29	159.83 / 125.44	108.89	91.75	75.51	48.73	30.44
1.0	251.58	242.90	219.21 / 155.79	139.48	120.38	100.77	82.35	52.42	32.33
0.5	280.33	270.27 / 179.06	166.86	148.96	128.09	106.77	86.87	54.81	33.55
0	290.91 / 187.88	183.80	170.80	152.32	130.81	108.89	88.46	55.65	33.97
x/b oder y/b or	0	0.5	1.0	1.5	2.0	2.5	3.0	4.0	5.0

z/b = 5.0

Tabelle 1.14b Spannungsbeiwert, $i(10^{-4})$ fuer z/b = 2.5; 3.0; 4.0; 5.0 [1]

Table 1.14b Stress coefficient, $i(10^{-4})$ for z/b = 2.5; 3.0; 4.0; 5.0 [1]

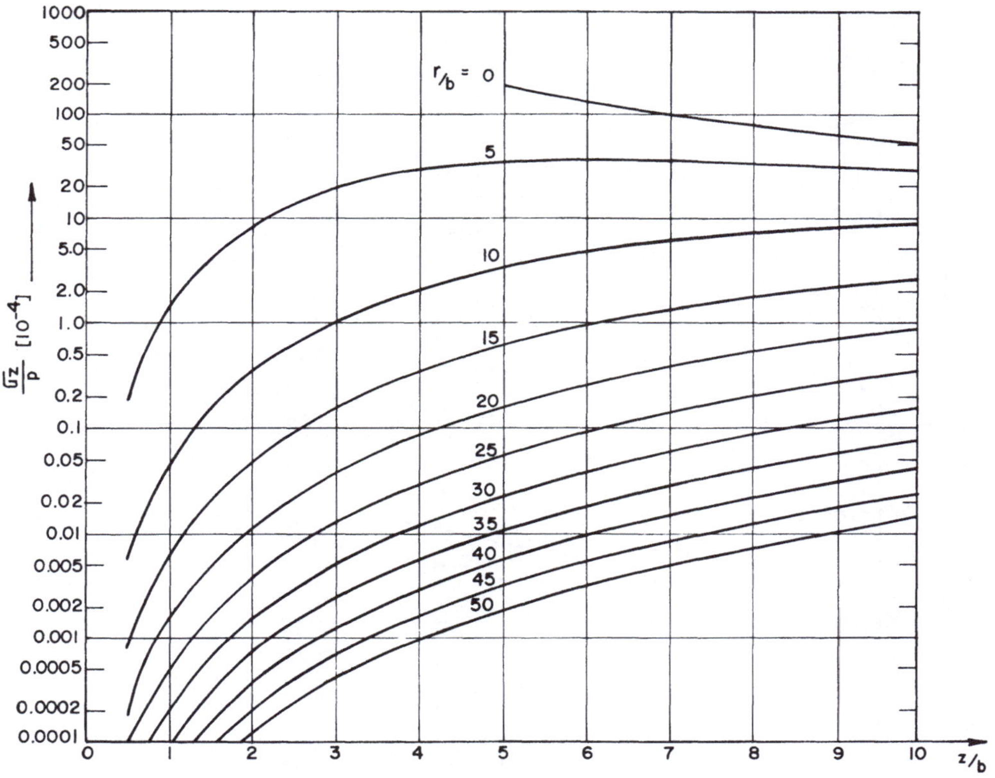

Fig. 1.27a Spannungsbeiwerte σ_z/p (10^{-4}) fuer $0<z/b\leq 10$ [1]

Stress coefficients σ_z/p (10^{-4}) for $0<z/b\leq 10$ [1]

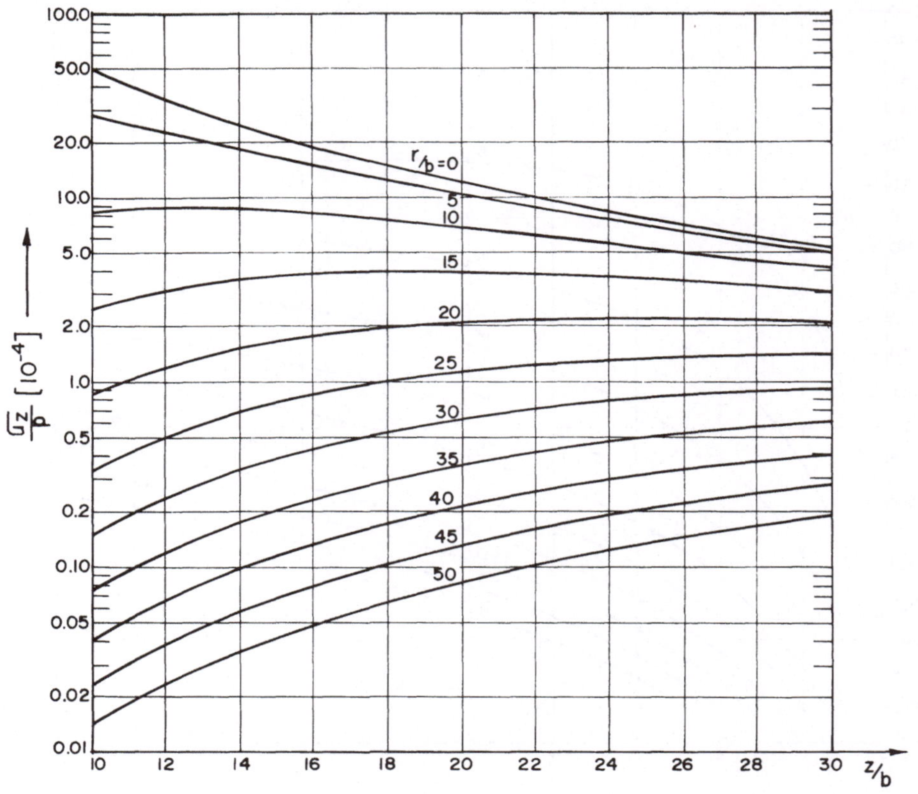

Fig. 1.27b Spannungsbeiwerte σ_z/p (10^{-4}) fuer $10 \le z/b \le 30$ [1]
Stress coefficients σ_z/p (10^{-4}) for $10 \le z/b \le 30$ [1]

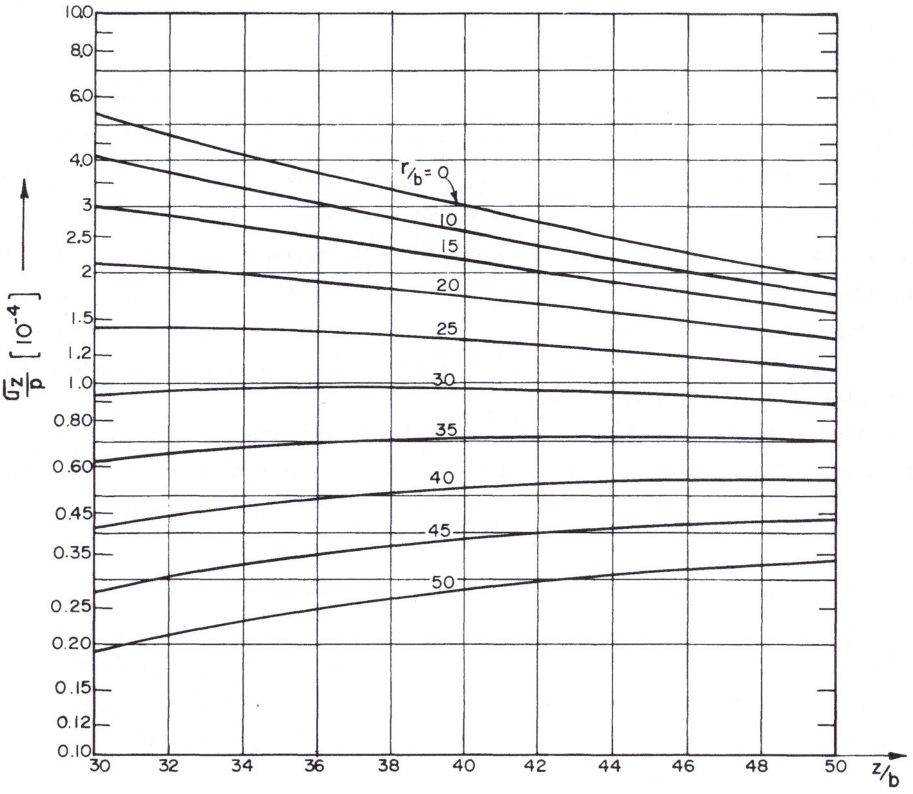

Fig. 1.27c Spannungsbeiwerte σ_z/p (10^{-4}) fuer $30 \leq z/b \leq 50$ [1]

Stress coefficients σ_z/p (10^{-4}) for $30 \leq z/b \leq 50$ [1]

(ii) Erlaeuterung

Es wurde empfohlen, dass man Tabelle 1.14b zur Berechnung lotrechten Bodenspannungen unter der Last flaeche bzw. in der Naehe der Lastflaeche verwendet, waehrend Fig. 1.27a bis Fig. 1.27c zur Berechnung lotrechten Bodenspannungen ausserhalb der Lastflaeche angewandet werden sollen.

Die Tabellen und Tafel koennen auch fuer Rechtecke angewandet werden, in dem die Rechtecke in Quadrate geteilt werden.

(ii) Explanatory Note

It is suggested to use Table 1.14a and Table 1.14b for determining vertical stresses in soil under the loaded area or close to the loaded area. Fig. 1.27a till Fig. 1.27c should be used for determining vertical stresses in soil due to loaded area located at some distance away from the point under consideration.

The Tables and Charts may also be used for rectangular loads by subdividing the rectangle into squares.

(iii) Anwendungsbeispiel

(iii)a *Gegeben*

Quadratische, gleichmaessig verteilte Flaechenlast $a = b = 2.0$ m

$p = 500$ kN/m^2

(iii) Example

(iii)a *Given*

Uniformly distributed square surface loading $a = b = 2.0$ m

$p = 500$ kN/m^2

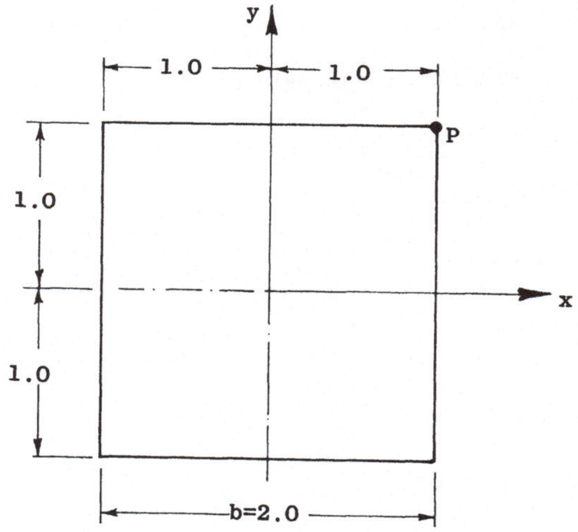

Gesucht

Lotrechte Bodenspannung in der Tiefe $z = 1.0$ m unter Punkt P.

Required

Vertical soil stress at the depth $z = 1.0$ m under Point P.

Loesung

Solution

$$\frac{z}{b} = \frac{1}{2} = 0.5$$

$$\frac{x}{b} = \frac{1}{2} = 0.5$$

$$\frac{y}{b} = \frac{1}{2} = 0.5$$

Aus Tabelle 1.14a ergibt sich

From Table 1.14a

$$\frac{\sigma z}{p} = (2325.28)10^{-4}$$

$$\sigma_z = p(2325.28)10^{-4} = 116.3\ kN/m^2$$

EineVergleichsberechnung nach Steinbrenner liefert identisches Ergebnis.

An identical result is obtained by using the table of Steinbrenner.

(iii)b *Gegeben*

Rechteckige, gleichmaessig verteilte Flaechenlast.

(iii)b *Given*

Uniformly distributed rectangular surface loading.

a = 6.0 m

b = 4.0 m

$p = 500\ kN/m^2$

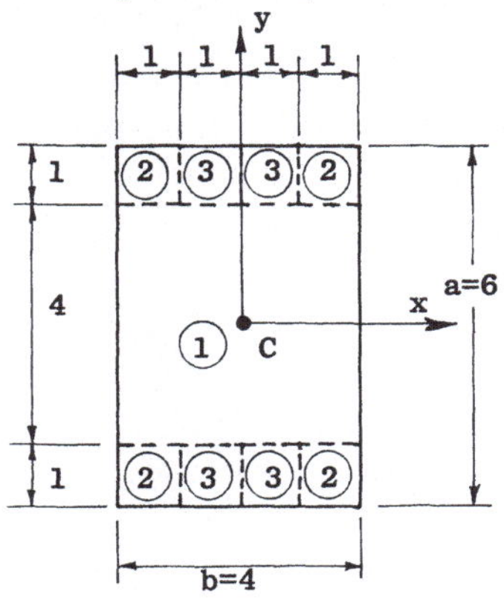

Gesucht

Lotrechte Bodenspannung unter dem Punkt C in einer Tiefe von z = 4.0 m

Required

Vertical soil stress under Point C at a depth of z = 4.0 m

Loesung

Die Rechteckflaeche wird in 9 quadratische Flaechen geteit.

Flaeche 1: $\frac{x}{b} = 0$; $\frac{y}{b} = 0$; $\frac{z}{b} = 1$

Aus Tabelle 1.14a,

$i_1 = 0.3369$

Flaeche 2: $\frac{x}{b} = 1.5$; $\frac{y}{b} = 2.5$; $\frac{z}{b} = 4$

Aus Tabelle 1.14a,

$i_2 = 0.0103$

Flaeche 3: $\frac{x}{b} = 0.5$; $\frac{y}{b} = 2.5$, $\frac{z}{b} = 4$

Aus Tabelle 1.14a,

$i_3 = 0.0127$

Solution

The rectangular area is divided into 9 squares.

Area 1: $\frac{x}{b} = 0$, $\frac{y}{b} = 0$; $\frac{z}{b} = 1$

From Table 1.14a,

$i_1 = 0.3369$

Area 2: $\frac{x}{b} = 1.5$; $\frac{y}{b} = 2.5$; $\frac{z}{b} = 4$

From Table 1.14a,

$i_2 = 0.0103$

Area 3: $\frac{x}{b} = 0.5$; $\frac{y}{b} = 2.5$; $\frac{z}{b} = 4$

From Table 1.14a

$i_3 = 0.0127$

$$\sigma_z = p(i_1 + 4 \cdot i_2 + 4 \cdot i_3)$$

$$= 500\left[(0.3369 + 4(0.0103) + 4(0.0127)\right]$$

$$= 500(0.4289) = 214.4 \text{ kN/m}^2$$

(iv) Quelle

Reference

[1] Kondner/Krizek (1965): Calculation of the vertical stress distribution beneath groups of footings
Geotechnique 13, PP 396

a.4 Lotrechte und Waagerechte Spannungen

(i) Tafeln

a.4 Vertical and horizontal stresses

(i) Charts

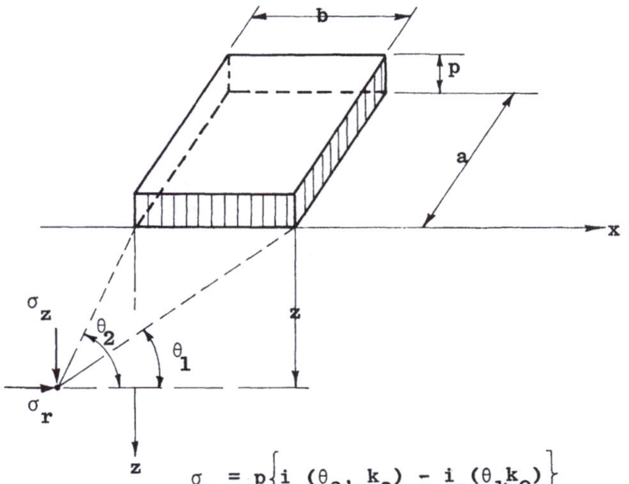

$$\sigma_z = p\left\{i_z(\theta_2,\ k_0) - i_z(\theta_1, k_0)\right\}$$

$$\sigma_x = p\left\{i_x(\theta_2,\ k_0) - i_x(\theta_1,\ k_0)\right\}$$

$$k_0 = a/_z$$

(ii) Erlaeuterung

Nach Fig. 1.28 kann man die Spannungen unter jeden Punkt einer rechteckigen, gleichmaessig verteilten Flaechenlast berechnen. Man zerlegt die Flaeche in zwei Teilrechtecke parallel zur den Aussenkanten. Fuer die 2 aneinanderstossenden Punkte dieser Teiflaechen greift man entsprechend den Verhaeltnissen a/z und θ_1 bzw θ_2 der Teilflaechen die Einflussbeiwerte i/z bzw. i/x ab. Diese 2 i/z bzw. i/x-Werte werden addiert und σ_z bzw. σ_x errechnet.

(ii) Explanatory note

The stress under any point of a uniformly distributed rectangular suface loading may be determined using Fig. 1.28. The rectangular area is divided into two segments having sides parallel to the outer edges. For the two adjoining rectangles - whose common side includes the point under consideration - one calculates the respective values of a/z, θ_1 and θ_2 to obtain the appropriate influence factors i_z and i_x. The influence factors thus obtained are then added to get σ_z and σ_x.

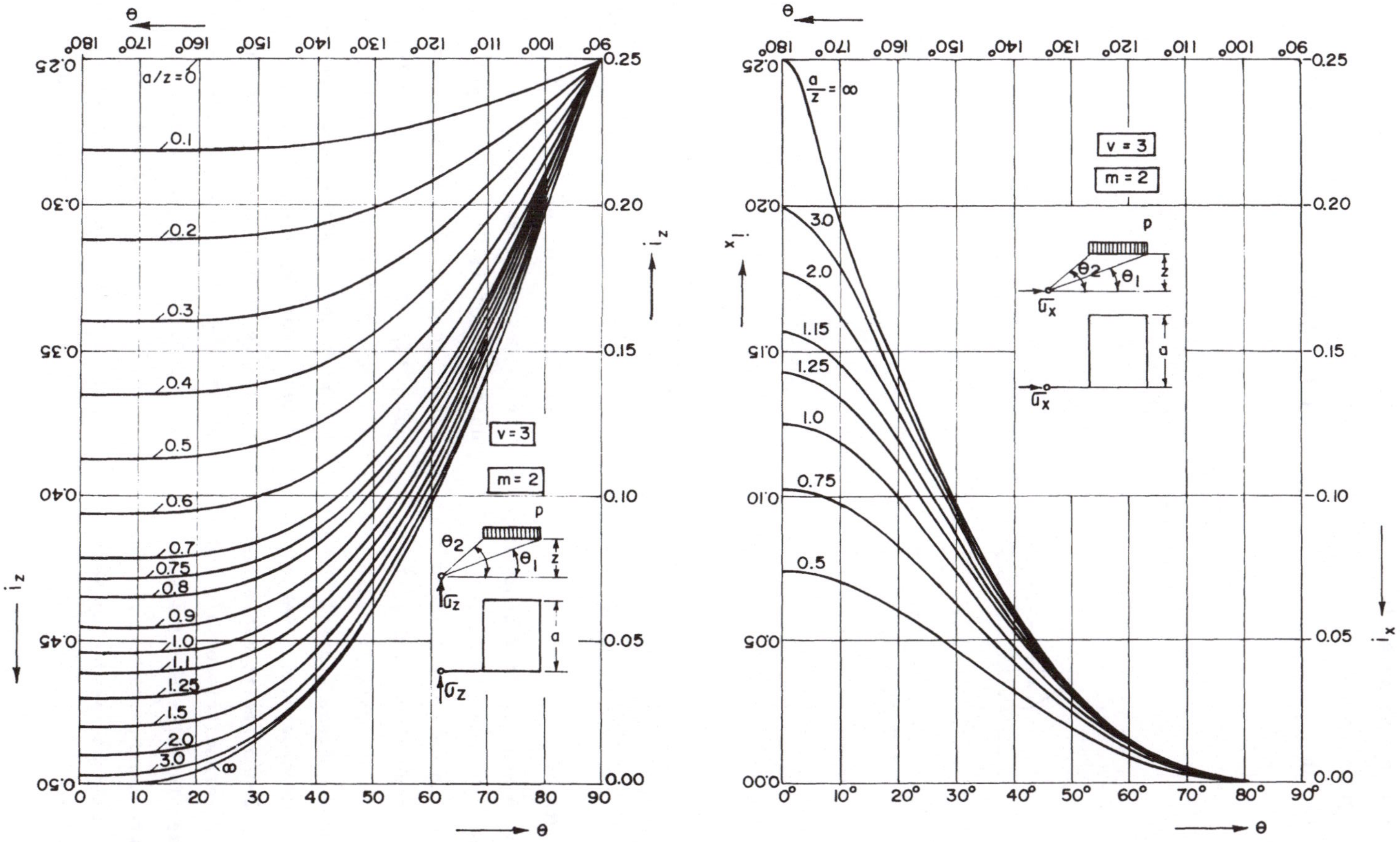

Fig. 1.28 Einflusslinien zur Ermittlung σ_z und σ_x [1]

Influence lines for the determination of σ_z and σ_x [1]

(iii) Anwendungsbeispiel

Gegeben

Rechteckige, gleichmassig verteilte Flaechenlast $a = 12.0$ m
$b = 8.0$ m
$p = 200$ kN/m^2

(iii) Example

Given

Uniformly distributed rectangular surface loading $a = 12.0$ m
$b = 8.0$ m
$p = 200$ kN/m^2

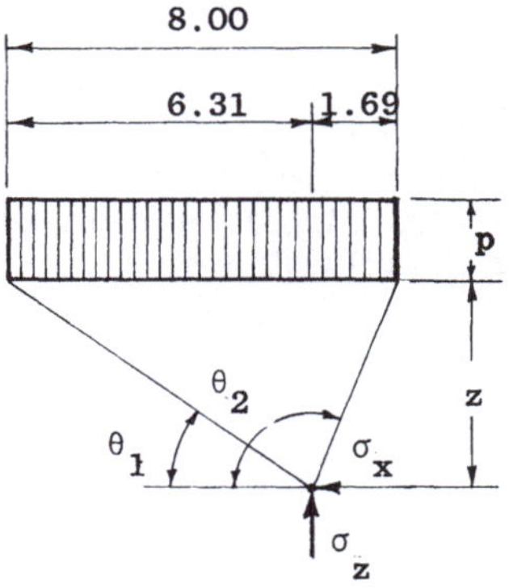

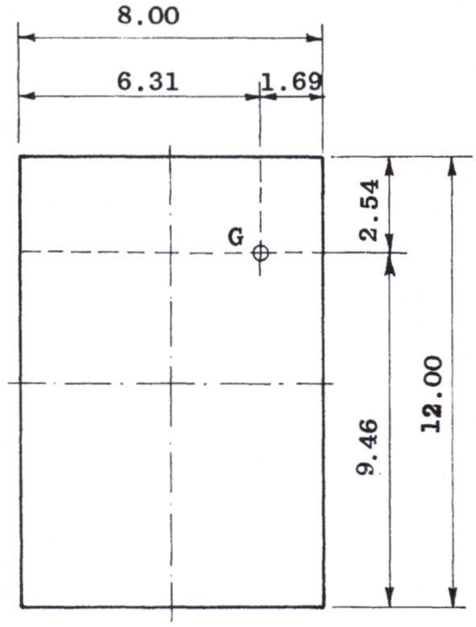

Gesucht

Lotrechte und waagerechte Bodenspannungen in verschiedenen Tiefen z unter einem beliebigen, z.B. Punkt G.

Required

Vertical and horizontal soil stresses at different depths, z, under any given point, say, Point G.

Loesung

Solution

$$k_{01} = 9.46/z$$

$$\theta_1 = \text{arccotan } 6.31/z$$

$$k_{02} = 2.54/z$$

$$\theta_2 = - \text{arccotan } 1.69/z$$

Zu diesen parametern liest man die Funktionswerte i_z und i_x aus Fig. 1.28 ab. Durch die oben angegebenen Subtraktionen und Additionen folgen daraus die Ergebnisse $\frac{\sigma_2}{p}$ und $\frac{\sigma_x}{p}$.

Using the above parameters one reads the influence coefficients i_z and i_x from Fig. 1.28. By carrying out the above indicated subtractions and additions, one obtains the results $\frac{\sigma_2}{p}$ and $\frac{\sigma_x}{p}$.

1	2	3	4	5	6	7	8	9
Tiefe unter Grundungs-Sohle Depth under Foundation	Winkel Angle		Parameter zur Beruecksichtigung der Bauwerkslaenge Parameters for the length of the structure		Aus Fig. 1.28 abgelesene Einflussbeiwerte Influence factors read from Fig. 1.28			
z	θ_1	θ_2	k_{01}	k_{02}	$i_z(\theta_2, k_{0_1})$	$i_z(\theta_2, k_{0_2})$	$i_z(\theta_1, k_{0_1})$	$i_z(\theta_1, k_{0_2})$
m	°	°	-	-	-	-	-	-
2.5	21.6	125.0	3.79	1.03	0.422	0.400	0.007	0.047
5.0	38.4	108.7	1.89	0.51	0.350	0.314	0.036	0.121
7.5	49.9	102.7	1.26	0.34	0.320	0.285	0.076	0.165
10.0	57.8	99.6	0.95	0.25	0.296	0.269	0.112	0.194
12.5	63.1	97.8	0.76	0.20	0.285	0.263	0.140	0.211

1	10	11	12	13
Tiefe unter Grundungs-Sohle Depth under Foundation	Spalte 6-8 Column 6-8	Spalte 7-9 Column 7-9	Spalte 10+11 Column 10+11	Spalte (12)(p) Column (12)(p)
z	-	-	$\sigma_{z/p}$	σ_z
m	-	-	-	kN/m^2
2.5	0.415	0.353	0.768	153.6
5.0	0.314	0.193	0.507	101.4
7.5	0.244	0.120	0.364	72.8
10.0	0.184	0.075	0.259	51.8
12.5	0.145	0.052	0.197	39.4

1	2	3	4	5	6	7	8	9
iefe unter rundungs- ohle	Winkel		Parameter zur Beruecksichtigung der Bauwerks laenge		Aus Fig. 1.28 abgelesene Einflussbeiwerte			
epth under oundation	Angle		Parameters for the length of the structure		Influence factors read from Fig. 1.28			
z	θ_1	θ_2	k_{01}	k_{02}	$i_x(\theta_2,k_{0_1})$	$i_x(\theta_2,k_{0_2})$	$i_x(\theta_1,k_{0_1})$	$i_x(\theta_1,k_{0_2})$
m	°	°	-	-	-	-	-	-
2.5	21.6	125.0	3.79	1.03	0.023	0.019	-0.133	-0.096
5.0	38.4	108.7	1.89	0.51	0.004	0.002	-0.063	-0.034
7.5	49.9	102.7	1.26	0.34	0.001	0.000	-0.029	-0.014
10.0	57.8	99.6	0.95	0.25	0.000	0.000	-0.015	-0.007
12.5	63.1	97.8	0.76	0.20	0.000	0.000	-0.008	-0.003

1	10	11	12	13
Tiefe unter Grundungs- sohle	Spalte 6-8	Spalte 7-9	Spalte 10+11	Spalte (12)(p)
Depth under Foundation	Column 6-8	Column 7-9	Column 10+11	Column (12)(p)
z	-	-	σ_x/p	σ_z
m	-	-	-	kN/m^2
2.5	0.156	0.115	0.271	54.2
5.0	0.067	0.036	0.103	20.6
7.5	0.030	0.014	0.044	8.8
10.5	0.015	0.007	0.022	4.4
12.5	0.008	0.003	0.011	2.2

(iv) Quelle

Reference

[1] Lehmann/Neuber (1955): Ein Beitrag zur Berechnung von Bodenspannungen fuer Setzungsberechnungen nach der Hypothese von Froehlich
Die Bautechnik 32, H.2

b) Halbraum nach Froehlich

(i) Tafeln

i)a Tafeln nach Kollbrunner [1]

b) Half-space according to Froehlich

(i) Charts

i)a Chart of Kollbrunner [1]

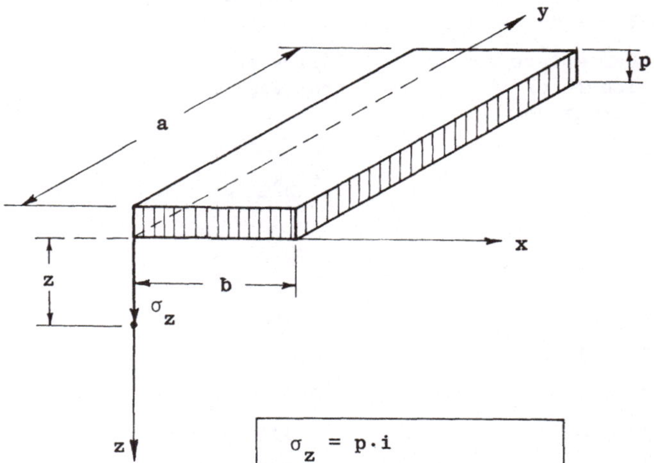

$$\sigma_z = p \cdot i$$

$$i = f(\nu,\ a/b,\ z/b)$$

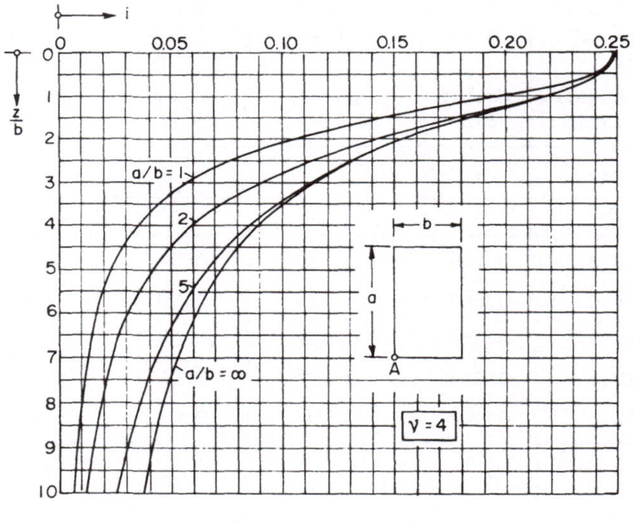

Fig. 1.29a Einflussbeiwert i [1]

Influence coefficient i [1]

(i)b Tafeln nach Lorenz [2]

(i)b Chart of Lorenz [2]

$\sigma_z = p \cdot i$

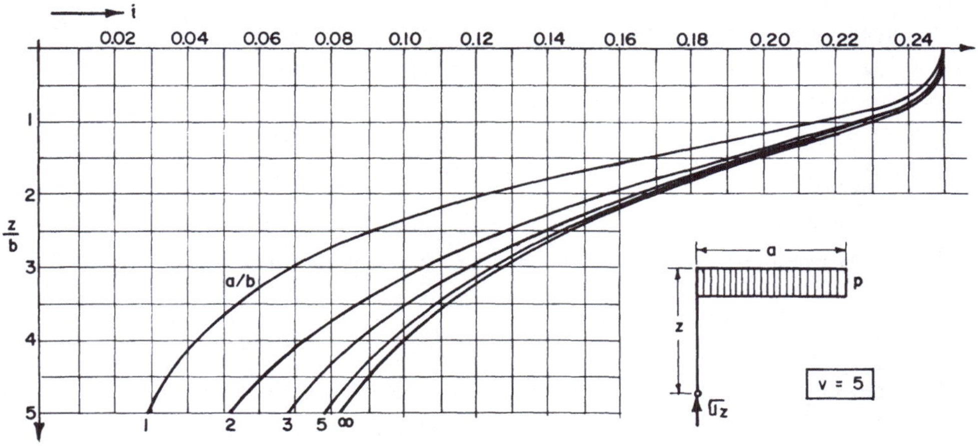

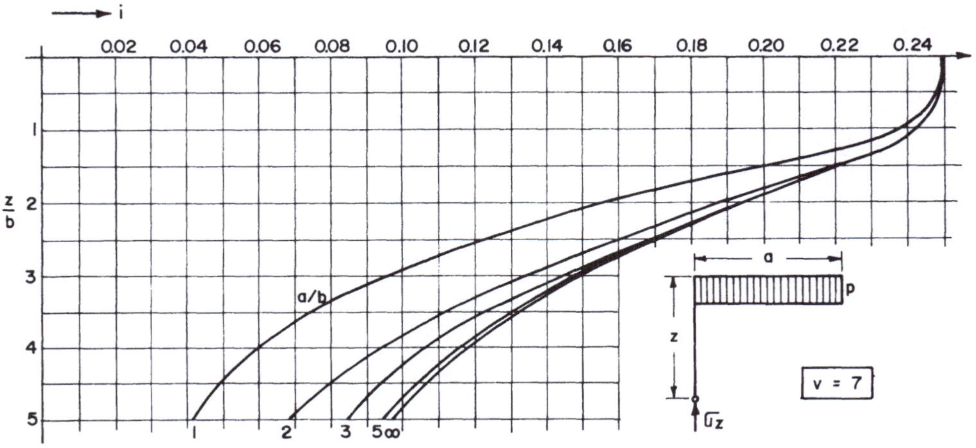

Fig. 1.29b Einflussbeiwert i [2]

Influence coefficient i [2]

(ii) Erlaeuterung

Die lotrechte Spannung wird unter dem Eckpunkt der rechteckigen Flaechenlast gerechnet.
Die Einflussbeiwerte, die in Fig. 1.29 angegeben sind, gelten fuer $\nu = 4$ bzw. $\nu = 5$ bzw. $\nu = 7$.

(ii) Explanatory note

The vertical stress is determined under the corner point of the rectangular surface loading. The influence coefficient given in Fig. 1.29 are valid for $\nu = 4$, $\nu = 5$ and $\nu = 7$ respectively.

(iii) Quellen
References

[1] Kollbrunner (1946): Fundation und Konsolidation
Bd. 1. S. 349.
Schweizer Druck und Verlagshaus, Zuerich.

[2] Lorenz (1949): Berechnung der Spannungsverteilung
unter rechtechfoermigen Lastkoerpern.
Bauplanung und Bautechnik 3, S. 109.

c) Halbraum nach Westergaard

(i) Formel/Tafel

c) Half-space according to Westergaard

(i) Formula/Chart

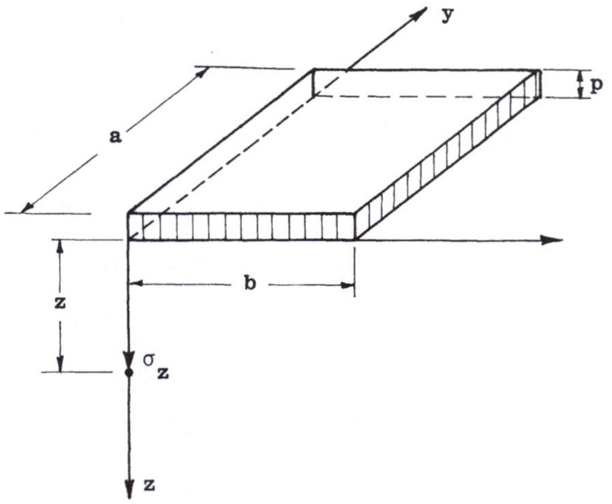

$$\sigma_z = \frac{p}{2\pi} \arctan \left\{ \frac{ba}{kz} \frac{1}{\sqrt{\left(a^2 + b^2 + k^2 z^2\right)}} \right\} = p \cdot i$$

$$k = \sqrt{\frac{1 - 2\mu}{2(1-\mu)}}$$

Fuer $\mu = 0$,

For $\mu = 0$, $k = \sqrt{\tfrac{1}{2}}$

(ii) Erlaeuterung

Die lotreche Spannung wird unter dem Eckpunkt der rechteckigen Flaechenlast gerechnet.

(ii) Explanatory note

The Vertical stress is determined under the corner point of the rectangular surface loading.

(iii) Quelle

Source

[1] Fandum (1948): Influence values for estimating stresses in elastic foundations. Proc. 2 ICSMFE, Rotterdam, Vol. 3, PP 77.

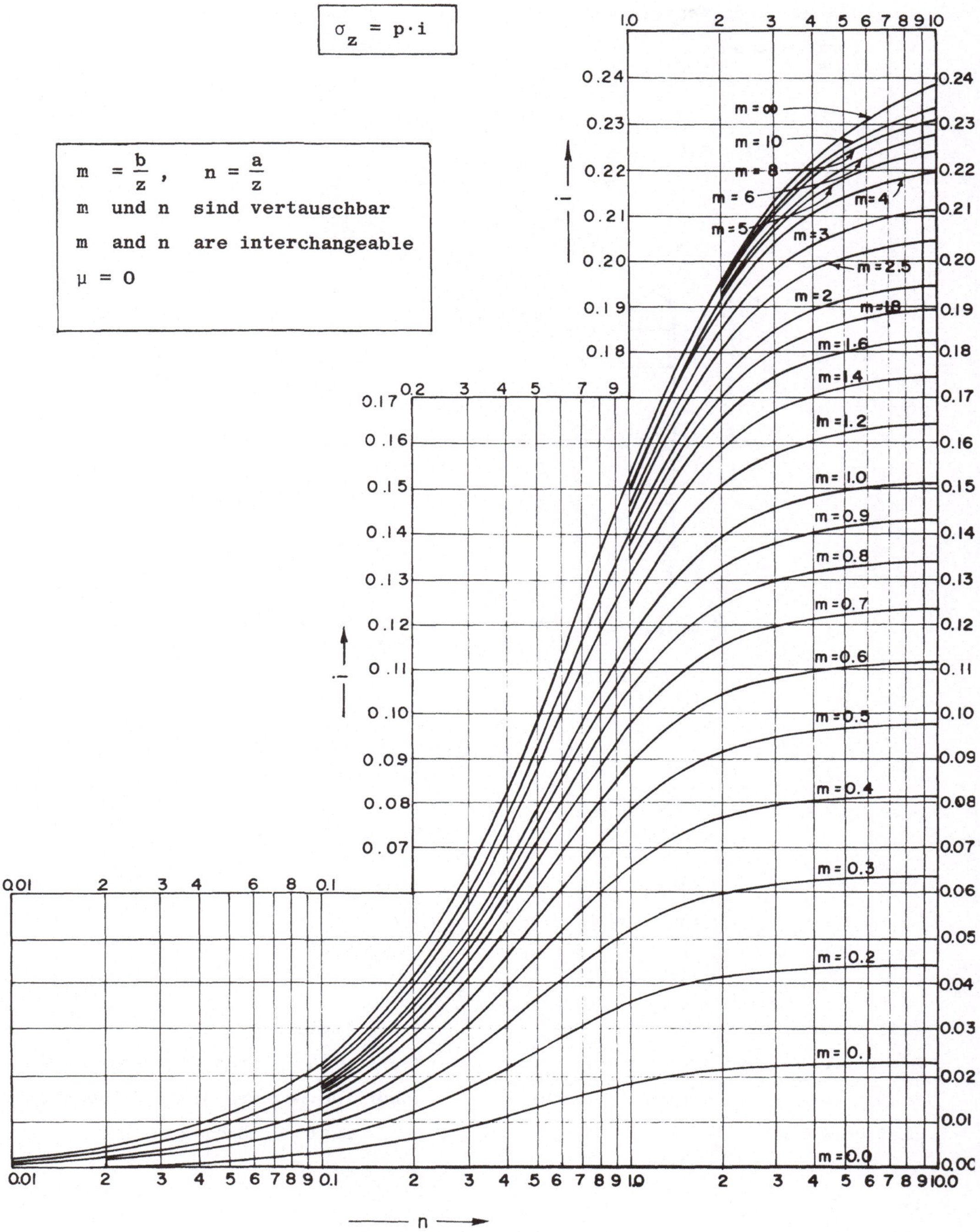

Fig. 1.30 Einflussbeiwert i [1]

Influence value i [1]

1.1.1.4.2 Dreieckige Belastung

a) Halbraum nach Boussinesq

(i) Formeln/Tafeln/Tabellen

1.1.1.4.2 Traiangular Loading

a) Half-space according to Boussinesq

(i) Formulae/Charts/Tables

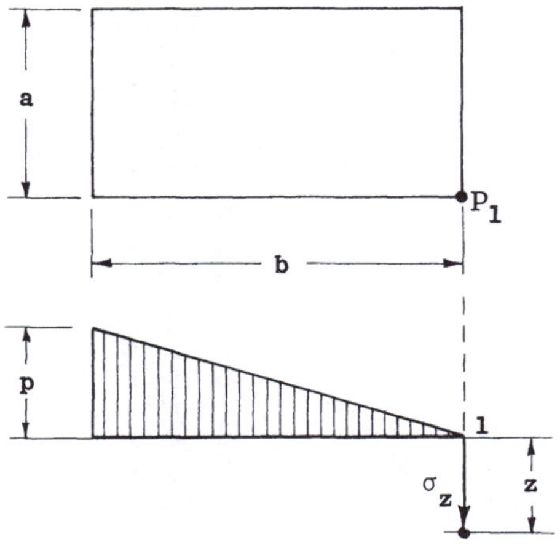

$$\sigma_z = \frac{p}{2\pi}\left[\frac{a \cdot b \cdot z}{R(z^2 + b^2)} + \frac{a \cdot z}{b \cdot R} \cdot \frac{R - \sqrt{a^2 + z^2}}{\sqrt{a^2 + z^2}}\right]$$

$$R = \sqrt{a^2 + b^2 + z^2}$$

$$\boxed{\sigma_z = p \cdot i_1}$$

	i_1											
a/b	$z/$Schmalseite b oder a					$z/$smaller side b or a						
	0.25	0.5	0.75	1.0	1.5	2	3	4	5	6	8	10
$\frac{1}{10}$	0.0039	0.0071	0.0095	0.0113	0.0132	0.0141	0.0147	0.0146	0.0142	0.0135	0.0119	0.0102
$\frac{1}{5}$	0.0077	0.0142	0.0190	0.0223	0.0257	0.0269	0.0259	0.0232	0.0201	0.0171	0.0122	0.0090
$\frac{1}{3}$	0.0128	0.0235	0.0311	0.0357	0.0396	0.0387	0.0310	0.0248	0.0190	0.0146	0.0093	0.0064
$\frac{1}{2}$	0.0191	0.0348	0.0446	0.0497	0.0502	0.0447	0.0313	0.0216	0.0154	0.0114	0.0068	0.0046
1.0	0.0370	0.0606	0.0687	0.0666	0.0522	0.0384	0.0214	0.0131	0.0089	0.0063	0.0036	0.0024
2.0	0.0374	0.0633	0.0753	0.0773	0.0682	0.0553	0.0352	0.0232	0.0161	0.0118	0.0069	0.0046
5.0	0.0374	0.0637	0.0765	0.0795	0.0732	0.0631	0.0463	0.0349	0.0269	0.0213	0.0139	0.0099

Tabelle 1.15 a Einflussbeiwert i_1 zur Berechnung der Bodenspannungen unter dem Eckpunkt 1 [1, 2]

Table 1.15 a Influence coefficient i_1 for determining the soil stresses under corner point 1 [1, 2]

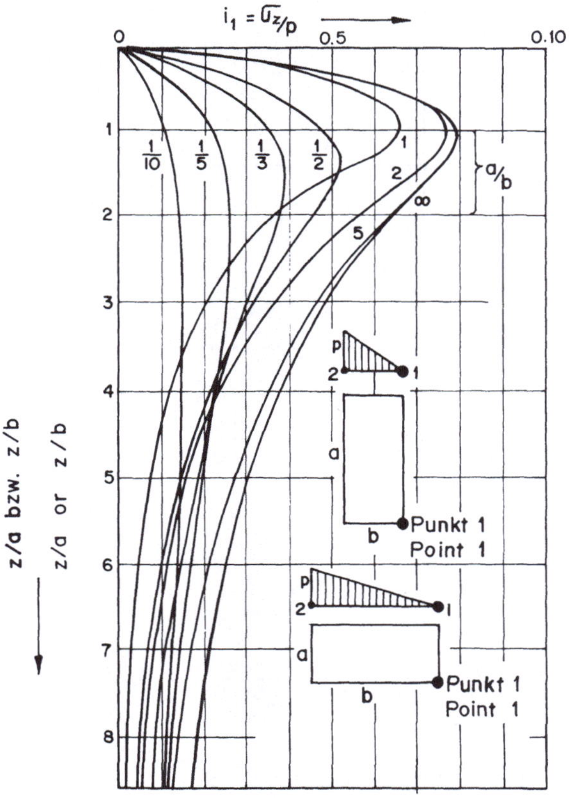

Fig. 1.31 a) Einflussbeiwert i_1 zur Berechnung der Bodenspannungen unter dem Eckpunkt 1 [1, 2]

Influence coefficient i_1 for determining the soil stresses under corner point 1 [1, 2]

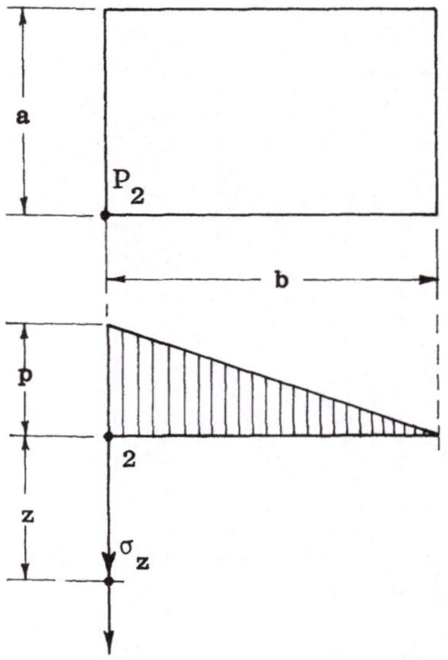

$$\sigma_z = \frac{p}{2\pi}\left[\arctan\frac{a\cdot b}{z\cdot R} + \frac{a\cdot z}{a^2+z^2}\cdot\frac{R-\sqrt{a^2+z^2}}{b}\right]$$

$$R = \sqrt{a^2 + b^2 + z^2}$$

$$\boxed{\sigma_z = p\cdot i_2}$$

i_2												
a/b	$z/$Schmalseite b oder a					$z/$smaller side b or a						
	0.25	0.5	0.75	1.0	1.5	2	3	4	5	6	8	10
1/10	0.2446	0.2328	0.2144	0.1933	0.1538	0.1233	0.0840	0.0612	0.0468	0.0370	0.0248	0.0177
1/5	0.2408	0.2256	0.2049	0.1821	0.1407	0.1094	0.0700	0.0480	0.0331	0.0260	0.0160	0.0108
1/3	0.2356	0.2162	0.1923	0.1675	0.1242	0.0929	0.0550	0.0356	0.0245	0.0177	0.0103	0.0065
1/2	0.2292	0.2043	0.1771	0.1502	0.1059	0.0755	0.0419	0.0259	0.0174	0.0124	0.0072	0.0046
1.0	0.2103	0.1719	0.1373	0.1086	0.0688	0.0456	0.0233	0.0139	0.0091	0.0064	0.0037	0.0024
2.0	0.2110	0.1758	0.1464	0.1226	0.0878	0.0648	0.0380	0.0243	0.0166	0.0120	0.0070	0.0046
5.0	0.2110	0.1762	0.1476	0.1250	0.0933	0.0732	0.0497	0.0364	0.0261	0.0218	0.0141	0.0099

Tabelle 1.15 b Einflussbeiwert i_2 zur Berechnung der Bodenspannungen unter dem Eckpunkt 2 [1, 2]

Table 1.15 b Influence coefficient i_2 for determining the soil stresses under corner point 2 [1,2]

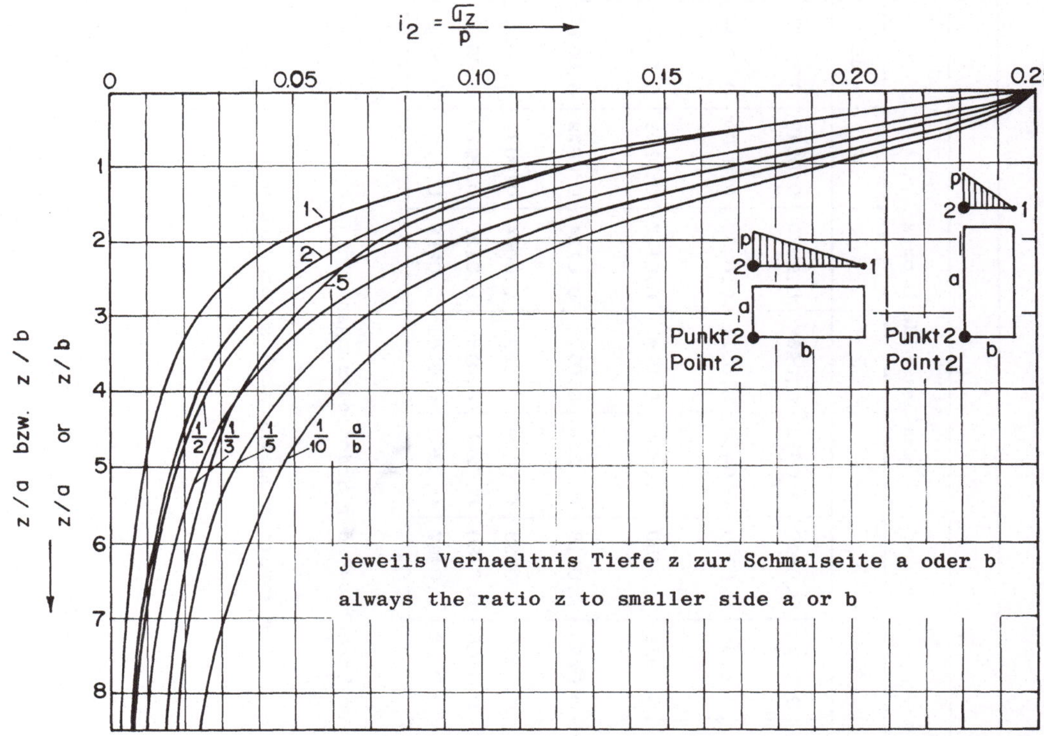

Fig. 1.31 b) Einflussbeiwert i_2 zur Berechnung der Bodenspannungen unter dem Eckpunkt 2 [2]

Influence coefficient i_2 for determining the soil stresses under corner point 2 [2]

(ii) Erlaeuterung

Die Tabellen und Tafeln eignen sich auch fuer die Spannungsberechnung laengs der Achse 1-1 und 2-2 (s. skizze) und zwar im Bereich inne - halb sowie ausserhalb der Belastung. Durch Superposition von verschiedenen Rechtecken koennen die Spannungen berechnet werden. z.B. fuer Punkt 3:

$\sigma_z = p \{i(b,a_1) + i(b,a_2)\}$

(ii) Explanatory note

The tables and charts may also be used to determine the vartical stresses along the axis 1-1 and 2-2 (Ref. sketch below) within or outside the loaded area. This is effected by superposing different rectangles. For point 3 for example :

$\sigma_z = p \{i(b,a_1) + i(b,a_2)\}$

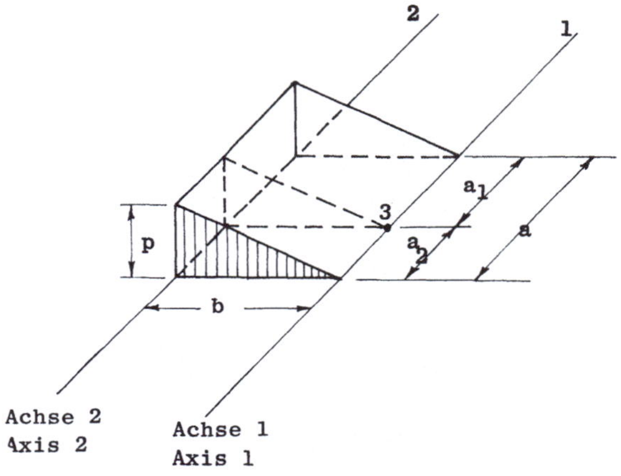

(iii) Quellen

References

[1] DIN 4019, Blatt 2: Setzungsberechnung bei schraeg und bei aussermittig wirkender Balastung (Verkantung) Feb. 1961.

[2] Jelinek (1949): Setzungsberechnung aussmittig belasteter Fundamente. Bauplannung und Bautechnik 3, S. 115.

b) Halbraum nach Froehlich

(i) Formeln/Tafeln

b) Half-space according to Froehlich

(i) Formulae/Chart

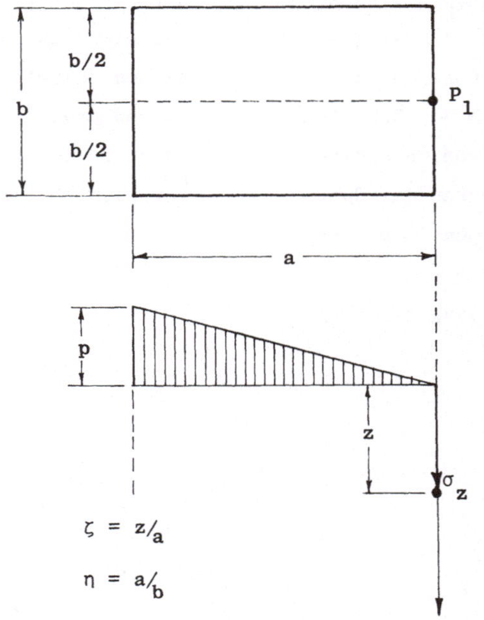

$\zeta = z/a$

$\eta = a/b$

$$\sigma_z = \frac{p}{2\pi}\left[\frac{\zeta}{\sqrt{\zeta^2\cdot\eta^2+\frac{1}{4}}}\left(1-\frac{1}{12(\zeta^2\cdot\eta^2+\frac{1}{4})}\right)-\frac{\zeta^5}{(\zeta^2+1)^2\sqrt{\zeta^2\cdot\eta^2+\eta^2+\frac{1}{4}}}\cdot\left(1-\frac{1}{12(\zeta^2\cdot\eta^2+\eta^2+\frac{1}{4})}\right)\right] = p\cdot i_1$$

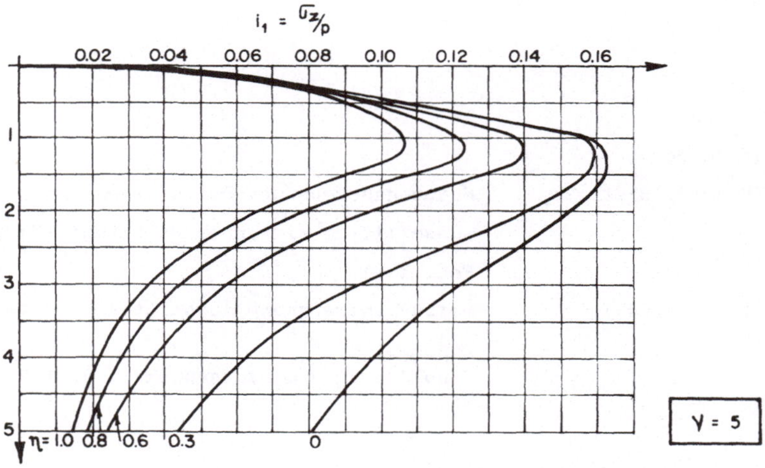

Fig. 1.32a Einflussbeiwerte i_1 zur Berechnung der Bodenspannungen unter dem Eckpunkt 1 [1]

Influence coefficients i_1 for determining the soil stresses under corner point 1 [1]

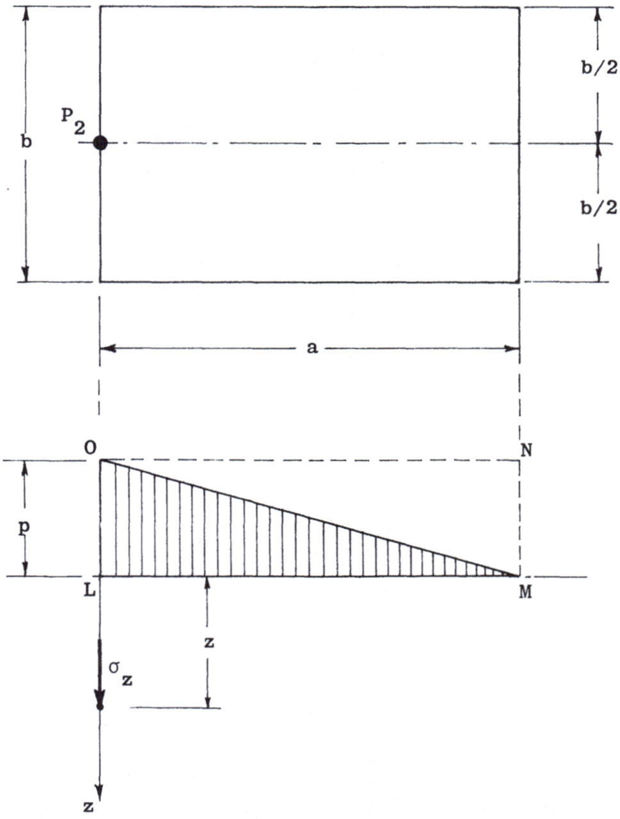

σ_z infolge gesamter rechteckiger Belastung (OLMN) - σ_{z1}

σ_z due to total rectangular loading (OLMN) - σ_{z1}

Beispiel fuer a/b = 0 — Example for a/b = 0

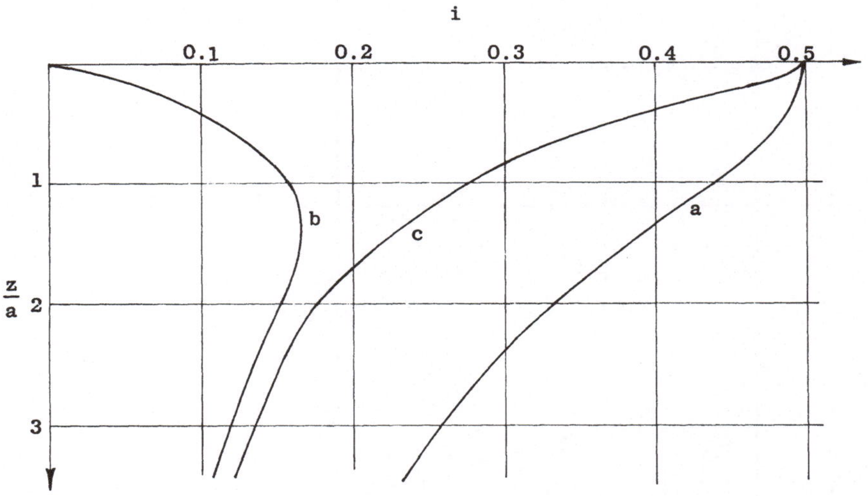

a) Spannungen infolge rechteckiger Belastung OLMN

b) Spannungen infolge dreieckiger Belastung MN

c) Gesuchte Spannungen infolge dreieckiger Belastung LMO

a) Stresses due to the rectangular loading OLMN

b) Stresses due to the triangular loading MNO

c) Required stresses due to the Triangular loading LMO

(ii) Erlaeuterung

Formeln und Tafeln sind fuer lotrechte Spannungen unter dem Kantenmittelpunkt fuer $\nu = 5$ abgeleitet.

(ii) Explanatory Note

The formulae and chars are derived for vertical stresses under the middpoint of the edge for $\nu = 5$.

(iii) Quelle

Reference

Lorenz (1949): Berechnung der Spannungsverteilung unter rechteckfoermigen Lastkoerpern. Bauplanung und Bautechnik 3, PP 109.

1.1.1.5 Kreisfoermiger Grundriss

1.1.1.5.1 Gleichmaessige Belastung

a) Halbraum nach Boussinesq

(i) Tafeln/Tabelle

1.1.1.5 Circular Plan Area

1.1.1.5.1 Uniformly distributed loading

a) Half-space according to Boussinesq

(i) Charts/Table

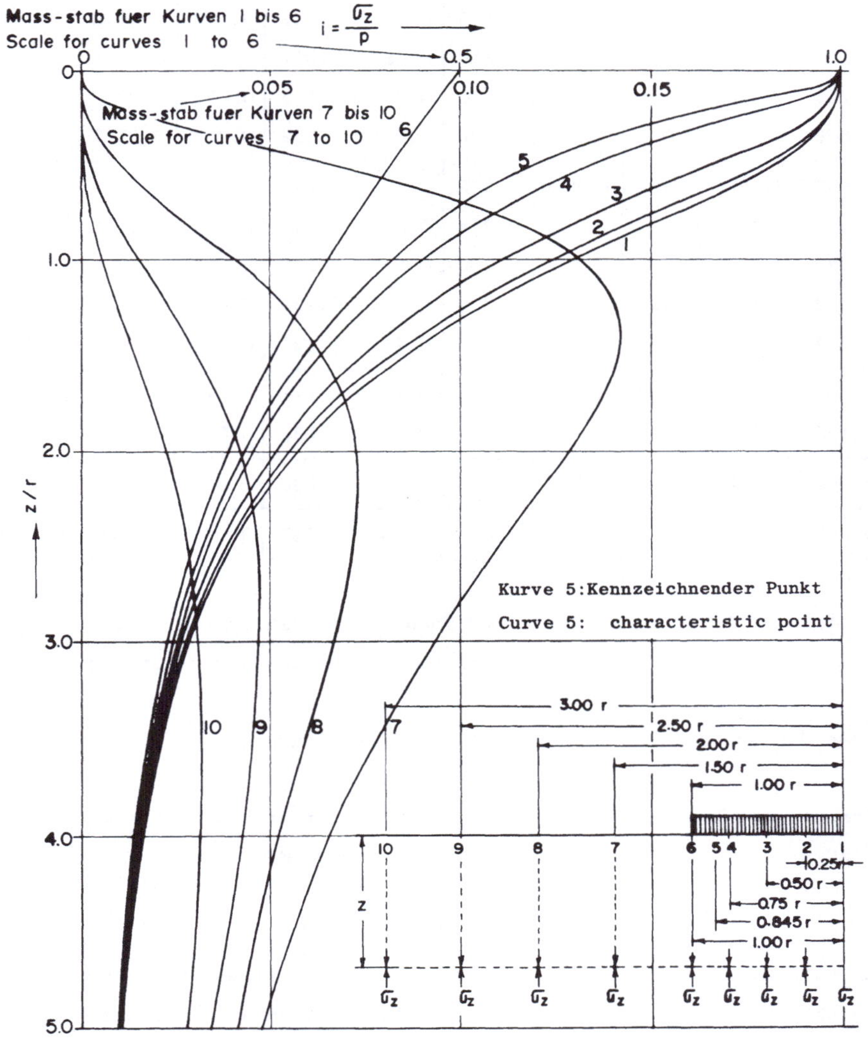

Fig. 1.33 Lotrechte Spannungen in verschiedenen Punkten und Tiefen unter schlaffen, gleichmaessig verteilter Kreislast [1]

Vertical stresses at different points and depths under uniformly distributed circular surface loading [1]

$$i = \frac{\sigma_z}{p}$$

$\frac{z}{r}$	Pkt 1	Pkt 2	Pkt 3	Pkt 4	Pkt 5	Pkt 6	Pkt 7	Pkt 8	Pkt 9	Pkt 10
0.2	0.992	0.990	0.977	0.898	0.817	0.465	0.011	0.001	0.0002	0.0001
0.4	0.949	0.936	0.885	0.735	0.650	0.430	0.047	0.006	0.0016	0.0006
0.6	0.864	0.840	0.766	0.615	0.546	0.397	0.087	0.016	0.0048	0.0017
0.8	0.756	0.727	0.652	0.523	0.470	0.363	0.115	0.028	0.0097	0.0037
1.0	0.646	0.619	0.553	0.449	0.409	0.330	0.132	0.041	0.0157	0.0064
1.2	0.547	0.523	0.469	0.388	0.358	0.298	0.140	0.052	0.0222	0.0097
1.4	0.460	0.442	0.400	0.337	0.314	0.269	0.142	0.061	0.0283	0.0132
1.6	0.390	0.374	0.342	0.294	0.276	0.241	0.140	0.067	0.0337	0.0167
1.8	0.332	0.319	0.295	0.258	0.244	0.217	0.135	0.071	0.0383	0.0200
2.0	0.284	0.274	0.256	0.227	0.216	0.195	0.129	0.073	0.0418	0.0230
2.5	0.200	0.193	0.184	0.168	0.162	0.150	0.111	0.072	0.0466	0.0286
3.0	0.146	0.142	0.137	0.128	0.124	0.118	0.093	0.067	0.0471	0.0315
4.0	0.087	0.085	0.084	0.080	0.078	0.076	0.066	0.052	0.0419	0.0316
5.0	0.057	0.056	0.056	0.054	0.053	0.052	0.047	0.041	0.0346	0.0282

Tabelle 1.16 Zahlenwerte der Spannungskurven in Fig. 1.33 [1]
Numerical values for the stress curves of Fig. 1.33 [1]

(ii) Anwendungsbeispiel

Gegeben

Kreisflaeche mit r = 4.0 m,
$p = 200\ kN/m^2$

Gesucht

Spannungsverlauf in einem horizontalen schnitt bei z = 4.0 m

(ii) Example

Given

Circular area with r = 4.0 m,
$p = 200\ kN/m^2$

Required

Stress distribution along a horizontal section at a depth z = 4.0 m

$z/_r = 4/_4 = 1.0$

Aus Tabelle 1.16

From Table 1.16

Punkt Point	1	2	3	4	5	6	7	8	9	10
i	0.646	0.619	0.553	0.449	0.409	0.330	0.132	0.041	0.0157	0.0064
$\sigma_z = p \cdot i$ kN/m^2	129.2	123.8	110.6	89.8	81.8	66.0	26.4	8.2	3.1	1.3

(iii) Quelle

Reference

[1] Grasshoff (1954): Der Einfluss der Schichtstaerke auf die Sohldruckverteilung und die Biegemomente einer kreisfoermigen Gruendungsplatte. Bautechnik 31, S. 330.

b) Halbraum nach Froehlich

(i) Formeln/Tafeln

b) Half-space according to Froehlich

(i) Formulae/Charts

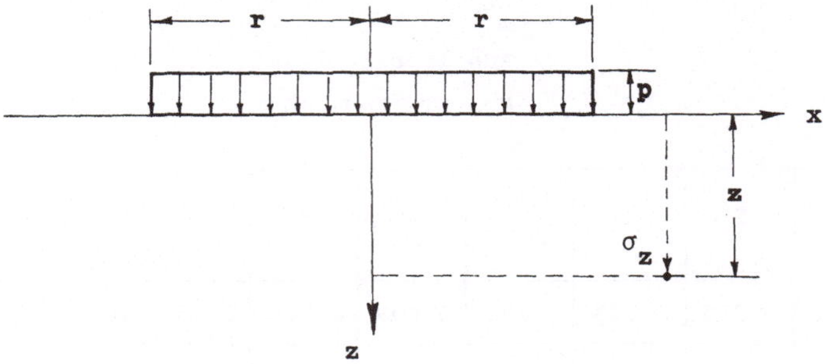

Fuer $\frac{x}{r} = 0$

For $\frac{x}{r} = 0$

$$\sigma_z = p \left\{ 1 - \left(\frac{z/_r}{\sqrt{1+(\frac{z}{r})^2}} \right)^{\nu} \right\}$$

Fuer $\nu = 2$

For $\nu = 2$

$$\sigma_z = \frac{p}{2} \left\{ 1 \pm \sqrt{\frac{\left[(\frac{x}{r})^2+(\frac{z}{r})^2\right]^2 - 2\left[(\frac{x}{r})^2+(\frac{x}{r})^2\right] + 1}{\left[(\frac{x}{r})^2+(\frac{z}{r})^2\right]^2 - 2\left[(\frac{x}{r})^2-(\frac{z}{r})^2\right] + 1}} \right\} = p \cdot i$$

Fuer $\nu = 4$

For $\nu = 4$

$$\sigma_z = p \left\{ \frac{1}{2} \pm \frac{1}{2} \sqrt{\frac{\left[(\frac{z}{r})^2-(\frac{x}{r})^2\right]^2 - 2\left[(\frac{x}{r})^2-(\frac{z}{r})^2\right] + 1}{\left[(\frac{x}{r})^2+(\frac{z}{r})^2\right]^2 - 2\left[(\frac{x}{r})^2-(\frac{z}{r})^2\right] + 1}} \right.$$

$$\left. - (\frac{z}{r})^2 \frac{(\frac{z}{r})^2\left[(\frac{z}{r})^2+1\right] + 2(\frac{z}{r})^2(\frac{x}{r})^2 + (\frac{x}{r})^2\left[(\frac{x}{r})^2-1\right]}{\left[\left[(\frac{x}{r})^2+(\frac{z}{r})^2\right]^2 - 2\left[(\frac{x}{r})^2-(\frac{z}{r})^2\right] + 1\right]^{\frac{3}{2}}} \right\} = p \cdot i$$

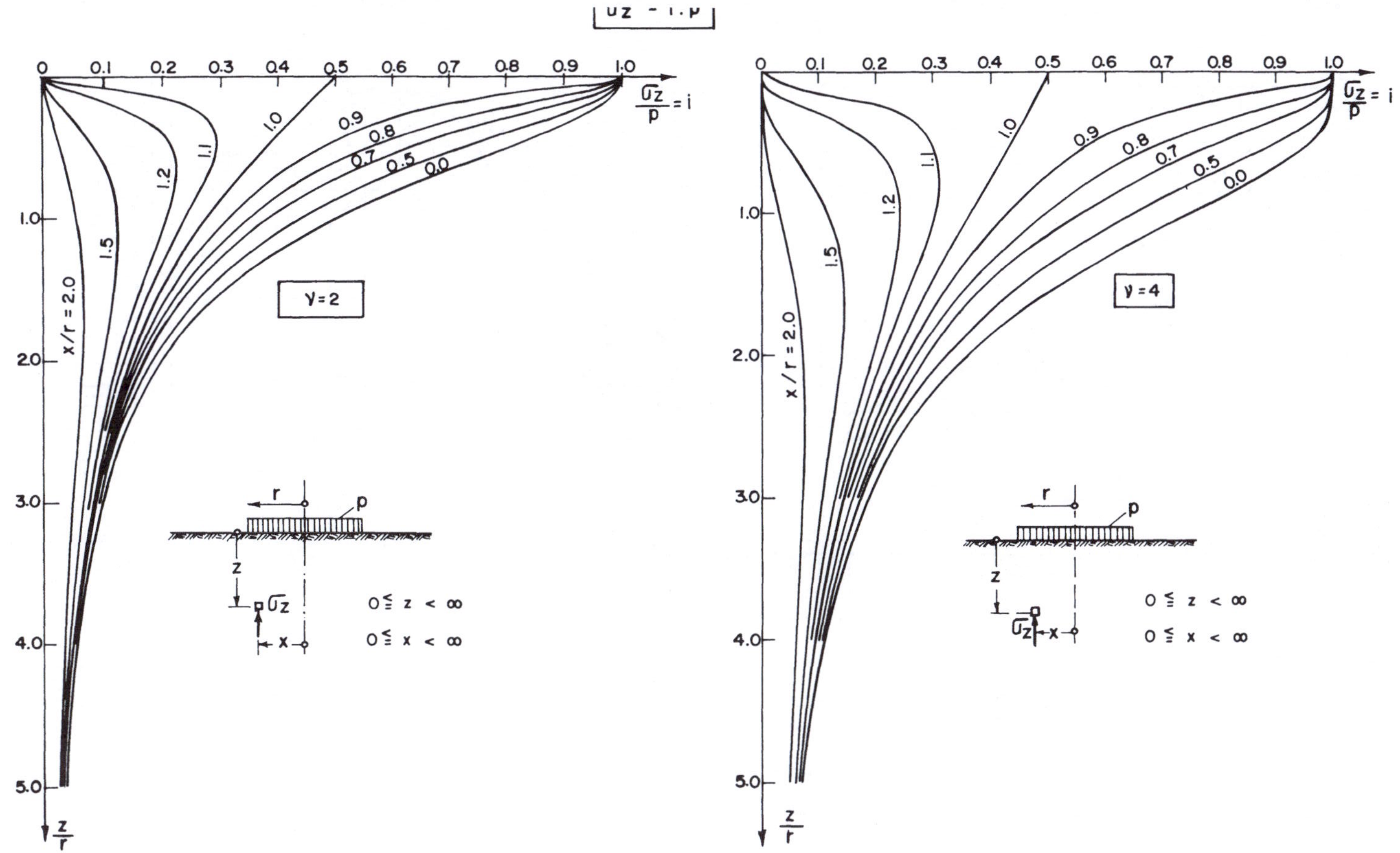

Fig. 1.34 Verlauf der lotrechten Spannungen fuer Konzentrationsfaktor $\nu = 2$ bzw. $\nu = 4$ [2]

Distribution of vertical stresses for concentration factor $\nu = 2$ and $\nu = 4$ [2]

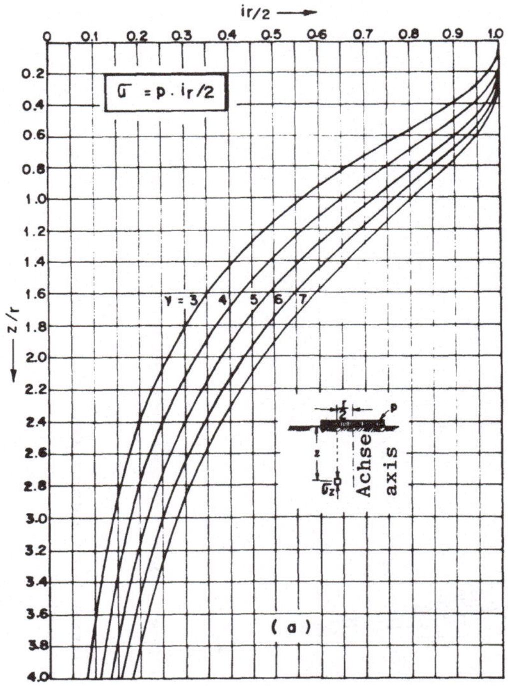

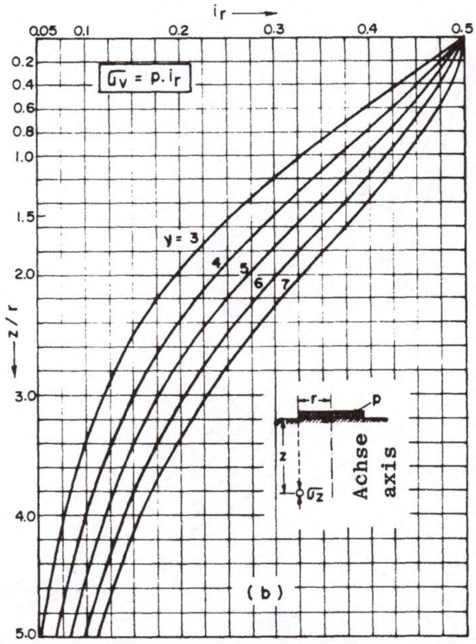

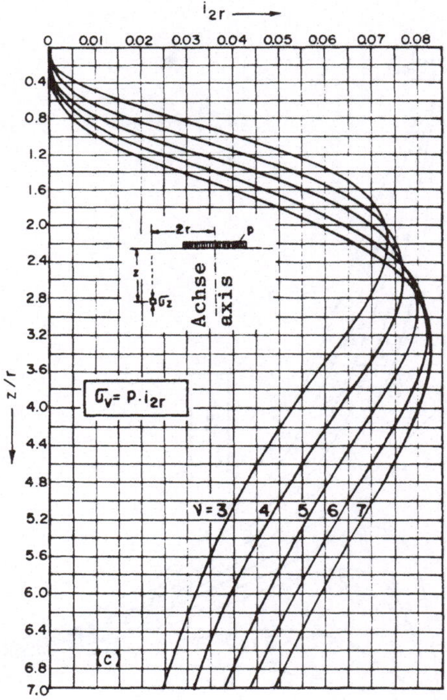

Fig. 1.35 Lotrechte Spannung, σ_v, fuer verschiedene Konzentrationsfaktoren unter:
Vertical stress, σ_v, for diffrent concentration factors under:

a) einem Punkt im Abstand r/2
   a point at a distance r/2
b) einem Randpunkt
   a point at the edge
c) einem Punkt im Abstand 2r
   a point at a distance 2r

Randpunkt, Naeherungsweise

At the edge,as approximation

$$\sigma_z = \frac{p}{2}\left[1-\frac{\cos^{\nu-1}\alpha}{\cos^{\nu-2}\beta}\right]$$

$$\tan\alpha = \frac{2r}{z}\ ;\ \tan\beta = \frac{\sqrt{2}.r}{z}$$

(ii) Vorausseitzung
Schlaffe gleichmaessig verteilte Kreislast

(ii) Assumption
Uniformly distributed circular surface loading

(iii) Quellen
References

[1] Lorenz/Neumener (1953): Spannungsberechnung infolge kreislasten unter beliebigen Punkten innerhalb und ausserhalb der Kreisflaeche. Bautechnik 30, S.127.

[2] Meissner (1958) Zur Berechnung der Spannungen unter beliebigen Punkten innerhalb und ausserhalb einer gleichmaessig belasteten schlaffen kreisflaeche. Bauingenieur 33, S. 342.

c) Halbraum nach Westergaard

c) Half-space according to Westergaard

(i) Formeln/Tabellen

(i) Formulae/Tables

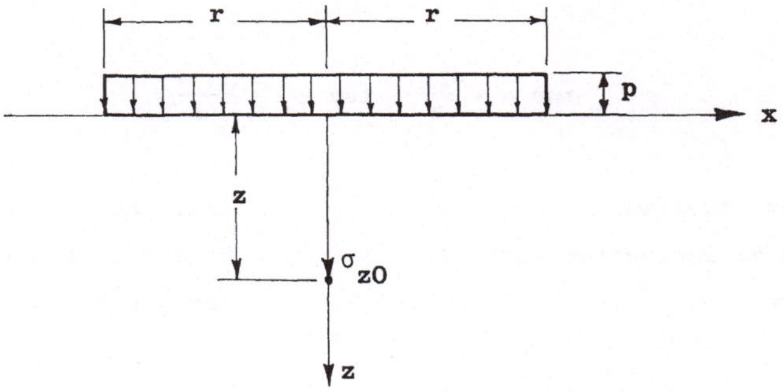

$$\sigma_{z0} = p\left\{1 - \frac{k}{\left[k^2 + \left(\frac{r}{z}\right)^2\right]^{\frac{1}{2}}}\right\}$$

$$k = \sqrt{\frac{1 - 2\mu}{2(1-\mu)}}$$

Fuer $\mu = 0$

For $\mu = 0$

$$\sigma_{z0} = p\left\{1 - \frac{1}{\sqrt{1 + 2\left(\frac{r}{z}\right)^2}}\right\}$$

$$= p \cdot i_{z0}$$

$\sigma_{z0} = p \cdot i_{z0}$

r/z	i_{z0}	r/z	i_{z0}	r/z	i_{z0}	r/z	i_{z0}	r/z	i_{z0}
0.00	0.0000	0.30	0.0794	0.60	0.2375	0.90	0.3822	1.20	0.4923
1	0.0001	1	0.0842	1	0.2428	1	0.3864	1	0.4955
2	0.0004	2	0.0890	2	0.2481	2	0.3906	2	0.4985
3	0.0009	3	0.0938	3	0.2534	3	0.3948	3	0.5016
4	0.0016	4	0.0988	4	0.2586	4	0.3989	4	0.5046
5	0.0025	5	0.1038	5	0.2638	5	0.4029	5	0.5076
6	0.0036	6	0.1089	6	0.2690	6	0.4069	6	0.5106
7	0.0049	7	0.1140	7	0.2741	7	0.4109	7	0.5135
8	0.0063	8	0.1191	8	0.2792	8	0.4149	8	0.5165
9	0.0080	9	0.1244	9	0.2843	9	0.4188	9	0.5193
0.10	0.0099	0.40	0.1296	0.70	0.2893	1.00	0.4227	1.30	0.5222
1	0.0119	1	0.1349	1	0.2943	1	0.4265	1	0.5250
2	0.0141	2	0.1402	2	0.2993	2	0.4303	2	0.5278
3	0.0165	3	0.1456	3	0.3043	3	0.4340	3	0.5306
4	0.0190	4	0.1510	4	0.3091	4	0.4377	4	0.5333
5	0.0218	5	0.1564	5	0.3140	5	0.4414	5	0.5360
6	0.0247	6	0.1618	6	0.3188	6	0.4451	6	0.5387
7	0.0277	7	0.1672	7	0.3236	7	0.4487	7	0.5414
8	0.0309	8	0.1726	8	0.3284	8	0.4522	8	0.5440
9	0.0343	9	0.1781	9	0.3331	9	0.4558	9	0.5466
0.20	0.0378	0.50	0.1835	0.80	0.3377	1.10	0.4593	1.40	0.5492
1	0.0414	1	0.1890	1	0.3424	1	0.4627	1	0.5517
2	0.0452	2	0.1944	2	0.3470	2	0.4662	2	0.5543
3	0.0490	3	0.1998	3	0.3515	3	0.4695	3	0.5568
4	0.0531	4	0.2053	4	0.3560	4	0.4729	4	0.5592
5	0.0572	5	0.2107	5	0.3605	5	0.4762	5	0.5617
6	0.0614	6	0.2161	6	0.3649	6	0.4795	6	0.5641
7	0.0658	7	0.2215	7	0.3693	7	0.4828	7	0.5665
8	0.0702	8	0.2268	8	0.3736	8	0.4860	8	0.5689
9	0.0748	9	0.2322	9	0.3779	9	0.4892	9	0.5713

r/z	i/z_0	r/z	i/z_0	r/z	i/z_0	i/z	i/z_0
1.50	0.5736	1.80	0.6344	2.30	0.7061	5.00	0.8600
1	0.5759	1	0.6361	.35	0.7119	.20	0.8653
2	0.5782	2	0.6379	.40	0.7174	.40	0.8702
3	0.5805	3	0.6396	.45	0.7227	.60	0.8747
4	0.5827	4	0.6413	.50	0.7278	.80	0.8790
5	0.5850	5	0.6430	.55	0.7328	6.00	0.8830
6	0.5872	6	0.6447	.60	0.7376	.50	0.8919
7	0.5893	7	0.6463	.65	0.7422	7.00	0.8995
8	0.5915	8	0.6480	.70	0.7467	.50	0.9061
9	0.5937	9	0.6496	.75	0.7510		
1.60	0.5958	1.90	0.6512	.80	0.7552	8.00	0.9120
1	0.5979	1	0.6528	.85	0.7592	9.00	0.9217
2	0.6000	2	0.6544	.90	0.7631	10.00	0.9295
3	0.6020	3	0.6560	.95	0.7669	12.00	0.9412
4	0.6041	4	0.6576	3.00	0.7706	14.00	0.9496
5	0.6061	5	0.6591	.10	0.7776	16.00	0.9559
6	0.6081	6	0.6606	.20	0.7842	18.00	0.9608
7	0.6101	7	0.6622	.30	0.7905	20.00	0.9647
8	0.6121	8	0.6637	.40	0.7964	25.00	0.9717
9	0.6140	9	0.6652	.50	0.8020	30.00	0.9764
1.70	0.6160	2.00	0.6667	.60	0.8073	40.00	0.9823
1	0.6179	2	0.6696	.70	0.8123	50.00	0.9859
2	0.6198	4	0.6725	.80	0.8171	100.00	0.9929
3	0.6217	6	0.6753	.90	0.8216	∞	1.0000
4	0.6235	8	0.6781	4.00	0.8259	$\mu = 0$	
5	0.6254			.20	0.8340	$m = \infty$	
6	0.6272	2.10	0.6809	.40	0.8413		
7	0.6290	.15	0.6876	.60	0.8481		
8	0.6308	.20	0.6940	.80	0.8543		
9	0.6326	.25	0.7002				

Tabelle 1.17 Einflussbeiwerte zur Berechnung lotrechten Spannungen in der Achse einer gleichmaessig belasteten Kreisflaeche [1]

Table 1.17 Influence values for vertical stresses beneath the centre of a uniformly loaded circular area [1]

(ii) Anwendungsbeispiel

Gegeben

Schlaffe kreislastflaeche mit
$r = 5.0$ m; $p = 200$ kN/m^2

Gesucht

Vergleich der lotrechten Spannungen in der Achse dieser Lastflaeche fuer eine Tiefe von $z = 5$ m nach den Verfahren von Boussinesq, Froehlich und Westergaard.

(ii) Example

Given

Uniform circular loading with
$r = 5.0$ m; $p = 200$ kN/m^2

Required

Comparison of the vertical stresses under the centre of a uniformly loaded circular area at the depths of $z = 5.0$ m using the methods of Boussinesq, Froehlich and Westergaard.

Loesung

a) Boussinesq

$\frac{z}{r} = \frac{5}{5} = 1$

Aus Tabelle 1.16 : $i = 0.646$

Solution

a) Boussinesq

$\frac{z}{r} = \frac{5}{5} = 1$

From Table 1.16 : $i = 0.646$

$$\sigma_z = p \cdot i = (200)(0.646) = 129.2 \text{ kN/m}^2$$

b) Froehlich

Fuer $\frac{x}{r} = 0$

b) Froehlich

For $\frac{x}{r} = 0$

$$\sigma_z = p\left\{1 - \left[\frac{\frac{z}{r}}{1+(\frac{z}{r})^2}\right]^{\nu}\right\}$$

$$\frac{z}{r} = \frac{5}{5} = 1$$

$$\sigma_z = p\left[1 - \left(\frac{1}{\sqrt{2}}\right)^{\nu}\right] = p\left(1 - \frac{1}{2^{\nu/2}}\right)$$

$\nu = 2$: $\quad \sigma_z = 200(1 - \frac{1}{2}) = 100.0$ kN/m^2

$\nu = 3$: $\quad \sigma_z = 200(1 - \frac{1}{2^{3/2}}) = 129.2$ kN/m^2

$\nu = 4$: $\quad \sigma_z = 200(1 - \frac{1}{2^2}) = 150.0$ kN/m^2

$\nu = 5$: $\quad \sigma_z = 200(1 - \frac{1}{2^{5/2}}) = 164.6$ kN/m^2

c) Westergaard

$\frac{r}{z} = \frac{5}{5} = 1.0$

Aus Tabelle 1.17, $i_{z0} = 0.4227$

c) Westergaard

$\frac{r}{z} = \frac{5}{5} = 1.0$

From Table 1.17, $i_{z0} = 0.4227$

$$\sigma_z = p \cdot i_{z0} = (200)(0.4227) = 84.5 \text{ kN/m}^2$$

(iii) Quelle

Reference

[1] Fadum (1948): Influence values for estimating stresses in elastic foundations.
Proc. 2. ICSMFE, Rotterdam, Vol. 3, PP 77

1.1.1.5.2 Gleichmaessige, conische und parabolische Belastung

(i) Formeln/Tafeln

1.1.1.5.2 Uniform, conical and parabolic loading

(i) Formuae/Charts

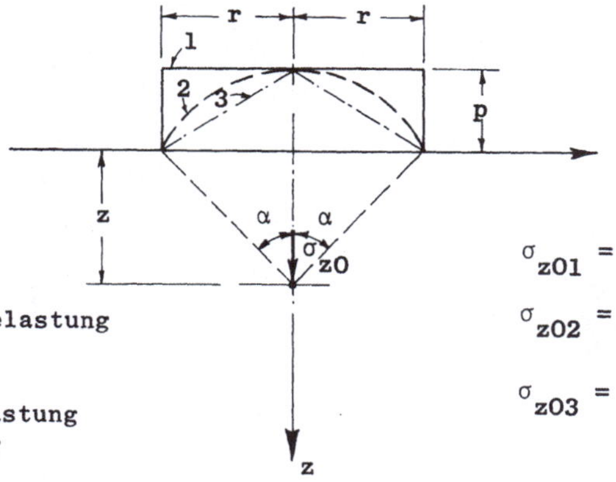

1. gleichmaessige Belastung
   uniform loading
2. parabolische Belastung
   parabolic loading
3. conische Belastung
   conical loading

$$\sigma_{z01} = p(1-\cos^3\alpha)$$

$$\sigma_{z02} = p\left\{1-2\ \mathrm{ctg}^2\alpha(1-\cos\alpha)\right\}$$

$$\sigma_{z03} = p(1-\cos\alpha)$$

$$\boxed{\sigma_{z0} = p \cdot i}$$

$$i = \frac{\sigma_i}{p}$$

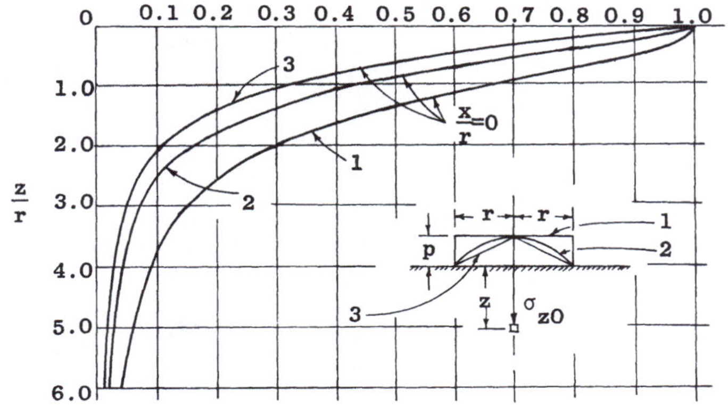

Fig. 1.36 a) Lotrechte Spannungen unter der Achse einer Kreisflaeche infolge gleichmaessiger, bzw. conischer bzw. parabolischer Belastung [1,2]

Vertical stresses under the centre of circular area acted upon by uniform, conical and parabolic loading [1, 2]

$\boxed{\sigma_z = i \cdot p}$

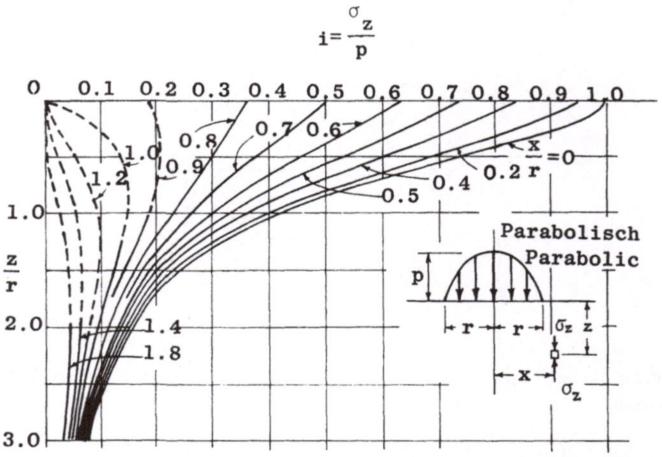

Fig. 1.36 b) Lotrechte Spannungen in verschiedenen Punkten und Tiefen bei parabolischer Belastung [1, 2]

Vertical stresses at different points and depths due to parabolic surface loading [1, 2]

(ii) Voraussetzung

Beliebige Poissonzahl m

(ii) Assumption

Any value of Poisson's number m

(iii) Quelle

References

[1] Harr/Lovell (1963): Vertical stresses under certain axisymmetrical loading. Highway Res. Board Record, 39.

[2] Szechy (1963): Der Grundbau 1. Band, S. 268 Springer Verlag Wien.

1.1.1.5.3 **Starres Kreisfundament**

(i) Formeln/Tafeln/Tabellen

A. Symmetrische Belatung

1.1.1.5.3 **Rigid Circular Foundation**

(i) Formulae/Charts/Tables

A. Symmetrical Loading

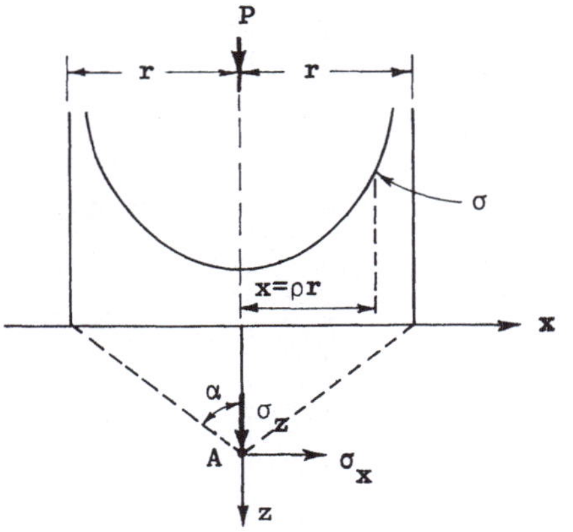

mit
with $p = \dfrac{P}{\pi r^2}$

$$\sigma = \frac{1}{2}\,\frac{p\cdot r}{\sqrt{r^2 - x^2}} = \frac{p}{2\sqrt{1-\rho^2}}$$

$$\sigma_z = p\left[\frac{1}{2}(1+\mu)\cdot\sin^2\alpha(1-2\mu + 2\cos^2\alpha)\right] = p\cdot i_z$$

$$\sigma_z = \sigma_y = p\left[\frac{1}{4}\cdot\sin^2\alpha(1 + 2\mu - 2\cos^2\alpha)\right] = p\cdot i_x$$

$\sigma_z = p \cdot i_z$

$\frac{z}{r}$	i_z			
	μ = 0.50	μ = 0.33	μ = 0.25	μ = 0
0	0	0.22	0.31	0.50
0.2	0.06	0.27	0.35	0.52
0.4	0.18	0.35	0.42	0.55
0.6	0.29	0.42	0.47	0.56
0.8	0.36	0.45	0.49	0.54
1.0	0.38	0.44	0.47	0.50
1.2	0.36	0.41	0.43	0.45
1.4	0.34	0.37	0.39	0.39
1.6	0.30	0.33	0.34	0.34
1.8	0.27	0.29	0.30	0.30
2.0	0.24	0.26	0.26	0.26
2.2	0.21	0.23	0.23	0.23
2.4	0.19	0.20	0.20	0.20
2.6	0.17	0.18	0.18	0.18
2.8	0.15	0.16	0.16	0.16
3.0	0.14	0.14	0.14	0.14
3.2	0.12	0.13	0.13	0.13
3.4	0.11	0.12	0.12	0.11
3.6	0.10	0.11	0.11	0.10
3.8	0.09	0.10	0.10	0.09
4.0	0.08	0.09	0.09	0.09

Tabelle 1.18 **Einflussbeiwert i_z zur Berechnung lotrechter Spannungen unter der Achse eines starren mittig balasteten Kreisfundaments[1]**

Table 1.18 **Influence value i_z for determining vertical stresses under the centre line of a rigid centrically loaded circular foundation [1]**

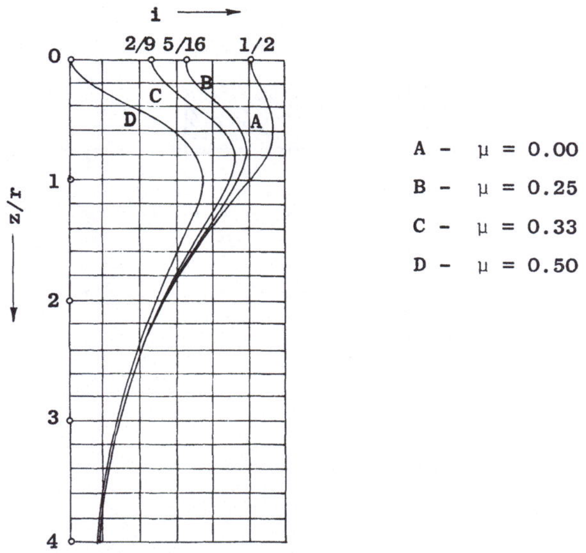

Fig. 1.37 Verlauf der Lotrechten Spannungen, σ_z, unter der Achse eines starren mittig belasteten Kreisfundaments [1]

Distribution of the vertical stresses, σ_z, under the centre line of a rigid centrically loaded circular foundation [1]

(ii) Voraussetzung

Beliebige Poissonzahl m

(ii) Assumption

Any value of Poisson's number m

(iii) Erlaeuterung

Die Virtuelle Spannung, σ_z, wird zu Setzungsberechnung unter starren Kreisfundamenten benuzt, nach der allgemeinen Formel:

(iii) Explanatory note

The virtual stress, σ_z, is used to calculate the settlement of a rigid circular footing using the general formula:

$$S = \frac{1}{E}\int_0^z \sigma_z \cdot d_z$$

wobei E = Elastizitaetsmodul des Halbraum

where E = modulus of elasticity of the half-space

B. Asymmetrische Belatung

B. Asymmetric loading

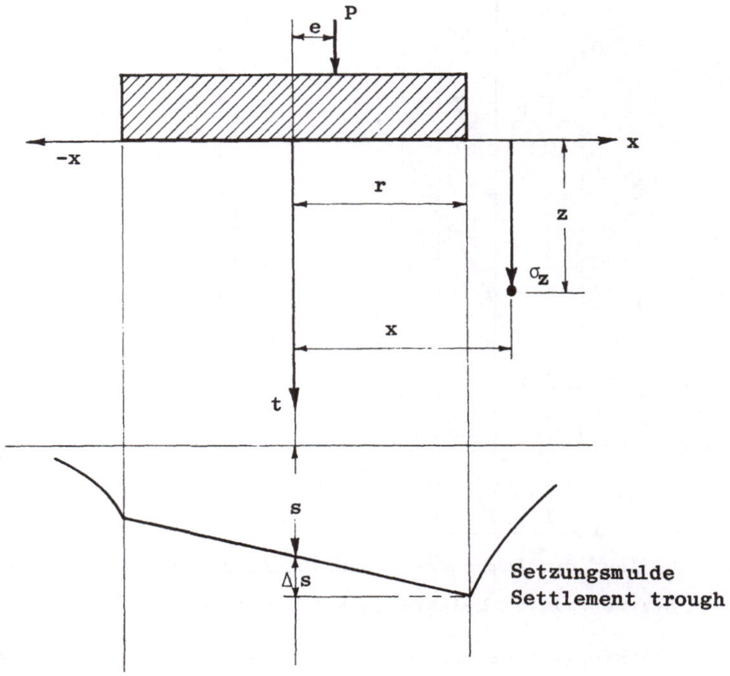

$$\sigma_z = p \cdot i \quad \text{wobei} \quad p = \frac{P}{\pi \cdot r^2}$$

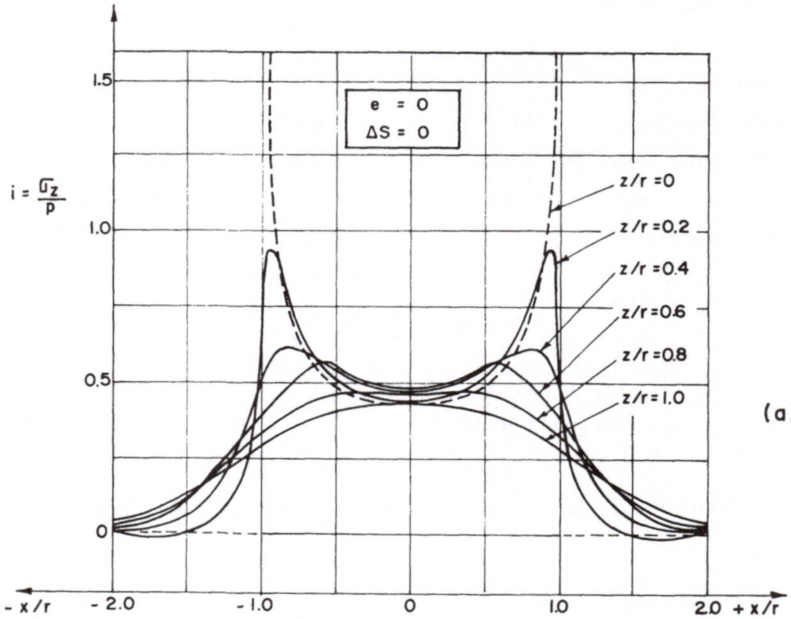

(a)

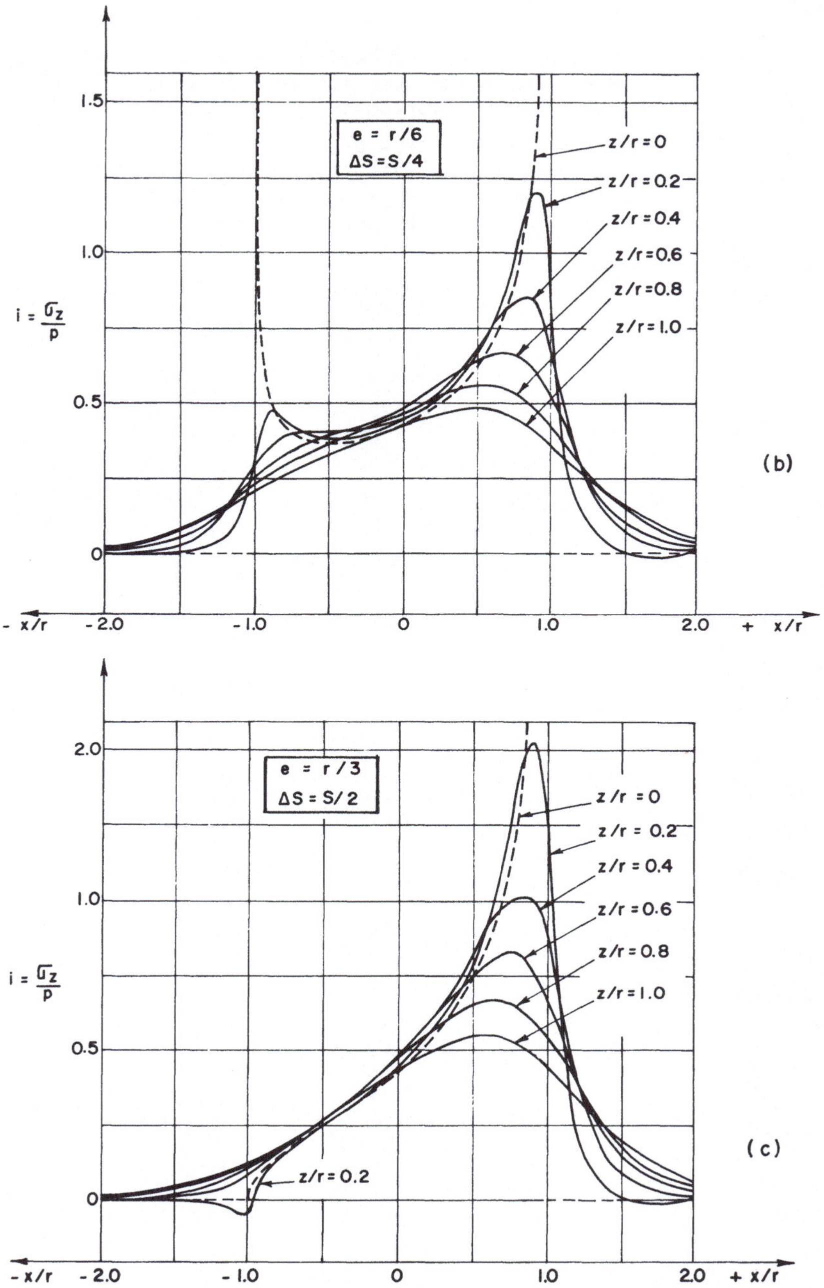

Fig. 1.38 Lotrechte Spannungen im elastisch-isotropen homogenen Halbraum unter einem starren Kreisfundament fuer verschiedene Schiefstellung in verschiedenen Tiefen und Punkten [2]

Vertical stresses in elastic isotropic homogeneous half-space under a rigid circular foundation for different inclination in different depths and points [2]

(ii) Quellen
References

[1] Fischer (1957): Zur Berechnung der Setzung einer starren, mittig belasteten Kreisplatte auf geschichteter Unterlage.
Beton-und Stahlbetonbau, 52, H. 10.

[2] Maki (1964): Asymmetric problems of the theory of elasticity for a semi-infinite solid and a thick plate.
Progress in solid mechanics Vol. 1, PP. 401
North-Holland Publishing Company.

1.1.1.6 Beliebiger Grundriss mit beliebiger Belastung

a) Halbraum nach Boussinesq

(i) Formeln/Tabelle

1.1.1.6 Any given area with different loading

a) Half space according to Boussinesq

(i) Formulae/Table

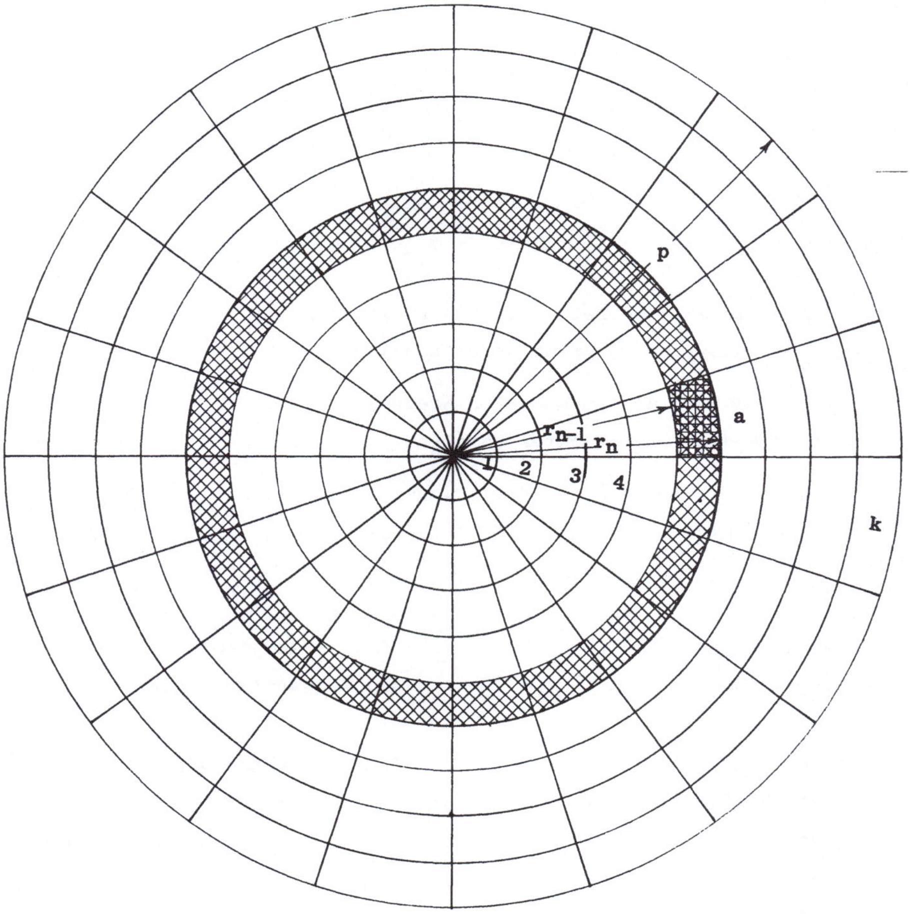

Verhaeltnis der Halbmessern der Konzentrischen Kreisen

Ratio of the radii of the concentric circles

1 : 2 : 3 : 4 .....k = 10

Die lotrechte Spannung, σ_{zn}, unter dem Kreisnetzmittelpunkt infolge eines belasteten Kreisringausschnittes [1]:

The vertical stress, σ_{zn} under the midpoint of the circular mesh due to a ring loading segment [1]

$$\sigma_{zn} = \frac{p}{a}\left[\left(\frac{(n-1)^2R^2}{k^2z^2} + 1\right)^{-1.5} - \left(\frac{n^2R^2}{k^2z^2} + 1\right)^{-1.5}\right]$$

$$\boxed{\sigma_{zn} = \frac{p}{a}\cdot i_1}$$

n = Nummer des betrachteten Kreisringes
 = number of the ring under consideration

R = Radius des aeussersten Kreises
 = Radius of the outer circle

a = Radialteilung der Kreisringe
 = Radial subdivision of the rings

(ii) Voraussetzung
Beliebige Poissonzahl

(ii) Assumption
Any Poisson's number

(iii) Erlaeutrung
Die belastete Fundamentflaeche wird in einem passenden Massstab aufgetragen, so dass sie vollstaendig von dem Kreisnetz ueberdeckt wird. Es erscheint ausreichend, mit 10 Kreisringen (k=10) und einer Radialteilung von a = 40 zur arbeiten. Das Kreis-

(iii) Explanatory note
The loaded area should be drawn using a suitable scale so that the whole of the area falls within the circular net (mesh). It appears that a net with 10 rings and 40 radial segments are sufficient to work with. The circular mesh consists of concentric circles

$$\sigma_z = \frac{p}{a} \cdot i_1$$

r/z	n = 1	2	3	4	5	6	7	8	9	10
0.05	0.0000	0.0001	0.0002	0.0002	0.0004	0.0004	0.0005	0.0005	0.0007	0.0007
0.10	0.0002	0.0005	0.0008	0.0011	0.0013	0.0016	0.0019	0.0022	0.0025	0.0028
0.167	0.0004	0.0012	0.0021	0.0029	0.0037	0.0045	0.0053	0.0060	0.0067	0.0074
0.20	0.0006	0.0018	0.0030	0.0041	0.0053	0.0064	0.0075	0.0085	0.0095	0.0104
0.25	0.00093	0.00280	0.00463	0.00643	0.00818	0.00984	0.01141	0.01290	0.01427	0.01551
0.3	0.00133	0.00401	0.00664	0.00921	0.01161	0.01387	0.01598	0.01788	0.01961	0.02109
0.4	0.00250	0.00711	0.01171	0.01600	0.01992	0.02342	0.02648	0.02900	0.03102	0.03254
0.5	0.00375	0.01106	0.01803	0.02431	0.02978	0.03434	0.03689	0.04045	0.04206	0.04280
0.6	0.00537	0.01584	0.02549	0.03386	0.04070	0.04580	0.04922	0.05103	0.05144	0.05075
0.7	0.00732	0.02134	0.03403	0.04446	0.05211	0.05713	0.05958	0.05992	0.05854	0.05586
0.8	0.00953	0.02770	0.04334	0.05548	0.06356	0.06771	0.06847	0.06671	0.06305	0.05831
0.9	0.01200	0.03466	0.05347	0.06689	0.07460	0.07709	0.07557	0.07122	0.06611	0.05769
1.0	0.01480	0.04234	0.06412	0.07834	0.08486	0.08504	0.08068	0.07367	0.06549	0.05711
1.2	0.02120	0.05935	0.08650	0.10025	0.10219	0.09604	0.08553	0.07351	0.06181	0.05123
1.4	0.02868	0.07839	0.10924	0.11950	0.11440	0.10089	0.08462	0.06888	0.05519	0.04387
1.6	0.03722	0.09882	0.13127	0.13518	0.12136	0.10073	0.07999	0.06209	0.04775	0.03671
1.8	0.04670	0.12036	0.15169	0.14679	0.12380	0.09705	0.07337	0.05465	0.04067	0.03038
2.0	0.05714	0.14246	0.16990	0.15435	0.12260	0.09117	0.06601	0.04749	0.03434	0.02510

r/z	n = 1	2	3	4	5	6	7	8	9	10
2.5	0.08691	0.19754	0.20341	0.15858	0.10977	0.06698	0.05639	0.03097	0.02245	0.01576
3.0	0.12127	0.24823	0.21984	0.14828	0.08558	0.06226	0.03507	0.02257	0.01500	0.01028
4.0	0.19960	0.32425	0.21377	0.11350	0.05944	0.03254	0.01885	0.01151	0.00737	0.00490
5.0	0.28446	0.36199	0.17675	0.08736	0.03821	0.01961	0.01089	0.00646	0.00406	0.00267
10.0	0.64645	0.26411	0.05782	0.01735	0.00754	0.00310	0.00161	0.00092	0.00056	0.00036
14.29	0.8114	0.1525	0.0243	0.0066	0.0025	0.0011	0.0006	0.0003	0.0002	0.0001
20	0.9106	0.0752	0.0098	0.0025	0.0009	0.0004	0.0002	0.0001	0.0001	0.0000
25	0.9488	0.0437	0.0052	0.0013	0.0005	0.0002	0.0001	0.0001	0.0000	0.0000
33.33	0.9763	0.0205	0.0023	0.0006	0.0002	0.0001	0.0000	0.0000	0.0000	0.0000
50	0.09925	0.0066	0.0007	0.0002	0.0001	0.0000	0.0000	0.0000	0.0000	0.0000
100	0.9990	0.0009	0.0001	0.0000	0.0000	0.0000	0.0000	0.0000	0.0000	0.0000
∞	1.0000	0.0000	0.0000	0.0000	0.0000	0.0000	0.0000	0.0000	0.0000	0.0000

Tabelle 1.19 Einflussbeiwert i_1 [1,2]

Table 1.19 Influence value i_1 [1,2]

netz besteht aus konzentrischen Kreisen mit Halbmessern in Verhaeltnis 1, 2, 3, 4, 5, 6, 7, 8, 9 und 10.

Der Mittlepunkt der Kreise wird auf den Punkt gelegt, unter dem die lotrechte Spannung bestimmt werden soll. Aus der Zeichung ist durch einfaches Abzaehlen zu entnehmen welcher Anteil des vollen Kreisringes tatsaechlich belastet ist. Bei diesem Abzaehlen kann auch eine unterschiedliche Belastung beruecksichtigt werden. Die belasteten Teilflaechen auf den einzelnen Kreisringen brauchen nur einmal ausgezaehlt zu werden, das Ergebnis der Auszaehlung ist dann je nach dem Tiefenverhaeltnis r:z mit der entsprechenden Einflusszahl (Tabelle 1.19) zu multiplizieren. Allgemein gilt damit fuer die Spannung im Kreisnetzmittelpunkt infolge Teilbelastung des n-ten Kreisringes:

$$\sigma_{zn} = (\frac{1}{a}\sum \text{belastete Teilflaechen} \cdot p)i$$

Die gesamt spannung is

that have ratio of their radii as: 1, 2, 3, 4, 5, 6, 7, 8, 9 and 10.

The sketch of the loaded area should be placed on the mesh in such a way that the centre of the mesh coincides with the point below which it is desired to find the vertical stress.

From the mesh one counts the loaded segment for a given ring. Different counts are made for different magnitude of loading. The loaded segments are to be counted only once. Having determined the load within each ring, one multiplies it by the influence value i (Table 1.19). The vertical stress in the middle of the nth ring may be expressed as:

$$\sigma_{zn} = (\frac{1}{a}\sum \text{loaded area} \cdot p)i$$

The total stress under the point will then be

$$\sigma_n = \Sigma\sigma_{zn}$$

(iv) Anwendumgsbeispiel

Gegeben

Ein Gebaeudegrundriss mit drei verschiedenen Belastungen

(iv) Example

Given

A plan of a structure with three different loadings

Gesucht

Lotrechte Spannungen unter einer Bauwerkskante infolge der Belastung p_{01} = 100 kN/m^2; p_{02} = 200 kN/m^2 und p_{03} = 300 kN/m^2 in den Tiefen z = 20.0 m, 10 m, 6.67 m, und 5.0 m.

Required

Vertical stresses under an edge of the structure due to the loading p_{01} = 100 kN/m^2; p_{02} = 200 kN/m^2 and p_{03} = 300 kN/m^2 at the depths of z = 20,0 m; 10.0 m; 6.67 m; and 5.0 m.

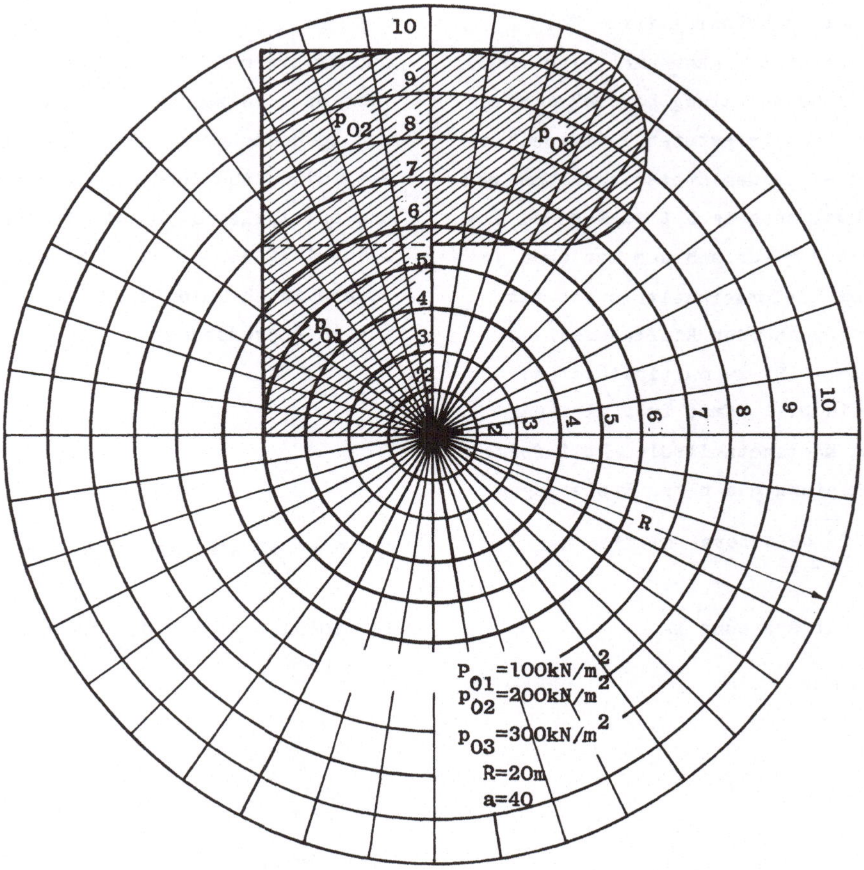

n		i				σ_{zn}			
		$r/_z = 1$	$r/_z = 2$	$r/_z = 3$	$r/_z = 4$	z = 20m	z = 10m	z=6.67m	z = 5m
1	$\frac{1}{40}(10)(100) =$ 25.0	0.01480	0.05714	0.12127	0.19960	0.3700	1.4285	3.0318	4.9900
2	$\frac{1}{40}(10)(100) =$ 25.0	0.04234	0.14246	0.24823	0.32425	1.0585	3.5615	6.2058	8.1063
3	$\frac{1}{40}(10)(100) =$ 25.0	0.06412	0.16990	0.16990	0.21984	1.6030	4.2475	5.4960	5.3443
4	$\frac{1}{40}(10)(100) =$ 25.0	0.07834	0.15435	0.14828	0.11350	2.8375	3.8588	3.7070	2.8375
5	$\frac{1}{40}(5.8)(100)+1.0(200)+1.0)(300) = \frac{1080}{40} =$ 27.0	0.08486	0.12260	0.08558	0.05944	2.2912	3.3102	2.3107	1.6022
6	$\frac{1}{40}(1.0)(100)+(4.0)(200)+(4.0)(300) = \frac{2100}{40} =$ 52.5	0.08504	0.09117	0.06226	0.03254	4.4646	4.7864	3.2687	1.7084
7	$\frac{1}{40}(4.25)(200)+(4.5)(300) = \frac{2200}{40} =$ 55.0	0.08068	0.06601	0.03507	0.01885	4.4374	3.6306	1.9289	1.0368
8	$\frac{1}{40}(3.5)(200)+4.5(300) = \frac{2050}{40}$ 51.25	0.07367	0.04749	0.02257	0.01151	3.7756	2.4339	1.1567	0.5899
9	$\frac{1}{40}(3.2)(200)+4.2(300) = \frac{1900}{40} =$ 47.5	0.06549	0.03434	0.01500	0.00737	3.1108	1.6312	0.7125	0.3501
10	$\frac{1}{40}$ $(0.4+0.25)200+(0.25+0.5+0.33)(300) = \frac{454}{40} = 11.35$	0.05711	0.02510	0.01028	0.00490	0.6482	0.2850	0.1167	0.0556
					$\sigma_z = \Sigma\sigma_{zn}$:	24.5968	29.1736	27.9348	26.6231

$\sigma_{z20} = 24.5 \text{ kN/m}^2$

$\sigma_{z10} = 29.2 \text{ kN/m}^2$

$\sigma_{z6.67} = 27.9 \text{ kN/m}^2$

$\sigma_{z5} = 26.5 \text{ kN/m}^2$

(v) Quellen

References

[1] Lorenz/Grewe (1959): Spannungsberechnung bei beliebiger Lastflaechenform und Belastung.
Bauingenieur 33, S.24.

[2] Salas (1948): Soil pressure computation: A modification of the Newmark's method.
Proc. 2. ICSMFE, Roterdam, Vol. 7, PP 30.

b) Halbraum nach Froehlich

(i) Formeln/Tabelle

Der Rechengang ist wie unter 1.1.1.6a. Die Spannungen haengen allgemein von ν zusammen in der Form [1]

b) Half-space according to Froehlich

(i) Formulae/Table

The calculation is similar to 1.1.1.6a. The stresses generally depend on ν and have the following form [1]

$$\sigma_{zn} = \frac{p}{a}\left[\left\{\frac{(n-1)^2 R^2}{k^2 z^2} + 1\right\}^{-\nu/2} - \left\{\frac{n^2 R^2}{k^2 z^2} + 1\right\}^{-\nu/2}\right]$$

$$= \frac{p}{a} \cdot i_z$$

Fuer $\nu = 5$ ergeben sich folgende Werte (Tabelle 1.20).

For $\nu = 5$, the following values are given (Table 1.20).

(ii) Quelle (S.S. 136)
Reference (Ref. PP. 136)

$\nu = 5$

R/z	n = 1	2	3	4	5	6	7	8	9	10
0.25	0.00156	0.00467	0.00770	0.01063	0.01348	0.01608	0.01852	0.02076	0.02273	0.02446
0.3	0.00225	0.00669	0.01104	0.01512	0.01902	0.02250	0.02566	0.02836	0.03068	0.03250
0.4	0.00400	0.01183	0.01927	0.02613	0.03217	0.03724	0.04132	0.04434	0.04630	0.04740
0.5	0.00623	0.01833	0.02956	0.03928	0.04722	0.05320	0.05711	0.05907	0.05937	0.05819
0.6	0.00894	0.02616	0.04152	0.05402	0.06318	0.06878	0.07120	0.07071	0.06804	0.06384
0.7	0.01215	0.03521	0.05492	0.06968	0.07897	0.08287	0.08226	0.07829	0.07207	0.06459
0.8	0.01583	0.04550	0.06941	0.08566	0.09370	0.09451	0.08984	0.08176	0.07191	0.06165
0.9	0.01998	0.05664	0.08470	0.10128	0.10677	0.10318	0.09387	0.08160	0.06951	0.05559
1.0	0.02456	0.06884	0.10042	0.11618	0.11756	0.10883	0.09462	0.07866	0.06345	0.05010
1.2	0.03510	0.09554	0.13196	0.14191	0.13188	0.11162	0.08879	0.06783	0.05061	0.03723
1.4	0.04736	0.12460	0.16184	0.16055	0.13666	0.10579	0.07737	0.05478	0.03821	0.02650
1.6	0.05123	0.15507	0.18821	0.17160	0.13356	0.09496	0.06432	0.04261	0.02806	0.01856
1.8	0.07662	0.18598	0.20995	0.17547	0.12510	0.08212	0.05192	0.03246	0.02039	0.01298
2.0	0.09340	0.21660	0.22639	0.17328	0.11355	0.06925	0.04119	0.02452	0.01481	0.00912
2.5	0.14062	0.28594	0.24463	0.15103	0.08165	0.04261	0.02246	0.01217	0.00684	0.00398
3.0	0.19382	0.34257	0.23673	0.11935	0.05501	0.02551	0.01232	0.00627	0.00337	0.00189
4.0	0.31000	0.39967	0.18280	0.06571	0.02393	0.00947	0.00412	0.00194	0.00099	0.00053
5.0	0.57244	0.39566	0.12426	0.03463	0.01082	0.00391	0.00160	0.00072	0.00036	0.00019
10.0	0.82322	0.09889	0.01473	0.00232	0.00055	0.00017	0.00006	0.00003	0.00001	0.00001
∞	1.00000	0	0	0	0	0	0	0	0	0

Tabelle 1.20 Einflussbeiwerte i_2 [1]

Table 1.20 Influence value i_2 [1]

1.1.2 Waagerechte Lasten

1.1.2.1 Einzellast

a) Halbraum nach Boussinesq

(i) Formeln

1.1.2 Horizontal Loading

1.1.2.1 Point Load

a) Half-space according to Boussinesq

(i) Formulae

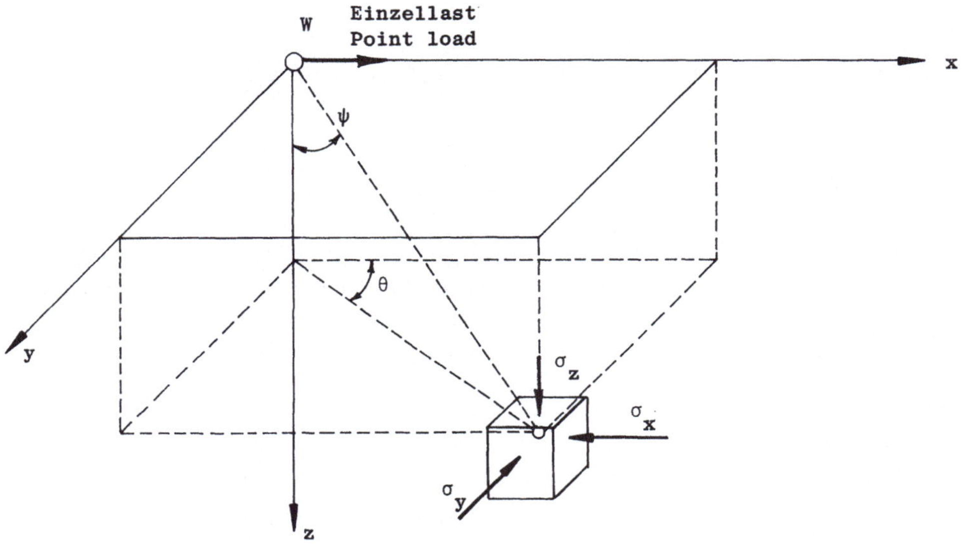

$$\sigma_z = \frac{3W}{z\pi z^2} \cdot \cos\theta \cdot \sin\psi \cdot \cos^4\psi$$

$$\sigma_x = \frac{W}{2\pi z^2} \cdot \cos\theta \cdot \sin\psi \cdot \cos^2\psi \left[3\cos^2\theta \sin^2\psi - \frac{m-2}{m} \left\{ 1 - \frac{3}{(1+\cos\psi)^2} + \frac{\cos^2\theta \sin^2\psi (3+\cos\psi)}{(1+\cos\psi)^3} \right\} \right]$$

$$\sigma_y = \frac{W}{2\pi z^2} \cos\theta \cdot \sin\psi \cdot \cos^2\psi \left[3 \sin^2\theta \cdot \sin^2\psi - \frac{m-2}{m} \left\{ 1 - \frac{1}{(1+\cos\psi)^2} + \frac{\sin^2\theta \sin^2\psi (3+\cos\psi)}{(1+\cos\psi)^3} \right\} \right]$$

$$\tau_{zx} = \frac{3W}{2\pi z^2} \cdot \cos^2\theta \cdot \sin^2\psi \cdot \cos^3\psi$$

$$\tau_{zy} = \frac{3}{2\pi z^2} \cdot \cos\theta \cdot \sin\theta \cdot \sin^2\psi \cdot \cos^3\psi$$

$$\tau_{xy} = \frac{W}{2\pi z^2} \sin\theta \cdot \sin\psi \cdot \cos^2\psi \left[3\cos^2\theta \sin^2\psi + \right.$$

$$\left. \frac{m-2}{m} \cdot \frac{1}{(1+\cos\psi)^2} \left\{ 1-\cos^2\theta \sin^2\psi \cdot \frac{3+\cos\psi}{1+\cos\psi} \right\} \right]$$

Fig. 1.39 Gleichungen der Spannungen infolge waagerechter Einzellast [1, 2]

Equations of stresses due to horizontal point load [1, 2]

(ii) Quellen

References

[1] Cerruti (1888): Sulla deformazione di un corpo elastico isotropo per alsune speciali condizioni ai limiti. Reale accademica dei lincei. Rendiconti, Roma, pp 785.

[2] Ohde (1939): Zur Theorie der Druckverteilung im Bangrund. Bauingenieur 20, PP 451.

b) **Halbraum nach Froehlich**

(i) **Formeln**

(i)a Gleichungen nach Froehlich [1]

b) **Half-space according to Froehlich**

(i) **Formulae**

(i)a **Equations according to Froehlich [1]**

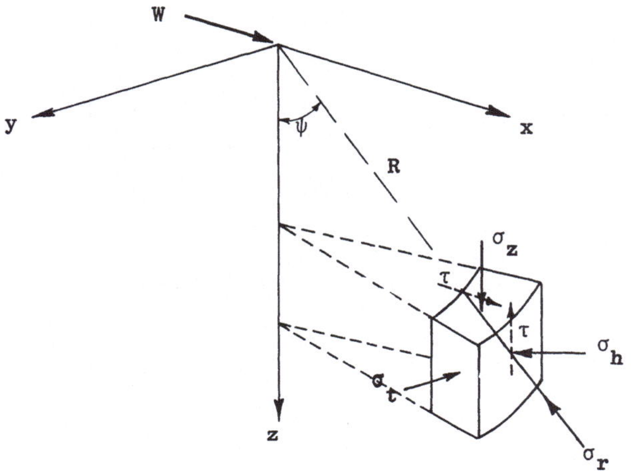

$$\sigma_r = \frac{\nu \cdot W}{2\pi \cdot R^2} \cdot \sin^{\nu-2}\psi \qquad = \frac{\nu \cdot W}{2\pi \cdot z^2} \cdot \sin^{\nu-2}\psi \cdot \cos^2\psi$$

$$\sigma_z = \frac{\nu \cdot W}{2\pi \cdot R^2} \cdot \sin^{\nu-2} \cdot \cos^2\psi = \frac{\nu \cdot W}{2\pi \cdot z^2} \cdot \sin^{\nu-2} \cdot \cos^4\psi$$

$$\sigma_h = \frac{\nu \cdot W}{2\pi \cdot R^2} \cdot \sin^{\nu}\psi \qquad = \frac{\nu \cdot W}{2\pi \cdot z^2} \cdot \sin^{\nu}\psi \cdot \cos^2\psi$$

$$\tau = \frac{\nu \cdot W}{2\pi \cdot R^2} \cdot \sin^{\nu-1}\psi \cdot \cos\psi = \frac{\nu \cdot W}{2\pi \cdot z^2} \cdot \sin^{\nu-1}\psi \cdot \cos^3\psi$$

Fig. 1.40 **Gleichungen der Spannungen infolge waagerechter Einzellast [1]**

Equations of stresses due to horizontal point load [1]

(i)b Gleichungen nach Ohde [2] (i)b Equations according to Ohde [2]

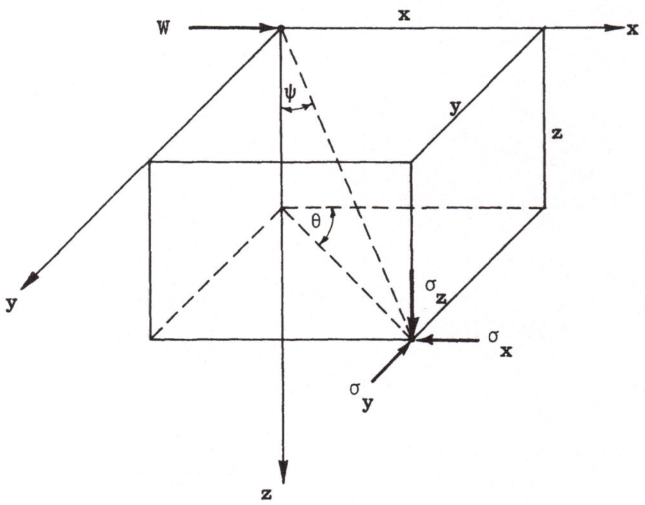

$$\sigma_z = (\nu-2)\cdot\frac{\nu}{2\pi}\cdot\frac{W}{z^2}\cdot\cos\theta\cdot\sin\psi\cdot\cos^{\nu+1}\psi$$

$$\sigma_x = (\nu-2)\cdot\frac{\nu}{2\pi}\cdot\frac{W}{z^2}\cdot\cos^3\theta\cdot\sin^3\psi\cos^{\nu-1}\psi$$

$$\sigma_y = (\nu-2)\cdot\frac{\nu}{2\pi}\cdot\frac{W}{z^2}\cdot\sin^2\theta\cdot\cos\theta\cdot\sin^3\psi\cos^{\nu-1}\psi$$

$$\tau_{zx} = (\nu-2)\cdot\frac{\nu}{2\pi}\cdot\frac{W}{z^2}\cdot\cos^2\theta\cdot\sin^2\psi\cos^{\nu}\psi$$

$$\tau_{zy} = (\nu-2)\cdot\frac{\nu}{2\pi}\cdot\frac{W}{z^2}\cdot\sin\theta\cos\theta\sin^2\psi\cos^{\nu}\psi$$

$$\tau_{xy} = (\nu-2)\cdot\frac{\nu}{2\pi}\cdot\frac{W}{z^2}\cdot\sin\theta\cos^2\theta\sin^3\psi\cos^{\nu-1}\psi$$

Fig. 1.41 Gleichungen der Spannungen infolge waagrechter Einzellast [2]

Equations of stresses due to horizontal point load [2]

(ii) **Erlaeuterung**

Der Unterschied zwischen den Spannungen nach Froehlich und nach Ohde besteht darin, dass Ohde seine Gleichung unter der Voraussetzung abgeleitet hat, dass die Spannungen bei $\nu = 3$ den Spannungen des Boussinesqschen Halbraumes mit $m = 2$ entspreschen (wie Froehlich bei lotrechter Einzellast). Diese Bedingung wird bei Froehlich fuer eine waagerechte Einzellast nicht erfuellt.

(ii) **Explanatory Note**

The difference between the stresses according to Froehlich and Ohde lies in the assumptions made by the authors. Ohde derived his equation by assuming that for $\nu = 3$, the stresses correspond to those stresses in an elastic half-space of Boussinesq (as in the case for vertical point load according to Froehlich). The stresses according to Froehlich however, do not satisfy this condition for horizontal line load.

(iii) Anwendungsbeispiel

Gegeben

$W = 1000$ kN

Gesucht

Lotrechte Spannung im Punkt $x = 4$ m, $y = 4$ m $z = 6$ m fuer den Boussinesqschen Halbraum und Froehlichen Halbraum fuer $\nu = 3$ nach Froehlich und nach Ohde.

Loesung

$\tan\theta = 4/4 = 1;\quad \theta = 45^o$

$\tan\psi = \sqrt{4^2 + 4^2}/6 = 0.943;\ \psi = 43.3^o$

a) Boussinesqscher Halbraum

$$\sigma_z = \frac{3W}{2\pi z^2} \cdot \cos\theta \cdot \sin\psi \cdot \cos^4\psi$$

$$= \frac{(3)1000}{(2)\pi(36)}(0.707)(0.686)(0.728^4)$$

$$= 1.81 \text{ kN/m}^2$$

(iii) Example

Given

$W = 1000$ kN

Required

Vertical stress at a point $x = 4$ m, $y = 4$ m and $z = 6$ m for a half-space of Boussinesq and that of Froehlich with $\nu = 3$; according to the Formulae of Froehlich and Ohde.

Solution

$\tan\theta = 4/4 = 1;\quad \theta = 45^o$

$\tan\psi = \sqrt{4^2 + 4^2}/6 = 0.943;\ \psi = 43.3^o$

a) Boussinesq's Half-space

$$\sigma_z = \frac{3W}{2\pi z^2} \cdot \cos\theta \cdot \sin\psi \cdot \cos^4\psi$$

$$= \frac{(3)(1000)}{(2)\pi(36)}(0.707)(0.686)(0.728^4)$$

$$= 1.81 \text{ kN/m}^2$$

b) Froehlischer Halbraum

Nach den Formeln von Froehlich

$$\sigma_z = \frac{\nu \cdot W}{2\pi \cdot z^2} \cdot \sin^{\nu-2}\psi \cdot \cos^4\psi$$

$$= \frac{(3)(1000)}{(2)(\pi)(36)}(0.686)(0.728^4)$$

$$= 2.56 \text{ kN/m}^2$$

Nach den Formel von Ohde

$$\sigma_z = (\nu-2) \cdot \frac{\nu}{2\pi} \cdot \frac{W}{z^2} \cdot \cos\theta \cdot \sin\psi \cdot \cos^{\nu+1}\psi$$

$$= \frac{(3)(1000)}{2\pi \cdot 36}(0.707)(0.686)(0.728^4)$$

$$= 1.81 \text{ kN/m}^2$$

b) Froehlich's Half-space

According to the formulae of Froehlich

$$\sigma_z = \frac{\nu \cdot W}{2\pi \cdot z^2} \cdot \sin^{\nu-2}\psi \cdot \cos^4\psi$$

$$= \frac{(3)(1000)}{(2)(\pi)(36)}(0.686)(0.728^4)$$

$$= 2.56 \text{ kN/m}^2$$

According to the formulae of Ohde

$$\sigma_z = (\nu-2) \cdot \frac{\nu}{2\pi} \cdot \frac{W}{z^2} \cdot \cos\theta \cdot \sin\psi \cdot \cos^{\nu+1}\psi$$

$$= \frac{(3)(1000)}{2\pi \cdot 36}(0.707)(0.686)(0.728^4)$$

$$= 1.81 \text{ kN/m}^2$$

(iv) Quellen

References

[1] Froehlich (1934): Druckverteilung im Bangrunde
Wien. Verlag von Julius Springer

[2] Ohde (1939): Zur Theorie derDruckverteilung im
Baugrund.
Bauingenieur 20, S.451.

1.1.2.2 Linienlast

a) Halbraum nach Boussinesq

(i) Formeln

(i)a Unendlich lange linienlast

1.1.2.2 Line load

a) Half-space according to Boussinesq

(i) Formulae

(i)a Line load with infinite length

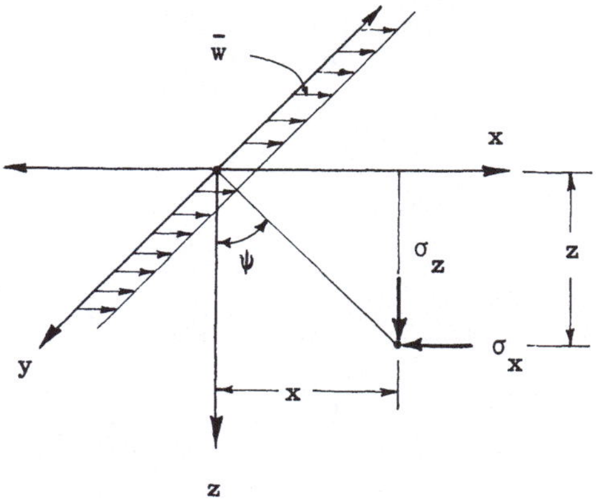

$$\sigma_z = \frac{2\bar{w}}{\pi z} \cdot \sin\psi\cos^3\psi$$

$$\sigma_x = \frac{2\bar{w}}{\pi z} \cdot \sin^3\psi\cos\psi$$

beliebge Poissonzahl m
any Poisson's number m

$$\tau = \frac{2\bar{w}}{\pi z} \cdot \sin^2\psi\cos^2\psi$$

Fig. 1.42 Gleichungen der Spannungen infolge unendlich langer waagerechter Linienlast [1, 2]

Equations for stresses due to horizontal line load with infinite length [1, 2] (i)b Endlich lange Linienlast

(i)b Line load with finite length

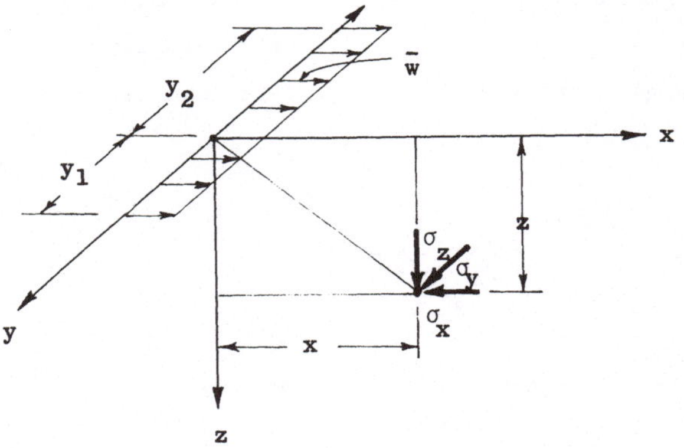

$$\bar{\sigma}_z = \frac{3\bar{w}}{2\pi} \cdot z^2 \cdot x \left[\frac{y_1}{3(z^2+x^2)(z^2+x^2+y_1^2)^{3/2}} + \frac{2}{3(z^2+x^2)} \cdot \frac{y_1}{(z^2+x^2)(z^2+x^2+y_1^2)^{1/2}} - \frac{y_2}{3(z^2+x^2)(z^2+x^2+y_2^2)^{3/2}} - \frac{2}{3(z^2+x^2)} \cdot \frac{y_2}{(z^2+x^2)(z^2+x^2+y_2^2)^{1/2}} \right]$$

$$\sigma_x = \frac{3\bar{w}}{2\pi} \cdot x^3 \left[\frac{y_1}{3(z^2+x^2)(z^2+x^2+y_1^2)^{3/2}} + \frac{2}{3(z^2+x^2)} \cdot \frac{y_1}{(z^2+x^2)(z^2+x^2+y_1^2)^{1/2}} - \frac{y_2}{3(z^2+x^2)(z^2+x^2+y_2^2)^{3/2}} - \frac{2}{3(z^2+x^2)} \cdot \frac{y_2}{(z^2+x^2)(z^2+x^2+y_2^2)^{1/2}} \right]$$

$$\sigma_y = \frac{3\bar{w}}{2\pi} \cdot x \left[\frac{y_1}{3(z^2+x^2+y_1^2)^{3/2}} + \frac{y_1}{3(z^2+x^2)(z^2+x^2+y_1^2)^{1/2}} + \frac{y_2}{3(z^2+x^2+y_2^2)^{3/2}} - \frac{y_2}{3(z^2+x^2)(z^2+x^2+y_2^2)^{1/2}} \right]$$

m = 2 fuer die waagerechte Spannungen und beliebig fuer die Lotrechte Spannung.

m = 2 for the horizontal stresses and any value for the vertical stress.

Fig. 1.43 Gleichungen der Spannungen infolge endlich langer waagerechter Linienlast [3]

Equations for stresses due to horizontal line load with a finite length [3]

(ii) Quellen

References

[1] Michell (1900): On some elementary distributions of stress in three dimensions and elementary distribution of plane stress. London Mathem. Soc. Proc. Vol XXXII

[2] Ohde (1939): Zur Theorie der Druckverteilung im Baugrund. Bauingenieur 20, S. 451

[3] Siemer (1967): Spannungen und Setzungen des Halbraums unter einfachen Flaechenlasten und unter starren Grundkoerpern aus waagerechter Beanspruchung. Mitt VGB, Aachen 41

b) Halbraum nach Froehlich

(i) Formeln

(i)a Unendlich lange Lininenlast nach Froehlich [1]

b) Half space according to Froehlich

(i) Formulae

Line load with infinite length according to Froehlich [1]

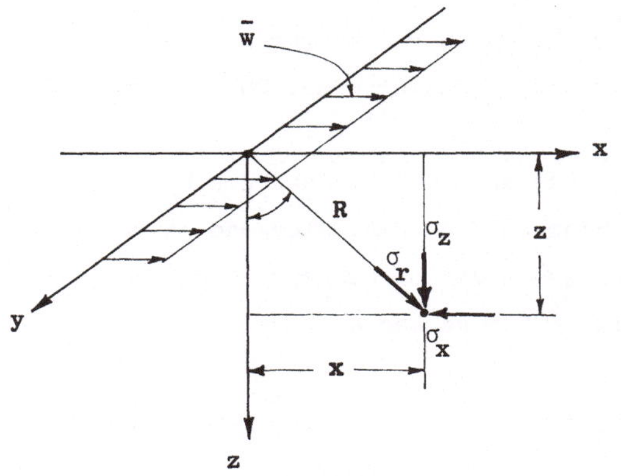

$$\sigma_r = f \cdot \frac{\bar{w}}{R} \cdot \sin^{\nu-2}\psi$$

$$\sigma_z = f \cdot \frac{\bar{w}}{R} \cdot \sin\psi^{\nu-2} \cdot \cos^2\psi$$

$$\sigma_x = f \cdot \frac{\bar{w}}{R} \cdot \sin^{\nu}\psi$$

$$\tau = f \cdot \frac{\bar{w}}{R} \cdot \sin\psi^{\nu-1} \cdot \cos\psi$$

Fuer ν = 1-6 ergeben sich dann folgende Spannungen:

For ν = 1-6 the following stresses are given;

ν	σ_r	σ_z	σ_x	τ	Faktoren fuer alle Spannungen Factors for all Stresses
1	$\frac{1}{\sin\psi}$	$\cos\psi\cot\psi$	$\sin\psi$	$\cos\psi$	$\frac{1}{\pi}\frac{\bar{w}}{R}$
2	1	$\cos^2\psi$	$\sin^2\psi$	$\sin\psi\cos\psi$	$\frac{1}{2}\frac{\bar{w}}{R}$
3	$\sin\psi$	$\cos^2\psi\sin\psi$	$\sin^3\psi$	$\sin^2\psi\cos\psi$	$\frac{2}{\pi}\frac{\bar{w}}{R}$
4	$\sin^2\psi$	$\cos^2\psi\sin^2\psi$	$\sin^4\psi$	$\sin^3\psi\cos\psi$	$\frac{3}{4}\frac{\bar{w}}{R}$
5	$\sin^3\psi$	$\cos^2\psi\sin^3\psi$	$\sin^5\psi$	$\sin^4\psi\cos\psi$	$\frac{8}{3}\frac{\bar{w}}{R}$
6	$\sin^4\psi$	$\cos^2\psi\sin^4\psi$	$\sin^6\psi$	$\sin^5\psi\cos\psi$	$\frac{15}{16}\frac{\bar{w}}{R}$

Fig. 1.44 Gleichungen der Spannungen infolge unendlich langer waagercheter Linienlast [1]

Equations of stresses due to horizontal line load of infinite length [1]

(i)b Unendlich lange Linienlast nach Ohde [2]

(i)b Line load with infinite length according to Ohde [2]

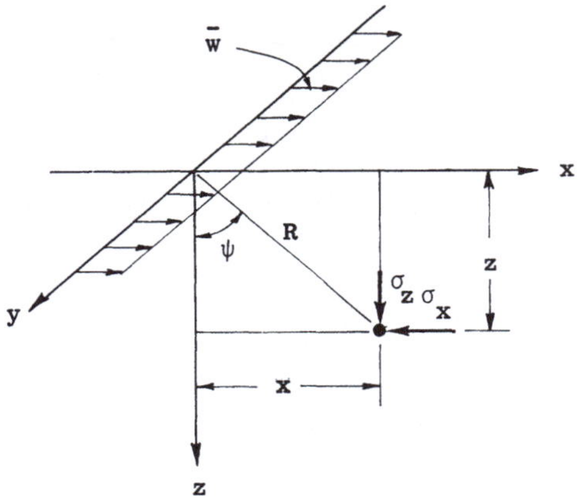

$$\sigma_z = f\cdot(\nu-2)\cdot\frac{\bar{w}}{z}\cdot\sin\psi\cdot\cos^{\nu}\psi \quad = f.(\nu-2)\cdot\frac{\bar{w}}{R}\cdot\sin\psi\cdot\cos^{\nu-1}\psi$$

$$\sigma_x = f\cdot(\nu-2)\cdot\frac{\bar{w}}{z}\cdot\sin^3\psi\cdot\cos^{\nu-2}\psi = f.(\nu-2)\cdot\frac{\bar{w}}{R}\cdot\sin^3\psi\cdot\cos^{\nu-3}\psi$$

$$\tau = f\cdot(\nu-2)\cdot\frac{\bar{w}}{z}\cdot\sin^2\psi\cdot\cos^{\nu-1}\psi = f.(\nu-2)\cdot\frac{\bar{w}}{R}\cdot\sin^2\psi\cdot\cos^{\nu-2}\psi$$

wobei f von ν folgendes zusammen haengt:

where f varies with ν as follows:

ν	1	2	3	4	5	6
f	$\frac{1}{\pi}$	$\frac{1}{2}$	$\frac{2}{\pi}$	$\frac{3}{4}$	$\frac{8}{3\pi}$	$\frac{15}{16}$

Fig. 1.45 Gleichungen der Spannungen infolge unendlich langer waagerechter Lininlast [2]

Equations of stresses due to horizontal line load of infinite length [2]

(ii) Quellen

References

[1] Froehlich (1934): Druckverteilung im Baugrunde. Wien. Verlag von Julius Springer.

[2] Ohde (1934): Zur Theorie der Druckverteilung im Baugrunde. Bauing. 20, S.451.

1.1.2.3 Streifenlast

1.1.2.3.1 Gleichmaessige Belastung

(i) Formeln/Tabelle

1.1.2.3 Strip loading

1.1.2.3.1 Uniformly distributed loading

(i) Formulae/Table

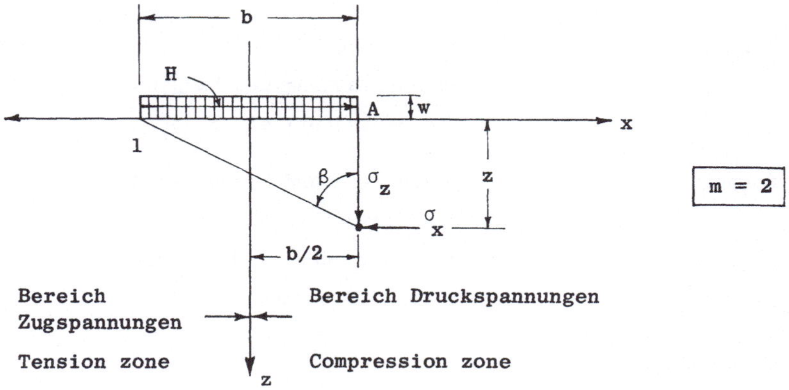

$$\sigma_z = \frac{w}{\pi} \cdot \frac{1}{1+\cot^2\beta} = \frac{w}{\pi} \cdot \frac{b^2}{z^2+b^2} = w \cdot i_{HE}$$

$$\sigma_x = \frac{w}{\pi} \left\{ - \frac{b^2}{z^2+b^2} + \ln \frac{z^2+b^2}{z^2} \right\}$$

Fig. 1.46 Gleichungen zur Bestimmung Spannungen unter dem Randpunkt A infolge waagerechter Streifen Last [1, 2, 3]

Equations for determining stresses under the corner point A due to strip horizontal load [1, 2, 3]

$\sigma_z = w \cdot i_{HE}$								
cotβ=z/b	0.00	0.25	0.50	0.75	1.00	1.50	2.00	3.00
i_{HE}	+0.3183	0.2996	0.2546	0.2037	0.1592	0.0979	0.0637	0.0318

$\sigma_z = w \cdot i_{HE}$								
cotβ=z/b	4.00	6.00	8.00	10.00	12.00	15.00	18.00	20.00
i_{HE}	0.0187	0.0086	0.0049	0.0032	0.0022	0.0014	0.0010	0.0080

Tabelle 1.21 Einflussbeiwert i_{HE} [1]

Table 1.21 Influence value i_{HE} [1]

(ii) Quellen

References

[1] DIN 4019 Blatt 2: Setzungsberechnungen bei schraeg und bei aussenmittig wirkender Belastung (Verkantung) Richtlinien.

[2] Kany (1964): Baugrundverformungen infolge waagerechter Schubbelastung der Baugrundoberflaeche Die Bautechnik 41, H. 41.

[3] Siemer (1967): Spannungen und Setzungen des Halbraums unter einfachen Flaechenlasten und unter starren Koerpern aus waagerechter Beanspruchung. Mitt. VGB Aachen, H. 41.

1.1.2.3.2 Dreiecksfoermige Belastung

(i) Formeln/Tabellen

1.1.2.3.2 Triangular loading

(i) Formulae/Tables

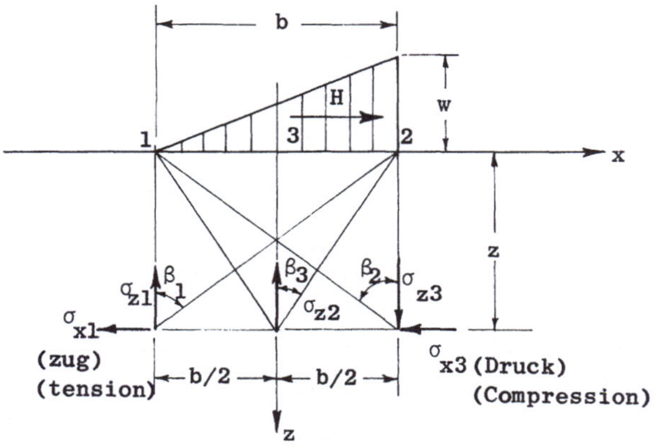

m = 2

$$\sigma_{z1} = \frac{w.z}{\pi b}\left\{\frac{b^3}{z^3+z\cdot b^2} - \frac{b}{z} + \operatorname{arctg}\frac{b}{z}\right\}$$

$$= \frac{w\cdot z}{\pi\cdot b}\left\{\beta_1 - \sin\beta_1\cdot\cos\beta_1\right\} = w\cdot i_1$$

$$\sigma_{x1} = \frac{w\cdot z}{\pi\cdot b}\left\{-\frac{b^3}{z^3+z\cdot b^2} + 3\frac{b}{z} - 3\operatorname{arctg}\frac{b}{z}\right\}$$

$$= \frac{w\cdot z}{\pi\cdot b}\left\{-\operatorname{tg}\beta_1\cdot\sin^2\beta_1 + 3\operatorname{tg}\beta_1 - 3\beta_1\right\}$$

$$\sigma_{z2} = \frac{w\cdot z}{\pi\cdot b}\left\{-\operatorname{arctg}\frac{b}{z} + \frac{b}{z}\right\} = \frac{w\cdot z}{\pi\cdot b}\left\{\operatorname{tg}\beta_2 - \beta_2\right\} = -w.i_2$$

$$\sigma_{x2} = \frac{w\cdot z}{\pi b}\left\{3\operatorname{tg}\beta_2 - 3\beta_2 + \operatorname{tg}\beta_2 \ln\frac{z^2}{z^2+b^2}\right\}$$

$$\sigma_{z3} = \frac{w\cdot z}{\pi b/2}\left\{\beta_3 - \sin\beta_3\cos\beta_3\right\} = w\cdot i_3$$

$$\sigma_{x3} = \frac{w\cdot z}{\pi b/2}\left\{-\operatorname{tg}\beta_3\sin^2\beta_3 + 3\operatorname{tg}\beta_3 - 3\beta_3\right\}$$

Fig. 1.47 Gleichungen zur Bestimmung Spannungen in verschiedenen Punkten[1, 2, 3]

Equations for determining stresses at different points [1, 2, 3]

σ_{z1}	=	$w \cdot i_1$
σ_{z2}	=	$w \cdot i_2$
σ_{z3}	=	$w \cdot i_3$

$\cot \beta = \frac{z}{b}$	Einflussbeiwerte Coefficients		
	i_1	i_2	i_3
0.00	0.0000	+0.3183	0.0000
0.25	+0.0868	+0.2128	+0.1125
0.50	+0.1125	+0.1421	+0.0908
0.75	+0.1070	+0.0967	+0.0604
1.00	+0.0908	+0.0683	+0.0405
1.50	+0.0604	+0.0376	+0.0208
2.00	+0.0405	+0.0231	+0.0123
3.00	+0.0208	+0.0110	+0.0057
4.00	+0.0123	+0.0064	+0.0033
6.00	+0.0057	+0.0029	+0.0014
8.00	+0.0033	+0.0016	+0.0008
10.00	+0.0021	+0.0011	+0.0005
12.00	+0.0014	+0.0007	+0.0003
15.00	+0.0011	+0.0003	+0.0003
18.00	+0.0007	+0.0003	+0.0002
20.00	+0.0005	+0.0003	+0.0002

Tabelle 1.22 Einflussbeiwerte i_1, i_2, und i_3 [1]

Table 1.22 Influence values i_1, i_2, and i_3 [1]

(ii) Quellen (s. Seite 152)

References (Ref. to PP. 152)

1.1.2.4 Rechteckiger Grundriss

1.1.2.4.1 Gleichmaessige Belatung

a) Halbraum nach Boussinesq

(i) Formelen/Tafeln/Tabellen

1.1.2.4 Rectangular plan

1.1.2.4.1 Uniform loading

a) Half space according to Boussinesq

(i) Formulae/Charts/Tables

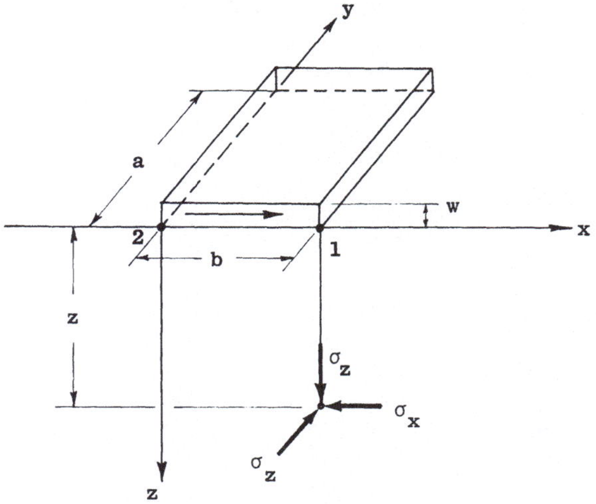

m = 2

$$\sigma_z = \frac{w}{2\pi}\left\{\frac{a}{\sqrt{z^2+a^2}} - \frac{z^2}{z^2+b^2}\cdot\frac{a}{\sqrt{z^2+b^2+a^2}}\right\} = w\cdot i_{R1}$$

$$\sigma_x = \frac{w}{2\pi}\left\{2\ln\left(\frac{\sqrt{z^2+b^2}}{z}\cdot\frac{a+\sqrt{z^2+a^2}}{a+\sqrt{z^2+b^2+a^2}}\right) - \frac{b^2}{z^2+b^2}\cdot\frac{a}{\sqrt{z^2+b^2+a^2}}\right\}$$

$$\sigma_y = \frac{w}{2\pi}\left\{\ln\left(\frac{\sqrt{z^2+b^2}}{z}\cdot\frac{a+\sqrt{z^2+a^2}}{a+\sqrt{z^2+b^2+z^2}}\right) - a\left(\frac{1}{\sqrt{z^2+a^2}} - \frac{1}{\sqrt{z^2+b^2+a^2}}\right)\right\}$$

Fig. 1.48a Gleichungen zur Bestimmung Spannungen unter dem Eckpunkt infolge rechteckig gleichmaessiger Belastung [2,3]

Equations for determining stresses under the corner point due to uniformly distributed rectangular loading [2, 3]

$$\sigma_z = 2 \cdot i_{R1}$$

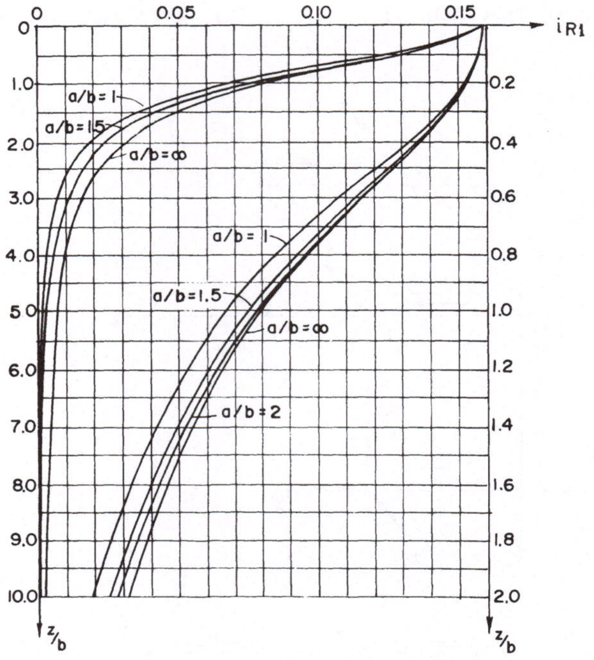

Fig. 1.48b Einflussbeiwert i_{R1} [3]

Influence value i_{R1} [3]

$\sigma_z = w \cdot i_{R1}$

z/b \ a/b	0.1	0.2	0.4	0.6	0.8	1.0	1.6	2.0
0	0.15915	0.15915	0.15915	0.15915	0.15915	0.15915	0.15915	0.15915
0.100	0.11238	0.14201	0.15382	0.15618	0.15694	0.15725	0.15751	0.15755
0.150	0.08794	0.12664	0.14773	0.15262	0.15426	0.15493	0.15550	0.15558
0.200	0.07058	0.11136	0.14012	0.14788	0.15062	0.15178	0.15276	0.15291
0.300	0.01908	0.08581	0.12262	0.13580	0.14103	0.14335	0.14542	0.14574
0.400	0.03657	0.06717	0.10490	0.12174	0.12926	0.13283	0.13619	0.13674
0.500	0.02838	0.05350	0.08870	0.10721	0.11644	0.12113	0.12582	0.12662
0.700	0.01823	0.03526	0.06266	0.08049	0.09109	0.09722	0.10420	0.10554
1.000	0.01022	0.02007	0.03745	0.05080	0.06024	0.06660	0.07534	0.07738
1.500	0.00448	0.00889	0.01714	0.02431	0.03020	0.03484	0.04297	0.04548
2.000	0.00226	0.00449	0.00879	0.01274	0.01622	0.01920	0.02533	0.02766
3.000	0.00077	0.00155	0.00306	0.00451	0.00588	0.00714	0.01023	0.01172
5.000	0.00018	0.00036	0.00072	0.00108	0.00143	0.00176	0.00269	0.00323
7.000	0.00007	0.00014	0.00027	0.00040	0.00054	0.00067	0.00104	0.00127
10.000	0.00002	0.00005	0.00009	0.00014	0.00019	0.00023	0.00037	0.00046
15.000	0.00001	0.00001	0.00003	0.00004	0.00006	0.00007	0.00011	0.00014
20.000	0.00000	0.00001	0.00001	0.00002	0.00002	0.00003	0.00005	0.00006
30.000	0.00000	0.00000	0.00000	0.00001	0.00001	0.00001	0.00001	0.00002
50.000	0.00000	0.00000	0.00000	0.00000	0.00000	0.00000	0.00000	0.00000
70.000	0.00000	0.00000	0.00000	0.00000	0.00000	0.00000	0.00000	0.00000
100.000	0.00000	0.00000	0.00000	0.00000	0.00000	0.00000	0.00000	0.00000

z/b \ a/b	2.5	3	5	10	20	50	100
0	0.15915	0.15915	0.15915	0.15915	0.15915	0.15915	0.15915
0.100	0.15757	0.15757	0.15758	0.15758	0.15758	0.15758	0.15758
0.150	0.15562	0.15564	0.15563	0.15565	0.15565	0.15565	0.15565
0.200	0.15298	0.15301	0.15303	0.15303	0.15303	0.15303	0.15303
0.300	0.14589	0.14595	0.14601	0.14601	0.14601	0.14601	0.14601
0.400	0.13700	0.13710	0.13719	0.13720	0.13720	0.13720	0.13720
0.500	0.12701	0.12716	0.12730	0.12732	0.12732	0.12732	0.12732
0.700	0.10623	0.10651	0.10677	0.10681	0.10682	0.10682	0.10682
1.000	0.07851	0.07901	0.07949	0.07957	0.07958	0.07958	0.07958
1.500	0.04710	0.04791	0.04879	0.04896	0.04897	0.04897	0.04897
2.000	0.02938	0.03034	0.03154	0.03181	0.03183	0.03183	0.03183
3.000	0.01306	0.01396	0.01541	0.01587	0.01591	0.01592	0.01592
5.000	0.00381	0.00428	0.00539	0.00602	0.00611	0.00612	0.00612
7.000	0.00154	0.00178	0.00246	0.00303	0.00317	0.00318	0.00318
10.000	0.00056	0.00066	0.00098	0.00139	0.00155	0.00157	0.00158
15.000	0.00017	0.00020	0.00032	0.00053	0.00066	0.00070	0.00070
20.000	0.00007	0.00009	0.00014	0.00025	0.00035	0.00039	0.00040
30.000	0.00002	0.00003	0.00004	0.00008	0.00013	0.00017	0.00018
50.000	0.00000	0.00001	0.00001	0.00002	0.00003	0.00006	0.00006
70.000	0.00000	0.00000	0.00000	0.00001	0.00001	0.00003	0.00003
100.000	0.00000	0.00000	0.00000	0.00000	0.00000	0.00001	0.00001

Tabelle 1.23 Zahlenwerte des Einfluzzbeiwertes i_1 (Fig. 1.47b) [1]

Table 1.23 Numerical values of the influence value (Fig. 1.47b) [1]

(ii) Erlaeuterung

Die Spannungen im Punkt 2 (Fig. 1.46) sind gleich so gross wie die im Punkt 1, nur mit umgekehrten Vorzeichen.

(ii) Explanatory note

The stresses in Point 2 (Fig. 1.46) are numerically equal to that in Point 1 with reversed sign.

(iii) Quellen

References

[1] Kany (1964): Baugrundverformungen infolge waagerechter schubbelastung der Baugrundoberflaeche. Veroeffentlichungen des Grundbauinstituts der Bayerischen Landesgewerbeanstalt. H. 6.

[2] Ohde (1939): Zur Theorie der Druckverteilung im Baugrund. Bauingenieur 20, S.451.

[3] Siemer (1967): Spannungen und Setzungen des Halbraums unter einfachen Flaechenlasten und unter starren Grundkorpern aus waagerechter Beanspruchung. Mitt. VGB Aachen, H. 41.

b) Halbraum nach Froehlich

b) Half-space according to Froehlich

(i) Formeln

(i) Formulae

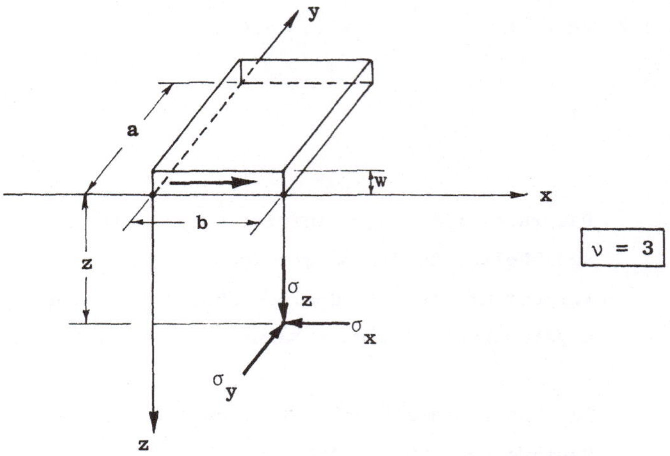

$$\sigma_z = \frac{w}{2\pi}\left[\arctan\frac{a}{z} - \frac{z^3}{(z^2+b^2)^{3/2}}\arctan\frac{a}{\sqrt{z^2+b^2}} + az\left(\frac{1}{z^2+a^2} - \frac{1}{z^2+b^2}\cdot\frac{z^2}{z^2+b^2+a^2}\right)\right]$$

$$\sigma_x = \frac{w}{\pi}\left[\arctan\frac{a}{z} - \left(1+\frac{1}{2}\cdot\frac{b^2}{z^2+b^2}\right)\cdot\frac{z}{\sqrt{z^2+b^2}}\arctan\frac{a}{\sqrt{z^2+b^2}} - \frac{1}{2}\cdot\frac{b^2}{z^2+b^2}\cdot\frac{a\cdot z}{z^2+b^2+a^2}\right]$$

$$\sigma_y = \frac{w}{2\pi}\left[\arctan\frac{a}{z} - \frac{z}{\sqrt{z^2+b^2}}\arctan\frac{a}{\sqrt{z^2+b^2}} - az\left(\frac{1}{z^2+a^2} - \frac{1}{z^2+b^2+a^2}\right)\right]$$

Fig. 1.49 Gleichungen zur Bestimmung Spannungen unter dem Eckpunkt infolge rechteckig gleichmaessig verteilter Belastung fuer $\nu = 3$ [1]

Equations for determining stresses under the corner Point due to uniformly distributed rectangular loading for $\nu = 3$ [1]

(ii) Quelle

Reference

Ohde (1939): Zur Theorie der Druckverteilung im Baugrund. Bauingenieur 20, S. 451.

1.1.2.4.2 Dreiecksfoermige Belastung

(i) Formeln/Tafeln/Tabellen

1.1.2.4.2 Triangular loading

(i) Formulae/Charts/Tables

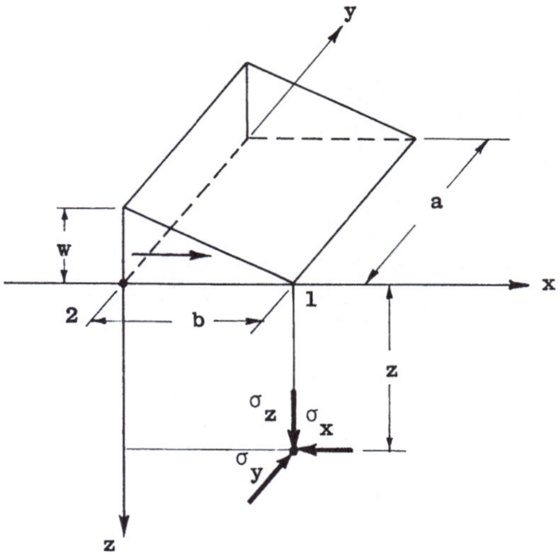

m = 2

$$\sigma_{z_1} = \frac{w}{2\pi}\left[\frac{\pi}{2}\cdot\frac{z}{b} - \frac{z^2}{z^2+b^2}\cdot\frac{a}{\sqrt{z^2+b^2+a^2}} - \frac{z}{b}\arctan\left(\frac{z}{ba}\sqrt{z^2+b^2+a^2}\right)\right]$$

$$= w.i_1$$

$$\sigma_{y_1} = \frac{w}{2\pi}\left[-\frac{\pi}{2}\cdot\frac{z}{b} + \frac{a}{\sqrt{z^2+b^2+a^2}} + \frac{z}{b}\arctan\left(\frac{z}{ba}\sqrt{z^2+b^2a^2}\right)\right]$$

$$\sigma_{x_1} = \frac{w}{2\pi}\left[\frac{b^2}{z^2+b^2}\cdot\frac{a}{\sqrt{z^2+b^2+a^2}} + 3\ \ln\frac{\sqrt{z^2+b^2}}{a+\sqrt{z^2+b^2+a^2}}\right.$$

$$-3\frac{a}{b}\ \ln\frac{\sqrt{z^2+a^2}}{b+\sqrt{z^2+b^2+a^2}} - 6\frac{z}{b}\left(\arctan\frac{b+\sqrt{z^2+b^2}}{z} - \frac{\pi}{4}\right)$$

$$\left.-\frac{3}{2}\frac{\sqrt{z^2+b^2}}{b}\cdot\frac{\left\{\frac{a}{\sqrt{z^2+b^2}}+\sqrt{\frac{z^2+b^2+a^2}{z^2+b^2}}\right\}^2 - 1}{\left\{\frac{a}{\sqrt{z^2+b^2}}+\sqrt{\frac{z^2+b^2+a^2}{z^2+b^2}}\right\}}\right]$$

$$+ \frac{3}{2} \cdot \frac{z}{b} \frac{\left\{ \frac{a}{z} + \sqrt{\frac{z^2 + a^2}{z^2}} \right\}^2 - 1}{\left\{ \frac{a}{b} + \sqrt{\frac{z^2 + a^2}{z^2}} \right\}}$$

$$+ \frac{3}{2} \cdot \ln \left\{ \frac{a}{\sqrt{z^2+b^2}} + \sqrt{\frac{z^2+b^2+a^2}{z^2+b^2}} \right\}^2$$

$$+ 6\frac{z}{b} \arctan \frac{\frac{a}{\sqrt{z^2+b^2}} + \sqrt{\frac{z^2+b^2+a^2}{z^2+b^2}} + \frac{b}{\sqrt{z^2+b^2}}}{\sqrt{\frac{z^2}{z^2+b^2}}}$$

$$-6\frac{z}{b} \arctan\left(\frac{a}{z} + \sqrt{\frac{z^2+a^2}{z^2}}\right) \Bigg]$$

Fig. 1.50a Gleichung zur Bestimmung Spannungen under den Eckpunkten infolge deiecksfoermiger Belastung [1]

Equations for determining stresses under the corner points due to triangular loading [1]

$\sigma_{z1} = w \cdot i_1$

$\sigma_{z2} = -w \cdot i_2$

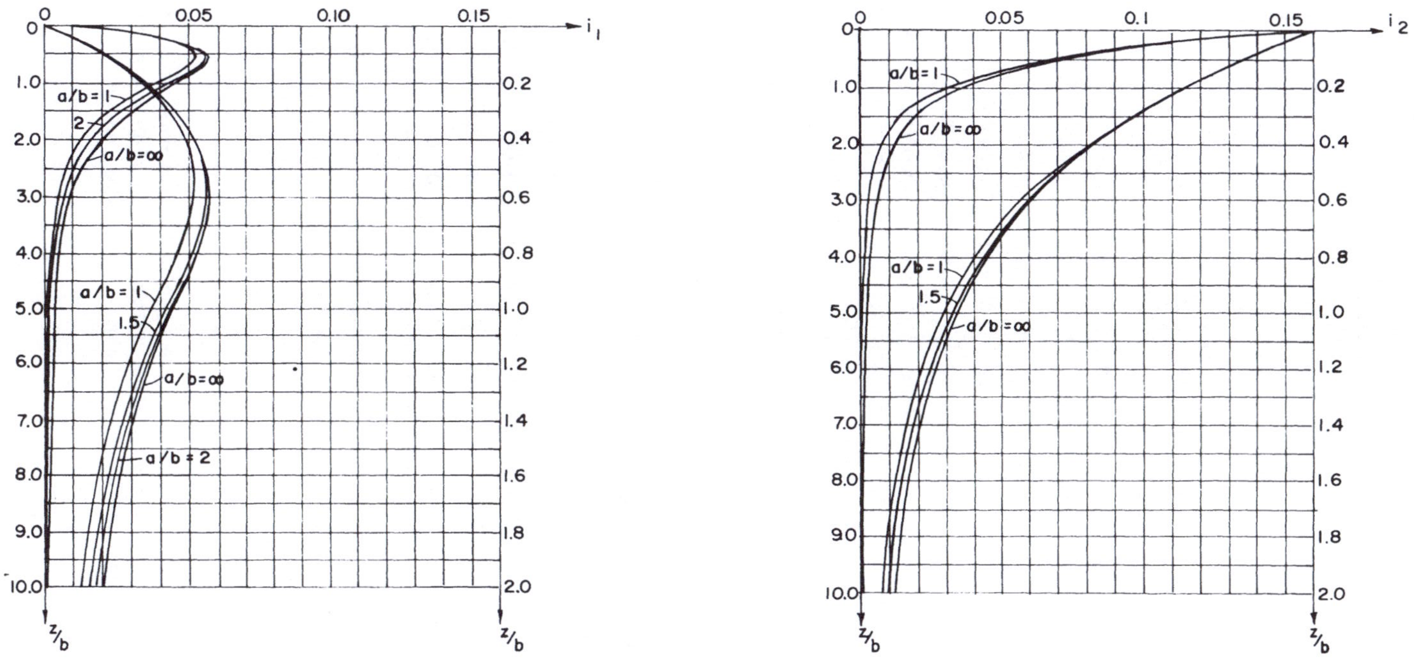

Fig. 1.50b Einflussbeiwerte i_1 und i_2

Influence values i_1 and i_2

$\sigma_{z1} = w \cdot i_1$

a/b / z/b	∞	10	5	2	1.5	1
0	0	0	0	0	0	0
0.1	0.0219	0.0218	0.0218	0.0218	0.0217	0.0216
0.25	0.0434	0.0434	0.0434	0.0435	0.0431	0.0423
0.5	0.0562	0.0563	0.0563	0.0558	0.0551	0.0526
1	0.0454	0.0454	0.0454	0.0440	0.0422	0.0374
2	0.0202	0.0202	0.0201	0.0175	0.0155	0.0121
5	0.0040	0.0040	0.0036	0.0021	0.0017	0.0012
10	0.0010	0.0009	0.0007	0.0003	0.0002	0.0002
20	0.0003	0.0002	0.0001	0.0000	0.0000	0.0000

$\sigma_{z2} = -w \cdot i_2$

z/b / z/b	∞	10	5	2	1.5	1
0	0.1592	0.1592	0.1592	0.1592	0.1592	0.1592
0.1	0.1357	0.1357	0.1357	0.1357	0.1357	0.1356
0.25	0.1064	0.1064	0.1064	0.1063	0.1062	0.1056
0.5	0.0711	0.0711	0.0710	0.0708	0.0704	0.0686
1	0.0342	0.0342	0.0341	0.0334	0.0323	0.0292
2	0.0116	0.0116	0.0115	0.0101	0.0090	0.0071
5	0.0021	0.0020	0.0018	0.0011	0.0008	0.0006
10	0.0006	0.0005	0.0003	0.0002	0.0001	0.0001
20	0.0001	0.0000	0.0000	0.0000	0.0000	0.0000

Tabelle 1.24 Zahlenwerte der Einflussbeiwerte i_1 bzw i_2 [1]
Numerical values of the Influence coeficients i_1 and i_2 [1]

(ii) Quelle
Reference

[1] Siemer (1967): Spannungen und Setzungen des Halbraums unter einfachen Flaechenlasten und unter starren Grundkoerpern aus waagerechter Beanspruchung. Mitt. VGB Aachen H.41

1.2 LASTEN IM HALBRAUM

1.2 LOADS IN HALF-SPACE

1.2.1 Lotrechte Last

1.2.1 Vertical Loads

1.2.1.1 Einzellast

1.2.1.1 Point load

(i) Formeln/Tabelle

(i) Formulae/Table

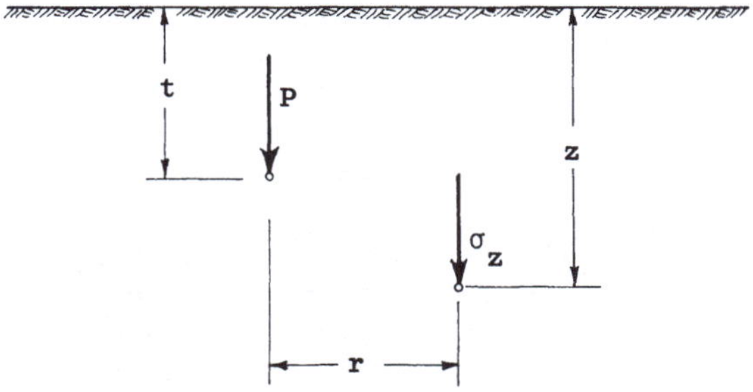

$$\sigma_z = \frac{3P}{8\pi \cdot z^2} \cdot \frac{m}{m-1} \left[\frac{m-2}{3m} \left\{ \frac{1}{\left[1+(\frac{r}{z})^2\right]^{3/2}} - \frac{1}{\left[(\frac{r}{z})^2+(1+2\frac{t}{z})^2\right]^{3/2}} \right\} \right.$$

$$+ \frac{3m-4}{m} \cdot \frac{(1+\frac{t}{z})(1+2\frac{t}{z})^2}{\left[(\frac{r}{z})^2+(\frac{2t}{z}+1)^2\right]^{5/2}} + \frac{1}{\left[1+(\frac{r}{z})^2\right]}$$

$$\left. - \frac{\frac{t}{z}(2\frac{t}{z}+1)(4\frac{t}{z}+5)}{\left[(\frac{r}{z})^2+(2\frac{t}{z}+1)^2\right]^{5/2}} + \frac{10\frac{t}{z}(\frac{t}{z}+1)(2\frac{t}{z}+1)^3}{\left[(\frac{r}{z})^2+(2\frac{t}{z}+1)^2\right]^{7/2}} \right]$$

Fuer r = 0 ergibt sich die Spannung σ_z unter der Lastachse zu:

For r = 0, the stress σ_z, under the axis becomes:

$$\sigma_{z0} = \frac{3P}{8\pi \cdot z^2} \cdot \frac{m}{m-1} \left[\frac{m-2}{3m} \left\{ 1 - \frac{1}{(2\frac{t}{z}+1)^3} \right\} + \frac{3m-4}{m} \cdot \frac{\frac{t}{z}+1}{(2\frac{t}{z}+1)^3} \right.$$

$$\left. +1 - \frac{\frac{t}{z}\left\{(4\frac{t}{z}+5)-10(\frac{t}{z}+1)\right\}}{(2\frac{t}{z}+1)^4} \right]$$

Fig. 1.51 **Gleichungen zur Berechnung den lötrechten Spannungen infolge Einzellast im Halbraum [1]**

Equations for the determination of vertical stresses due to point load in half-space [1]

Der Verlauf dieser Achsspannung σ_{z0} als Funktion von der Querdehnungszahl m und d Verhaeltnis t/z ist durch die folgende Abbildung gegeben:

The variation of σ_{z0} as a function of Poisson's number m and the ratio t/z is given in the following figure:

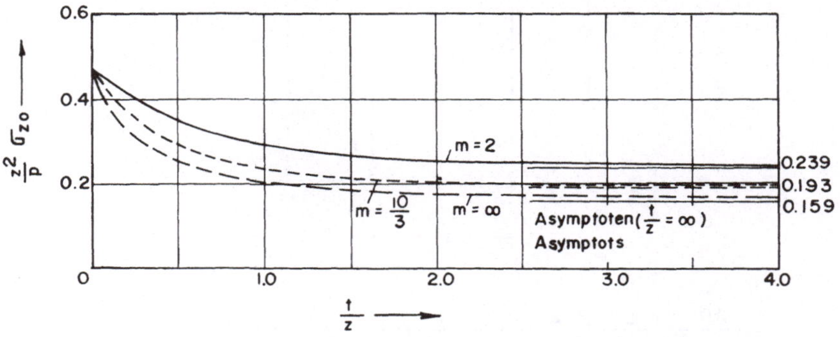

Fig. 1.52 Gleichungen zur Berechnung der Lotrechten Spannungen infolge Punktlast im Halbraum [1]

Equations for calculating the vertical stresses in half-space [1]

$$\sigma_z = \frac{2P}{t^2} \cdot i$$

z/t / r/t	0	0.1	0.2	0.3	0.5	0.75	1.0	1.5	2.0	2.5	3.0
0.1	0	0.01479	0.05277	0.11184	0.32992	1.30043	-0.05836	-0.46395	-0.13954	-0.06920	-0.04213
0.2	"	0.01187	0.04222	0.08786	114.04956	0.47471	-0.05675	-0.36124	-0.13113	-0.06737	-0.04150
0.3	"	0.00814	0.02886	0.05852	0.13286	0.13436	-0.05419	-0.25350	-0.11881	-0.06448	-0.04049
0.5	"	0.00185	0.00669	0.01264	0.01625	-0.02302	-0.04690	-0.11522	-0.08979	-0.05647	-0.03750
0.75	"	-0.00113	-0.00377	-0.00750	-0.01855	-0.03441	-0.03586	-0.05152	-0.05838	-0.04482	-0.03259
1.0	"	-0.00117	-0.00408	-0.00799	-0.01673	-0.02455	-0.02529	-0.02962	-0.03759	-0.03402	-0.02723
1.5	"	-0.00034	-0.00127	-0.00260	-0.00571	-0.00892	-0.01062	-0.01307	-0.01701	-0.01872	-0.01770
2.0	"	-0.00007	-0.00028	-0.00060	-0.00150	-0.00277	-0.00600	-0.00600	-0.00840	-0.01031	-0.01100
2.5	"	-0.00001	-0.00005	-0.00011	-0.00034	-0.00078	-0.00133	-0.00267	-0.00426	-0.00575	-0.00675
3.0	"	0.00000	0.00000	0.00000	-0.00005	-0.00018	-0.00042	-0.00116	-0.00217	-0.00325	-0.00414

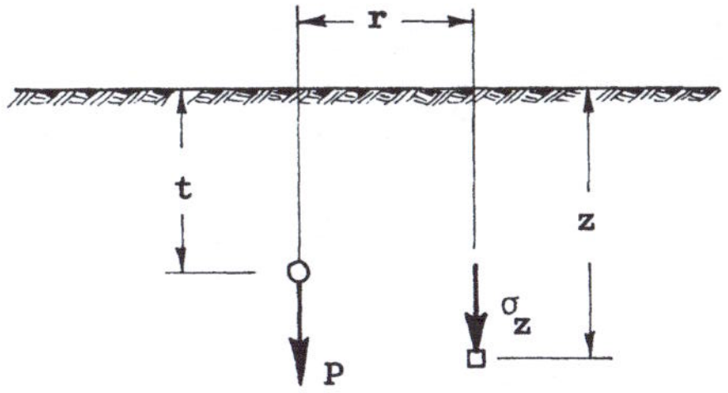

m = 2

Tabelle 1.25 Einflussbeiwerte zur Berechnung der Lotrechten Spannungen im Halbraum infolge Einzellast fuer m = 2 [2]

Table 1.25 Influence values for determining the vertical stresses in a half-space due to point load for m = 2 [2]

(ii) Voraussetzung:

Elastisch-isotroper Halbraum m = 2

(iii) Anwendungsbeispiel

Gegeben

P = 100 kN

t = 3.0 m

r = 0.3 m

z = 4.5 m

Gesucht

σ_z im Abstand r von P in der Tiefe z.

Loesung

$\frac{r}{t} = 0.1; \quad \frac{z}{t} = 1.5$

Aus Tabelle (1.25) i = -0.45395

(ii) **Assumption:**

Elastic-isotropic half-space m = 2

(iii) Example

Given

P = 100 kN

t = 3.0 m

r = 0.3 m

z = 4.5 m

Required

σ_z at a distance r from P at depth z.

Solution

$\frac{r}{t} = 0.1; \quad \frac{z}{t} = 1.5$

From Table (1.25) i =-0.45395

$$\sigma_z = \frac{(2)100.0}{(3.0)^2} \cdot (-0.46395) = -\ 10.2\ \text{kN/m}^2$$

(Druck)

(Compression)

(iv) Quellen

References

[1] Jelinek (1951): Der Einfluss von Gruendungstiefe und begrenzter Schichtmaechtigkeit auf die Druckausbreitung im Baugrund.
Bautechnik 28, S. 125.

[2] Mogami (1957): Numerical Tables for Calculation of Stress Components Induced in a Semi-Infinite Elastic Solid when a Force is Applied at a Point in the Interior of the Body.
Kajima Construction Technical Research Institute, PP 2.

1.2.1.2 Linienlast

(i) Formeln/Tafeln

1.2.1.2 Line Load

(i) Formulae/Charts

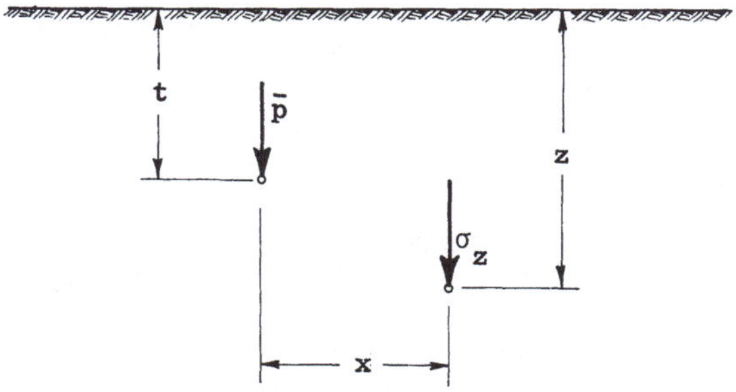

$$\sigma_z = \frac{\bar{p}}{\pi z}\left[\frac{1}{\left[1+(\frac{x}{z})^2\right]^2} + \frac{(1+2\frac{t}{z})^3}{\left[(1+2\frac{t}{z})^2+(\frac{x}{z})^2\right]^2}\right.$$

$$- \frac{m-1}{4m}\left\{\frac{1-(\frac{x}{z})^2}{\left[1+(\frac{x}{z})^2\right]^2} - \frac{(1+2\frac{t}{z})^2-(\frac{x}{z})^2}{\left[(1+2\frac{t}{z})^2+(\frac{x}{z})^2\right]^2}\right\}$$

$$+ \frac{m+1}{4m}\cdot\frac{t}{z}(1+\frac{t}{z})(1+2\frac{t}{z})\cdot\frac{(1+2\frac{t}{z})^2-3(\frac{x}{z})^2}{\left[(1+2\frac{t}{z})^2+(\frac{x}{z})^2\right]^3}$$

Fuer $x = 0$ ergibt sich die Spannung σ_z unter der Lastachse zu:

For $x = 0$, the stress σ_z under the axis becomes:

$$\sigma_{z0} = \frac{\bar{p}}{\pi z}\left[1+\frac{1}{(1+2\frac{t}{z})} - \frac{m-1}{4m}\left\{1-\frac{1}{(1+2\frac{t}{z})^2}\right\}\right.$$

$$\left. + \frac{m+1}{4m}\cdot\frac{\frac{t}{z}(1+\frac{t}{z})}{(1+2\frac{t}{z})^3}\right]$$

Fig. 1.53 Gleichungen zur Berechnung der lotrechten Spannungen infolge Linienlast im Halbraum [1]

Equations for the determination of vertical stress due to line load in half-space [1]

(ii) Voraussetzung:

Elastisch-isotroper Halbraum

(iii) Erlaueterung

Die Verteilung der Spannungen wird veraendert, wenn der Halbraum im unteren Teil aus einer unnachgibigen Unterlage besteht, wenn die Last nicht an seiner Oberkante, sondern in einem Schlitz (= Baugrube) angreift, oder wenn die Querdehnungszahl m anders angenommen wird. Die Spannungen Werden in solchen Faellen am einfachsten durch Einfuehrung einer Ersatztiefe k·z an Stelle der wirklichen Tiefe z berechnet. Die Tafeln fuer Linienlast und Einzellast (Fig. 1.52a, 1.52b) liefern die Reduktionsbeiwerte k, und zwar k_{z1} in Abhaengigkeit von z_1/z bei begrenzter Schichtmaechtigkeit bzw. k_t in Abhaengigkeit von t/z bei Last im Schlitz.

Fuer Flaechenlasten gibt es noch keine Tafeln. Daher muessen diese vorerst wie Linienlasten behandelt werden.

Bei anderem m sind die werte von k_m:

(ii) Assumption

Elastic-isotropic half-space

(iii) Explanatory note

The vertical stress distribution will change if the half-space is underlain by incompressible layer, or if the load acts in a slot or if the assumed Poisson's number ,m, differs from 2. The stresses in such cases will be calculated simply by considering an ideal depth k·z instead the true depth z. The charts for line load and point load (Fig. 1.52a, b) give the reduction coefficient k, i.e. k_{zl} as a function of zl/z for confined depth, or k_t as a function of t/z for loads in slot.

At the moment there are no charts provided for spread loadings. They should therefore be considered as line loads for the time being.

For different m values one may take the following k_m:

m	2	3	4	5
k_m	1	0.86	0.76	0.70

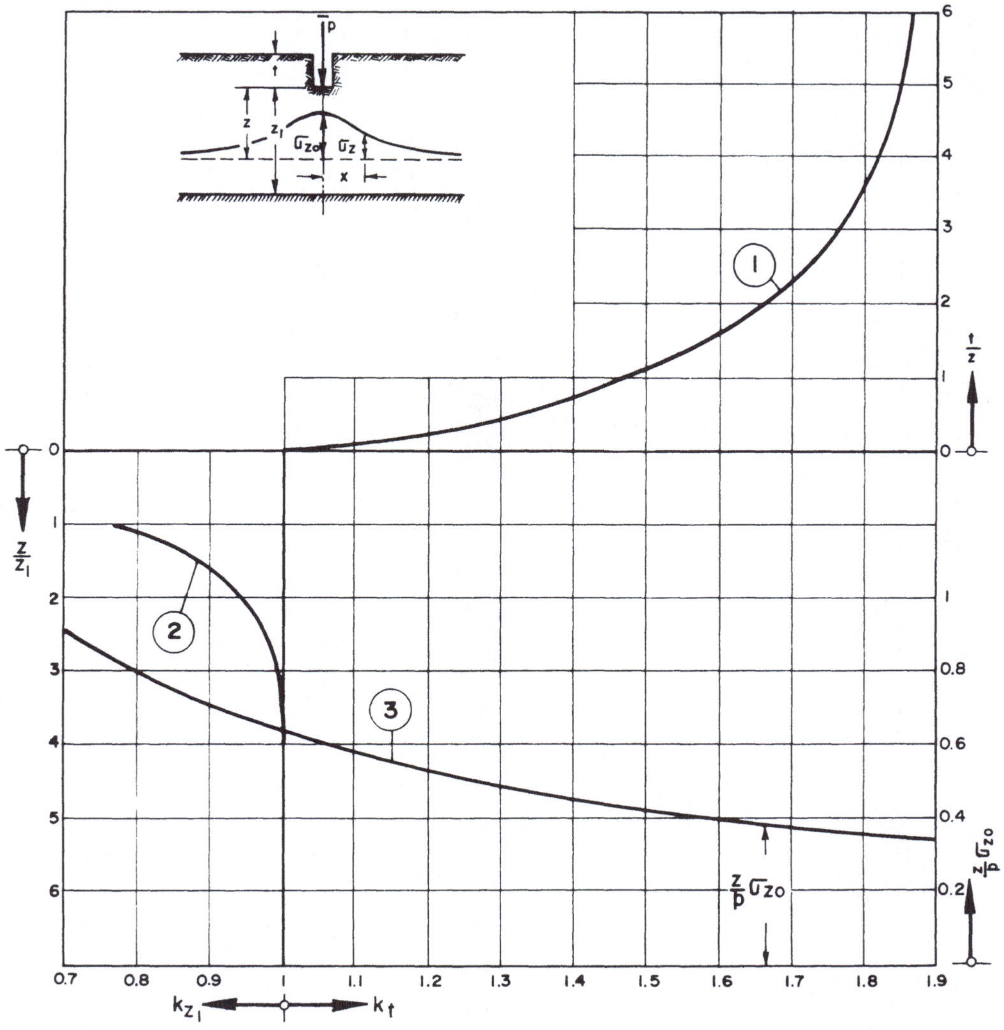

(1) Liefert k_t abhaengig von Gruendungstiefe t
Gives k_t as a function of t

(2) Liefert k_{z1} abhaengig von Schichtstaerke z_1
Gives k_{z1} as a function of layer depth z_1

(3) Liefert $\sigma_{z0}\ \frac{z}{p}$ abhaengig von k_t oder k_{z1} und damit die Achsspannung σ_{z0}
Gives $\sigma_{z0}\cdot\frac{z}{p}$ as a function of k_t or k_{z1} which ultimately gives the stress under the axis of the loading, σ_{z0}

Fig. 1.54a Einfluss von Gruendungstiefe, begrezter schichtmaechtigkeit und verschiedenen Querdehnungszahlen auf die Druckausbreitung im Baugrund bei Linienlast [1]

Influence of foundation depth, confined layer depth and different Poisson's number on stress propagation in soil foundation due to line load [1]

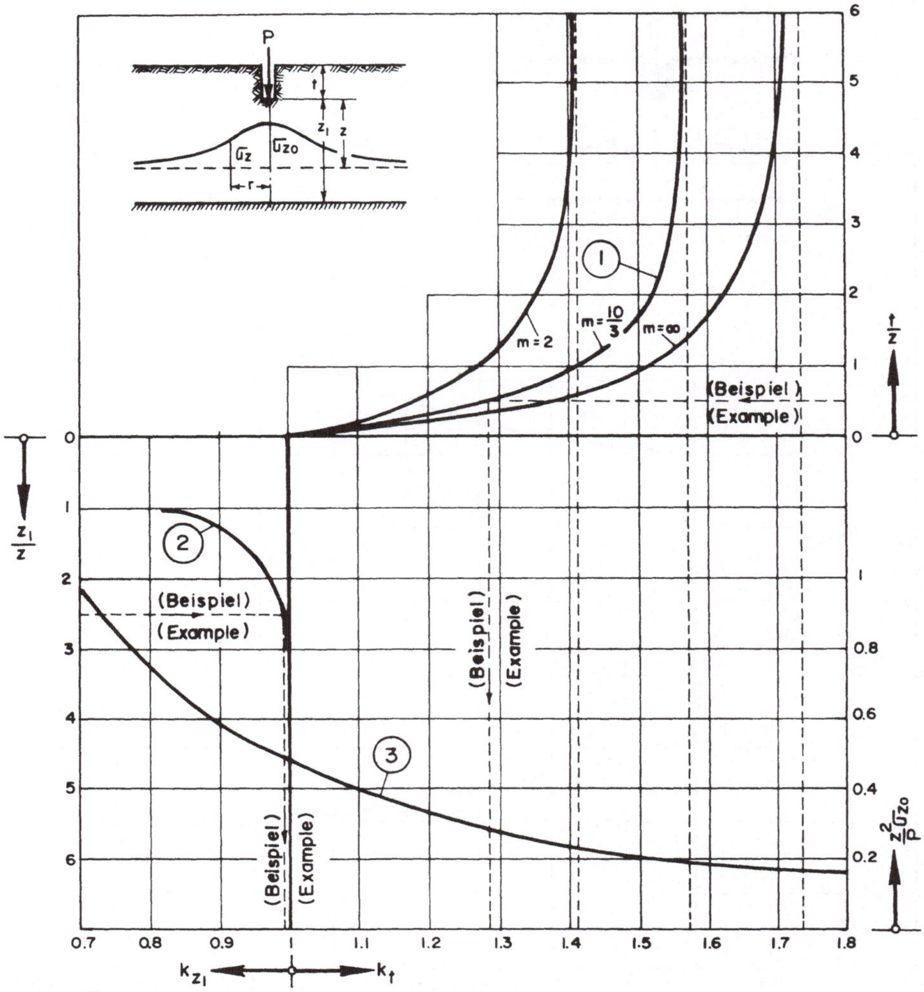

(1) Liefert k_t abhaengig von Gruendungstiefe t

Gives k_t as a function of t

(2) Liefert k_{z1} abhaengig von Schichtstaerke z_1

Gives k_{z1} as a function of layer depth z_1

(3) Liefert $\sigma_{z0}\frac{z^2}{P}$ abhaengig von k_t oder k_{z1} und damit die Achsspannung σ_{z0}

Gives $\sigma_{z0}\frac{z^2}{P}$ as a function of k_t or k_{z1} which ultimately gives the stress under the axis of the loading σ_{z0}

Fig. 1.54b Einfluss von Gruendungstiefe, begrezter schichtmaechtigkeit und verschiedenenQuerdehnungszahlen auf die Druckausbreitung im Baugrund bei Einzellast [1]

Influence of foundation depth confined layer depth and different Poisson's number on stress propagation in soil foundation due to point load [1]

Bei gleichzeitigem Vorkommen werden die k-Werte miteinander multipliziert. Die Spannungen σ_{z0} in der Bauwerksachse koennen aus den Tafeln ermittelt werden. Die Spannungen σ_z mit dem Abstand x (r) von der Achse (Angriffspunkt) muessen mit folgenden Formeln ausgerechnet werden:

The k-values are multiplied with each other if they occur simultaneously. The stresses σ_{z0} in the axis of the structure may be determined from charts. The stresses σ_z with a distance of x(r) from the line of action of the force should be calculated using the following equations:

Fuer Linienlast:

For line load:

$$\sigma_z = \frac{2\bar{p}}{\pi} \cdot \frac{k^3 z^3}{(x^2+k^2 \cdot z^2)^2}$$

Fuer Einzellast:

For point load:

$$\sigma_z = \frac{3p}{2\pi} \cdot \frac{k^3 z^3}{(r^2+k^2 \cdot z^2)^{5/2}}$$

(iv) Anwendungsbeispiel:

Gegeben

Einzellast P = 1000 kN

Angriffspunkt t = 2.0 m unter Gelaende

Schicktmaechtigkeit unter dem Angriffspunkt der Last z_1 = 10.0 m

Poissonzahl m = 3.33

Gesucht

Spannung σ_{z0} im z = 4.0 m Tiefe senkrecht unter der Last sowie σ_z in z = 4.0 m Tiefe und r = 1.0 m von der Last.

Loesung

Fuer m = 3.33 erhaelt man k_m durch Interpolation zu k_m = 0.83.

Fuer t/z = 2.0/4.0 = 0.5 und m = 3.33 ist nach Fig. 1.52b; k_t = 1.28.

Fuer z_1/z = 10.0/4.0 = 2.5 nach Fig. 1.52b: k_{z1} = 0.99.

Example:

Given

Point load P = 1000 kN

Point of action t = 2.0 below the ground surface under point of action of the load z_1 = 10.0 m

Poisson's number 3.33

Required

Stress σ_{z0} at z = 4.0 m depth below the load and also σ_z at z = 4.0 m depth and a distance of r = 1.0 m from the load.

Solution

For m = 3.33 one obtains k_m by interpolation k_m = 0.83.

For t/z = 2.0/4.0 = 0.5 and m = 3.33; from Fig. 1.52b; k_t = 1.28.

For z_1/z = 10.0/4.0 = 2.5; according to Fig. 1.52b k_{z1} = 0.99.

$$k = k_m \cdot k_t \cdot k_{z1} = (0.83)(1.28)(0.99) = 1.05$$

$$\sigma_{z0} = \frac{3\bar{p}}{2\pi} \cdot \frac{k^3 \cdot z^3}{(k \cdot z)^5} = \frac{(3)(1000)}{2 \cdot \pi} \frac{(1.05)^3 \ (4.0)^3}{\left[(1.05)\ (4.0)\right]^5}$$

$$= 27.1 \text{ kN/m}^2$$

$$\sigma_z = \frac{3\bar{p}}{2\pi} \cdot \frac{k^3 \cdot z^3}{(r^2 + k^2 \cdot z^2)^{5/2}}$$

$$= \frac{(3)1000}{2 \cdot \pi} \frac{(1.05)^3 \ (4.0)^3}{\left[(1.0^2 + 1.05^2 (4.0)^2\right]^{5/2}}$$

$$= 23.7 \text{ kN/m}^2$$

(v) Quelle

Reference

[1] Jelinek (1951): Der Einfluss von Gruendungstiefe und begrenzter Schichtmaechtigkeit auf die Druckausbreitung im Baugrund. Bautechnik 28, S. 125.

1.2.1.3 Gleichmaessige Streifenlast

(i) Formeln/Tafeln

1.2.1.3 Uniformly distributed strip loading

(i) Formulae/Charts

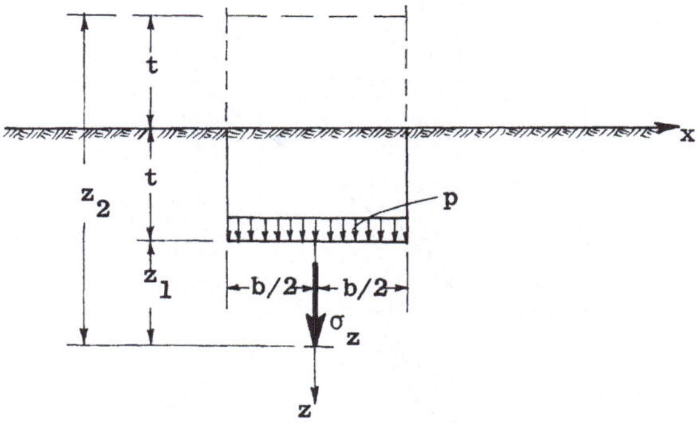

$$\sigma_z = \frac{p}{\pi}\left[\frac{(bz_1/2)}{z_1^2+b^2/4} + \arctan\frac{b}{2z_1} + \frac{bz_2/2}{z_2^2+b^2/4} + \mathrm{arc\ tan}\frac{b}{2z_2}\right.$$

$$- \frac{1-\mu}{2}\cdot z_1\left\{\frac{b/2}{z_1^2+b^2/4} - \frac{b/2}{z_2^2+b^2/4}\right\}$$

$$\left. + \frac{1+\mu}{2}\cdot\frac{z1.t}{z_2}\cdot\frac{b/2}{z_2^2+b^2/4}\right]$$

Bei

For

$$\mu = \frac{1}{3}, \quad \sigma_z = p\cdot i$$

Fig. 1.55a Gleichung zur Bestimmung der lotrechten Spannungen in der Achse einer streifenlast im Halbraum [1]

Equation for determining the vertical stresses along the axis of a strip loading in a half-space [1]

$$\sigma_z = p \cdot i$$

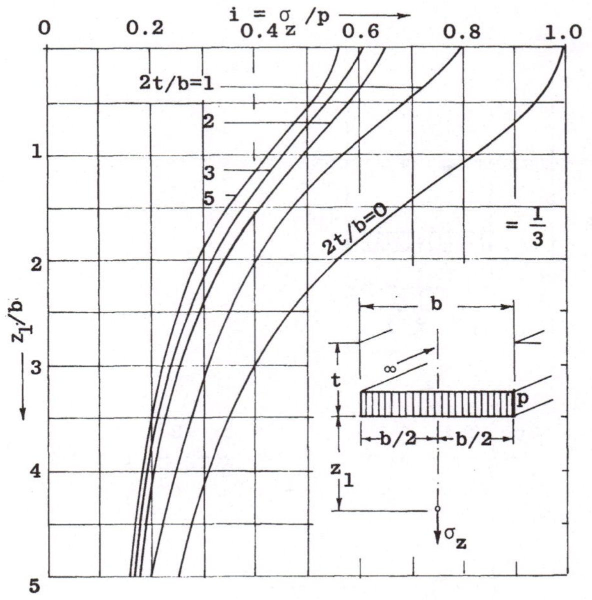

Fig. 1.55b Einflusszahlen der lotrechten Bodenspannungen in der Achse einer Streifenlast in Halbraum bei $\mu = 1/3$ [1]

Influence values of the vertical soil stresses along the axis of a strip loading in a half-space for $\mu = 1/3$[[1]

(ii) Quelle

Reference

[1] Kezdi (1985): Beitraege zur Berechnung der Spannungsverteilung im Boden. Bauingenieur 33, S. 54.

1.2.1.4 Gleichmaessige Kreislast

(i) Formeln/Tafeln

1.2.1.4 Uniformly distributed Circular loading

(i) Formulae/Charts

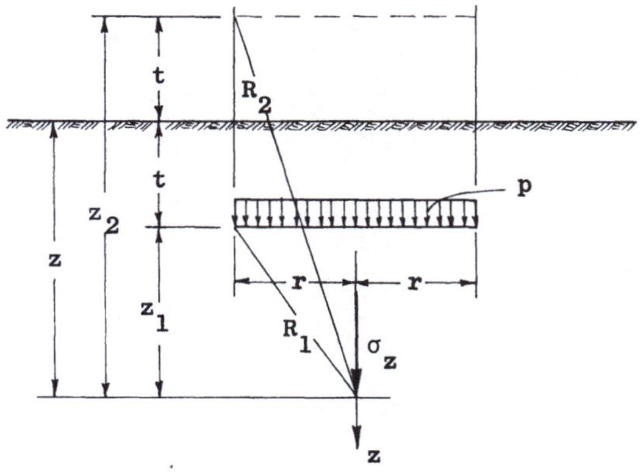

$$\sigma_z = -\frac{1}{1-\mu}\cdot\frac{3p}{2}\left[\frac{1-2\mu}{6}\cdot z_1\left(\frac{1}{R_2}-\frac{1}{R_1}-\frac{1}{z_2}-\frac{1}{z_1}\right)\right.$$

$$+\frac{1}{6}-\frac{z_1^3}{6R_1^3}+\left\{\frac{3-4\mu}{2}\cdot z_2-t(1-2\mu)\right\}\cdot\left(\frac{1}{3z_2}-\frac{z_2^2}{3R_2^3}\right)$$

$$\left.-\,t.z(z-t).\left(\frac{z_2^2}{R_2^5}-\frac{1}{R_2^3}\right)\right]$$

Bei $\mu = \frac{1}{3}$ ist $\sigma_z = p\cdot i$

For $\mu = \frac{1}{3}$, $\sigma_z = p\cdot i$

Fig. 1.56a Gleichung zur Bestimmung der lotrechten Spannungen in der Achse einer Kreisflaechenlast im Halbraum [1]

Equation for determining the vertical stresses along the axis of a spread circular loading in a half-space [1]

$$\sigma_z = i \cdot p$$

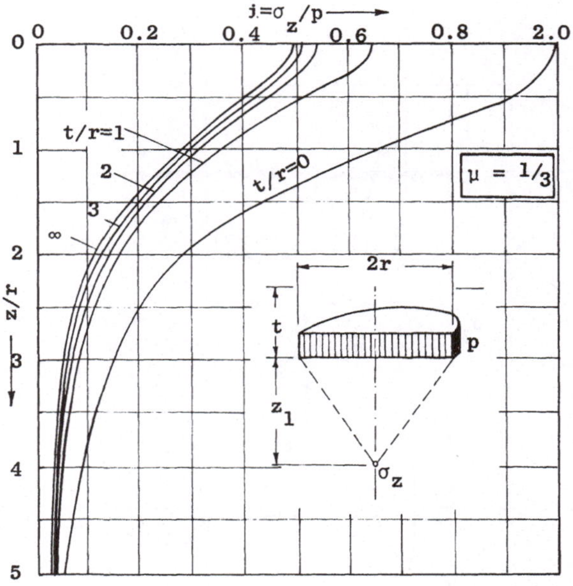

Fig. 1.56b Einflusszahlen der lotrechten Bodenspannungen in der Achse einer Kreisflaechenlast im Halbraum [1]

Equations for determining the vertical stresses along the axis of a spread circular loading in a half-space [1]

(ii) Quelle

Reference

[1] Kezdi (1958): Beitraege zur Berechnung der Spannungsverteilung im Boden. Bauingenieur 33, S. 54.

1.2.1.5 Pfahllasten

1.2.1.5.1 Punktfoermig

(i) Formeln/Tabellen

1.2.1.5 Pile loads

1.2.1.5.1 Point load

(i) Formulae/Tables

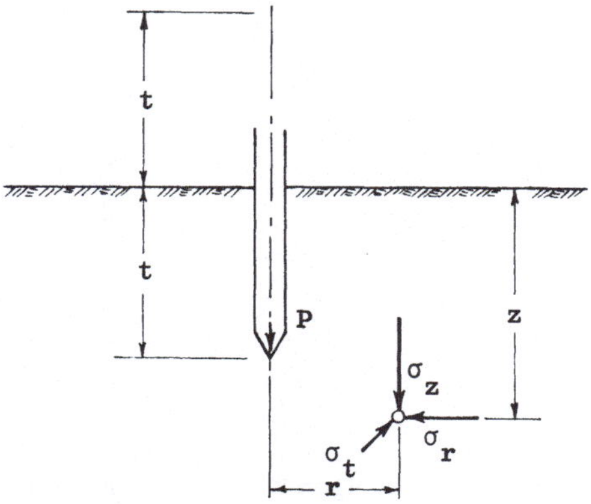

$$\frac{r}{t} = n; \quad \frac{z}{t} = m$$

$$A^2 = n^2 + (m-1)^2$$

$$B^2 = n^2 + (m+1)^2$$

$$i_z = \frac{\sigma_z \cdot t^2}{P} = \frac{1}{8\pi(1-\mu)} \left[- \frac{(1-2\mu)(m-1)}{A^3} + \frac{(1-2\mu)(m-1)}{B^3} \right.$$

$$\left. - \frac{3(m-1)^3}{A^5} - \left\{ \frac{3(3-4\mu)m(m+1)^2 - 3(m+1)(5m-1)}{B^5} \right\} - \frac{30m(m+1)^3}{B^7} \right]$$

$$i_r = \sigma_r \cdot \frac{t^2}{P} = \frac{1}{8\pi(1-\mu)} \left[\frac{(1-2\mu)(m-1)}{A^3} - \frac{(1-2\mu)(m+7)}{B^3} + \frac{4(1-\mu)(1-2\mu)}{B(B+m+1)} \right.$$

$$- \frac{3n^2(m-1)}{A^5} + \frac{6(1-2\mu)(m+1)^2 - 6(m+1) - 3(3-4\mu)n^2(m-1)}{B^2}$$

$$\left. - \frac{30n^2 m(m+1)}{B^7} \right]$$

$$i_t = \sigma_t \cdot \frac{t^2}{P} = \frac{1}{8\pi(1-\mu)} \left[\frac{(1-2\mu)(m-1)}{A^3} + \frac{(1-2\mu)(3-4\mu)(m+1) - 6(1-2\mu)}{B^3} \right.$$

$$\left. - \frac{4(1-2\mu)(1-\mu)}{B(B+m+1)} + \frac{6(1-2\mu)(m+1)^2 - 6(m+1)}{B^5} \right]$$

wobei: σ_z = Lotrechte Spannung

σ_r = Radiale Spannung

σ_t = Tangentiale Spannung

μ = Poissonziffer

where: σ_z = vertical stress

σ_r = radial stress

σ_t = tangential stress

μ = Poisson's ratio

Fig.1.57 Gleichungen fuer Spannungen infolge Punktfoermigen Pfahllasten im Halbraum [1]

Equations for stresses due to point load of a pile in half-space [1]

$$i_z = \sigma_z \cdot \frac{t^2}{P}$$

m \ n	0.00	0.02	0.04	0.06	0.08	0.10	0.12	0.16
0.1	+22	+22	+22	+21	+21	+20	+20	+18
0.2	+78	+78	+77	+76	+75	+73	+71	+66
0.3	+168	+168	+166	+164	+161	+156	+151	+139
0.4	+305	+304	+301	+296	+288	+279	+269	+244
0.5	+524	+522	+515	+503	+488	+469	+446	+396
0.6	+915	+909	+892	+864	+826	+781	+730	+618
0.7	+1744	+1725	+1670	+1584	+1474	+1347	+1211	+939
0.8	+4088	+3994	+3728	+3339	+2885	+2420	+1985	+1275
0.9	+16698	+15257	+11881	+8278	+5434	+3494	+2254	+982
1.0		93	93	93	92	92	92	91
1.1	16885	15444	12067	8464	5619	3679	2438	1165
1.2	4278	4183	3918	3528	3073	2608	2172	1460
1.3	1938	1919	1864	1778	1667	1539	1402	1129
1.4	1116	1110	1092	1064	1026	980	928	814
1.5	733	730	723	711	695	675	652	598
1.6	523	522	518	512	504	494	483	455
1.7	395	394	393	389	385	379	373	357
1.8	311	311	310	308	305	302	298	288
1.9	253	252	252	251	249	247	244	238
2.0	210	210	210	209	208	206	205	200
2.1	178	178	178	177	177	176	175	172
2.2	154	154	153	153	152	152	151	149
2.3	134	134	134	134	133	133	132	131
2.4	118	118	118	118	118	117	117	116
2.5	106	105	105	105	105	105	104	103
2.6	95	95	95	94	94	94	94	93
2.7	86	86	86	85	85	85	85	84
2.8	78	78	78	78	78	77	77	77
2.9	71	71	71	71	71	71	71	70
3.0	65	65	65	65	65	65	65	65

m \ n	0.20	0.30	0.40	0.50	0.60	0.80	1.00	1.20	1.60	2.00	2.50	3.00
0.1	+17	+12	+7	+3	+1	1	2	1	-	-	-	-
0.2	+59	+42	+25	+12	+3	5	5	4	2	1	-	-
0.3	+125	+86	+50	+23	+6	9	10	8	3	1	-	-
0.4	+216	+142	+79	+34	+7	14	16	12	5	2	1	-
0.5	+340	+205	+104	+40	+4	21	22	17	8	3	1	1
0.6	+505	+265	+115	+34	5	29	28	21	10	4	2	1
0.7	+697	+289	+97	+14	21	38	33	25	12	6	2	1
0.8	+792	+223	+44	18	40	45	37	28	14	7	3	1
0.9	+456	+60	26	51	56	51	41	31	16	8	3	2
1.0	90	86	81	76	69	56	44	33	18	9	4	2
1.1	636	233	137	101	82	61	46	35	19	10	5	2
1.2	975	398	208	134	100	67	49	37	21	11	5	3
1.3	884	467	264	168	119	74	52	39	22	12	6	3
1.4	697	447	285	190	135	81	56	41	23	13	7	3
1.5	539	393	277	197	145	87	59	43	24	14	7	4
1.6	422	335	256	193	147	92	62	45	26	15	8	4
1.7	338	284	229	182	144	94	64	47	27	16	9	5
1.8	276	241	204	168	138	94	66	48	28	17	9	5
1.9	230	207	180	154	130	92	66	49	29	18	10	6
2.0	195	179	160	140	121	89	66	49	29	18	10	6
2.1	168	156	142	127	112	85	65	49	30	19	11	6
2.2	146	138	127	115	103	81	63	49	30	19	11	7
2.3	129	122	114	105	95	77	61	48	31	20	12	7
2.4	114	109	103	96	88	73	59	47	31	20	12	7
2.5	102	98	93	88	81	68	56	46	31	20	12	8
2.6	92	89	85	80	75	64	54	45	30	21	13	8
2.7	84	81	78	74	70	61	52	43	30	21	13	8
2.8	76	74	72	68	65	57	49	42	30	21	13	8
2.9	70	68	66	63	60	54	47	40	29	21	13	9
3.0	64	63	61	59	56	51	45	39	28	20	13	9

Tabelle 1.26a Beiwerte i_z X 1000 fuer eine Punktlast $\mu = 0.1$ (+Zug, andernfalls Druck) [1]

Table 1.26a Coefficients i_z X 1000 for a point load $\mu = 0.1$ (+ tension, otherwise compression) [1]

$$i_z = \sigma_z \cdot \frac{t^2}{P}$$

m \ n	0.00	0.02	0.04	0.06	0.08	0.10	0.12	0.16
0.1	+26	+25	+25	+25	+24	+24	+23	+21
0.2	+91	+91	+90	+89	+87	+85	+83	+76
0.3	+196	+195	+194	+191	+187	+181	+175	+161
0.4	+354	+353	+349	+343	+334	+323	+311	+281
0.5	+606	+604	+595	+581	+563	+540	+514	+453
0.6	+1057	+1050	+1029	+995	+950	+896	+836	+703
0.7	+2010	+1988	+1922	+1820	+1688	+1536	+1375	+1054
0.8	+4707	+4594	+4277	+3813	+3273	+2724	+2213	+1388
0.9	+19214	+17491	+13474	+9227	+5922	+3707	+2319	+942
1.0		102	102	102	102	101	101	100
1.1	19419	17697	13679	9432	6126	3911	2522	1143
1.2	4917	4804	4486	4021	3481	2932	2419	1592
1.3	2225	2203	2137	2034	1901	1749	1587	1264
1.4	1279	1272	1251	1217	1172	1117	1055	920
1.5	839	836	827	813	794	770	742	678
1.6	598	596	592	585	575	563	549	516
1.7	451	450	448	444	438	432	424	405
1.8	354	354	352	350	347	343	338	327
1.9	287	287	286	284	282	280	277	269
2.0	238	238	238	237	235	234	232	227
2.1	202	202	201	201	200	199	197	194
2.2	174	173	173	173	172	171	170	168
2.3	151	151	151	151	150	149	149	147
2.4	133	133	133	133	132	132	131	130
2.5	118	118	118	118	118	117	117	116
2.6	106	106	106	106	106	105	105	104
2.7	96	96	96	95	95	95	95	94
2.8	87	87	87	87	87	86	86	86
2.9	79	79	79	79	79	79	79	78
3.0	73	73	73	73	72	72	72	72

m \ n	0.20	0.30	0.40	0.50	0.60	0.80	1.00	1.20	1.60	2.00	2.50	3.00
0.1	+19	+13	+8	+3	+1	2	2	1	1	-	-	-
0.2	+69	+48	+28	+13	+2	6	6	5	2	1	-	-
0.3	+144	+98	+56	+24	+4	12	12	9	4	1	-	-
0.4	+247	+160	+86	+34	+4	19	19	14	6	2	1	-
0.5	+387	+228	+110	+37	2	28	26	19	8	3	1	-
0.6	+569	+288	+116	+27	15	37	33	24	10	4	1	1
0.7	+771	+301	+89	+1	34	47	38	28	13	5	2	1
0.8	+836	+208	+23	36	54	54	42	31	15	7	2	1
0.9	+400	+23	50	67	68	58	45	33	16	8	3	1
1.0	99	95	89	83	75	61	46	34	18	9	4	2
1.1	599	213	129	99	83	63	48	36	19	10	4	2
1.2	1037	401	203	131	98	67	49	37	20	11	5	2
1.3	977	498	272	169	118	73	52	39	21	12	5	3
1.4	782	490	304	198	138	81	56	41	23	13	6	3
1.5	608	436	302	210	151	89	60	43	24	13	7	3
1.6	477	374	281	209	157	95	63	45	25	14	7	4
1.7	382	318	254	199	156	98	66	47	26	15	8	4
1.8	313	271	227	185	150	99	68	49	27	16	9	5
1.9	260	232	201	170	142	98	69	50	29	17	9	5
2.0	220	201	178	155	133	96	70	51	30	18	10	5
2.1	189	175	159	141	123	92	69	52	30	19	10	6
2.2	165	154	142	128	114	88	67	52	31	19	11	6
2.3	145	137	127	117	105	84	66	51	31	20	11	7
2.4	128	122	115	106	97	79	63	50	32	20	12	7
2.5	114	110	104	97	90	75	61	49	32	21	12	7
2.6	103	99	95	89	83	70	58	48	32	21	13	8
2.7	93	90	87	82	77	66	56	46	31	21	13	8
2.8	85	83	79	76	72	62	53	45	31	21	13	8
2.9	78	76	73	70	67	59	51	43	30	21	13	9
3.0	71	70	68	65	62	55	48	42	30	21	14	9

Tabelle 1.26b Beiwerte i_z x 1000 fuer eine Punktlast $\mu = 0.3$ (+ Zug, andernfalls Druck) [1]

Table 1.26b Coefficients i_z x 1000 for a point load $\mu = 0.3$ (+tension, otherwise compression) [1]

$$i_z = \sigma_z \cdot \frac{t^2}{P}$$

m \ n	0.00	0.02	0.04	0.06	0.08	0.10	0.12	0.16
0.1	+32	+32	+32	+31	+31	+30	+29	+27
0.2	+115	+115	+114	+112	+110	+107	+104	+95
0.3	+246	+245	+243	+239	+233	+227	+219	+200
0.4	+442	+440	+435	+427	+416	+402	+386	+347
0.5	+755	+751	+740	+722	+698	+669	+634	+556
0.6	+1311	+1302	+1275	+1232	+1174	+1104	+1026	+855
0.7	+2490	+2461	+2376	+2243	+2073	+1878	+1671	+1261
0.8	+5822	+5675	+5265	+4666	+3973	+3272	+2623	+1591
0.9	+23741	+21512	+16341	+10936	+6800	+4090	+2437	+870
1.0		119	119	119	119	118	118	117
1.1	23982	21752	16581	11176	7039	4328	2674	1104
1.2	6067	5920	5509	4910	4216	3514	2864	1829
1.3	2742	2713	2628	2494	2323	2127	1919	1507
1.4	1574	1565	1537	1493	1434	1364	1284	1110
1.5	1030	1026	1015	996	971	940	904	822
1.6	732	730	725	716	703	688	669	626
1.7	551	550	547	542	535	526	516	491
1.8	431	431	429	426	422	417	411	396
1.9	349	348	347	345	343	340	336	326
2.0	289	289	288	287	285	283	280	274
2.1	244	244	243	242	241	240	238	233
2.2	209	209	209	208	207	206	205	202
2.3	182	182	181	181	180	180	179	176
2.4	160	160	159	159	159	158	157	155
2.5	142	142	141	141	141	140	140	138
2.6	127	126	126	126	126	126	125	124
2.7	114	114	114	114	113	113	113	112
2.8	103	103	103	103	103	102	102	101
2.9	94	94	94	94	94	93	93	93
3.0	86	86	86	86	86	85	85	85

m \ n	0.20	0.30	0.40	0.50	0.60	0.80	1.00	1.20	1.60	2.00	2.50	3.00
0.1	+24	+17	+9	+4	-	3	2	2	1	-	-	-
0.2	+86	+59	+33	+14	+1	9	8	6	2	1	-	-
0.3	+178	+119	+65	+26	+1	17	16	11	4	1	-	-
0.4	+304	+192	+98	+34	3	27	25	17	6	2	-	-
0.5	+471	+269	+122	+33	13	39	34	23	9	3	1	-
0.6	+684	+330	+120	+14	32	52	42	29	11	4	1	-
0.7	+903	+324	+73	24	57	63	48	33	14	5	1	-
0.8	+915	+181	16	69	80	69	51	35	15	6	2	-
0.9	+300	45	92	96	90	71	52	37	17	7	2	1
1.0	115	110	103	95	86	68	51	37	18	8	3	1
1.1	532	176	115	95	83	66	50	37	19	9	3	1
1.2	1150	405	194	124	94	67	50	37	20	10	4	1
1.3	1145	553	286	171	117	72	52	38	20	10	4	2
1.4	934	566	338	212	144	82	55	40	22	11	5	2
1.5	732	514	345	234	163	92	60	42	23	12	5	2
1.6	577	445	328	238	174	101	65	45	24	13	6	3
1.7	462	380	299	230	176	107	69	48	26	14	7	3
1.8	378	325	268	216	172	110	73	51	27	15	7	4
1.9	314	278	238	199	164	110	75	53	29	16	8	4
2.0	266	241	212	182	154	108	76	55	30	17	9	4
2.1	228	210	189	166	144	105	76	56	31	18	9	5
2.2	198	185	168	151	133	101	75	56	32	19	10	5
2.3	173	163	151	137	123	96	74	56	33	20	11	6
2.4	153	146	136	125	114	91	71	55	33	20	11	6
2.5	137	131	123	114	105	86	69	54	34	21	12	7
2.6	123	118	112	105	97	81	66	53	34	21	12	7
2.7	111	107	102	96	90	77	64	52	34	22	13	7
2.8	101	98	94	89	84	72	61	50	34	22	13	8
2.9	92	89	86	82	78	68	58	49	33	22	13	8
3.0	84	82	79	76	72	64	55	47	33	22	14	8

Tabelle 1.26c Beiwerte i_z x 1000 fuer eine Punktlast $\mu = 0.5$ (+ Zug, andernfalls Druck) [1]

Table 1.26c Coefficients i_z x 1000 for a point load $\mu = 0.5$ (+ tension, otherwise compression) [1]

m	i_r und i_t (x 1000) and		
	μ = 0.1	μ = 0.3	μ = 0.5
1.1	+3530	+2258	-25
1.2	+879	+556	-20
1.3	+389	+242	-17
1.4	+218	+133	-14
1.5	+139	+84	-12
1.6	+97	+57	-10
1.7	+71	+41	-9
1.8	+55	+31	-8
1.9	+43	+24	-7
2.0	+35	+20	-6
2.1	+29	+16	-5
2.2	+25	+13	-5
2.3	+21	+11	-4
2.4	+19	+10	-4
2.5	+16	+9	-3
2.6	+14	+8	-3
2.7	+13	+7	-3
2.8	+12	+6	-2
2.9	+11	+5	-2
3.0	+10	+5	-2

Tabelle 1.26d Beiwerte i_r x 1000 und i_t x 1000 fuer eine Punktlast bei n = 0 (+ Zug; - Druck) [1]

Table 1.26d Coefficients i_r x 1000 and i_t x 1000 for a point load n = 0 (+tension, - Compression) [1]

(ii) Quelle

Reference

[1] Geddes (1966): Stresses in foundation soils due to vertical subsurface loading. Geotechnique XVI, PP 231.

1.2.1.5.2 Gelichmaessige Mantelreibung

(i) Formeln/Tabellen

1.2.1.5.2 Uniform skin friction

(i) Formulae/Tables

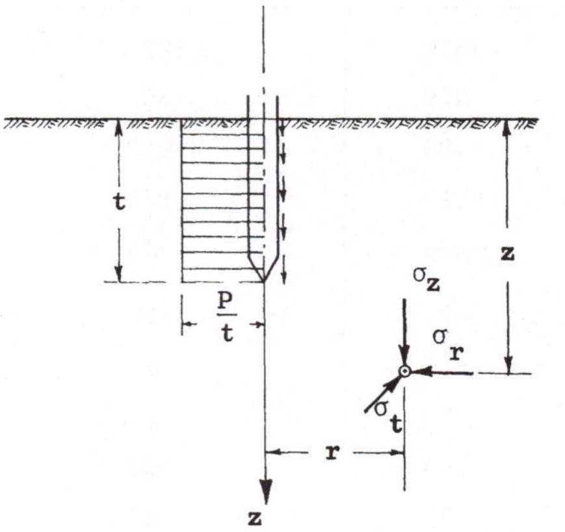

$$r/t = n, \quad z/t = m, \quad A^2 = n^2+(m-1)^2$$

$$F^2 = n^2+m^2, \quad B^2 = n^2+(m+1)^2$$

$$i_z = \frac{\sigma_z \cdot t^2}{P} = \frac{1}{8\pi(1-\mu)}\left[-\frac{2(2-\mu)}{A} + \frac{2(2-\mu) + 2(1-2\mu)\frac{m}{n}\left(\frac{m}{n}+\frac{1}{n}\right)}{B}\right.$$

$$-\frac{(1-2\mu)2\left(\frac{m}{n}\right)^2}{F} + \frac{n^2}{A^3} + \frac{4m^2-4(1+\mu)\left(\frac{m}{n}\right)^2 m^2}{F^3}$$

$$+\frac{4m(1+\mu)(m+1)\left(\frac{m}{n}+\frac{1}{n}\right)^2 - (4m^2+n^2)}{B^3}$$

$$\left.+\frac{6m^2\left(\frac{m^4-n^4}{n^2}\right)}{F^5} + \frac{6m\left(mn^2-\frac{1}{n^2}[m+1]^5\right)}{B^5}\right]$$

$$i_r = \sigma_r\,\frac{t^2}{P} = \frac{1}{8\pi(1-\mu)}\left[\frac{(1-2\mu)}{A} + \frac{(7-2\mu) - 12(1-\mu)\frac{m}{n}\left(\frac{m}{n}+\frac{1}{n}\right)}{B}\right.$$

$$-\left\{\frac{4(2-\mu)-12(1-\mu)\left(\frac{m}{n}\right)}{F}\right\} - \frac{n^2}{A^3} + \frac{4n^2-2m^2+2(1+2\mu)\left(\frac{m}{n}\right)^2 m^2}{F^3}$$

$$- \left\{ \frac{3n^2 - 2m^2 + 2(1+2\mu)\frac{m}{n}(m+1)^2\left(\frac{m}{n} + \frac{1}{n}\right)}{B^3} \right\}$$

$$+ \frac{6\left[n^2m^2 - m^4\left(\frac{m}{n}\right)^2\right]}{F^5} + \frac{6\left[\frac{m}{n}(m+1)^4\left(\frac{m}{n}+\frac{1}{n}\right) - m^2n^2\right]}{B^5}$$

$$+ 4(1-\mu)(1-2\mu)\left\{\frac{1}{F+m} - \frac{1}{B+m+1}\right\} \Bigg]$$

$$i_t = \sigma_t \cdot \frac{t^2}{p} = \frac{1}{8\pi(1-\mu)} \Bigg[\frac{(1-2\mu)}{A} + \frac{6-(1-2\mu)(3-4\mu)+6(1-2\mu)\frac{m}{n}\left(\frac{m}{n} + \frac{1}{n}\right)}{B}$$

$$+ \frac{2(1-2\mu)^2 - 6(1-2\mu)\left(\frac{m}{n}\right)^2 - 6}{F} + \frac{2m^2 - 4\mu n^2 + 2(1+2\mu)\frac{m}{n}(m+1)^2\left(\frac{m}{n} + \frac{1}{n}\right)}{B^3}$$

$$+ \frac{4\mu n^2 - 2m^2 - 2(1+2\mu)m^2\left(\frac{m}{n}\right)^2}{F^3}$$

$$- 4(1-\mu)(1-2\mu)\left\{\frac{1}{F+m} - \frac{1}{B+m+1}\right\} \Bigg]$$

Fig. 1.58 Gleichungen fuer Spannungen infolge gleichmaessiger Mantelreibung im Halbraum [1]

Equations for stresses due to a uniform skin friction in half-space [1]

$$i_z = \sigma_z \cdot \frac{t^2}{P}$$

m \ n	0.00	0.02	0.04	0.06	0.08	0.10	0.12	0.16
0.1		1508	1281	1011	765	570	425	244
0.2		811	778	726	664	596	528	404
0.3		585	575	558	535	509	480	417
0.4		498	493	485	474	460	445	409
0.5		480	476	471	463	454	443	416
0.6		516	512	506	498	488	477	449
0.7		623	616	606	593	576	557	515
0.8		876	857	828	791	750	707	621
0.9		1656	1528	1365	1201	1055	930	741
1.0		6203	3108	2076	1559	1249	1042	782
1.1	1684	1635	1507	1344	1180	1034	911	723
1.2	837	831	812	784	748	708	666	583
1.3	551	549	544	535	523	508	491	455
1.4	407	406	404	400	394	388	380	361
1.5	320	319	318	316	313	309	305	295
1.6	261	261	260	259	257	255	252	246
1.7	219	219	218	218	217	215	213	209
1.8	187	187	187	186	186	185	184	181
1.9	163	163	163	162	162	161	160	158
2.0	143	143	143	143	142	142	141	140
2.1	127	127	127	127	126	126	126	124
2.2	114	114	114	114	113	113	113	112
2.3	103	103	103	102	102	102	102	101
2.4	93	93	93	93	93	93	92	92
2.5	85	85	85	85	85	84	84	84
2.6	78	78	78	78	78	77	77	77
2.7	72	72	72	71	71	71	71	71
2.8	66	66	66	66	66	66	66	65
2.9	61	61	61	61	61	61	61	61
3.0	57	57	57	57	57	57	57	56

m \ n	0.20	0.30	0.40	0.50	0.60	0.80	1.00	1.20	1.60	2.00	2.50	3.00
0.1	149	56	27	15	9	4	2	1	-	-	-	-
0.2	305	155	86	52	33	14	7	3	1	-	-	-
0.3	355	228	146	95	63	29	14	7	2	1	-	-
0.4	369	272	193	134	94	46	23	12	4	1	-	-
0.5	386	303	228	166	120	62	33	18	6	2	1	-
0.6	417	333	255	191	141	77	42	24	8	3	1	-
0.7	471	364	276	208	157	89	51	30	11	4	2	1
0.8	543	393	291	220	167	98	58	35	14	6	2	1
0.9	609	410	298	225	173	105	64	40	16	7	3	1
1.0	625	411	298	227	176	110	69	44	19	9	4	2
1.1	592	398	292	224	176	112	73	48	21	10	4	2
1.2	509	370	280	218	174	114	76	51	24	12	5	2
1.3	417	329	260	208	169	114	78	53	26	13	6	3
1.4	341	286	237	195	162	112	78	55	28	14	7	3
1.5	283	248	213	181	153	109	78	56	29	16	7	4
1.6	238	215	190	165	143	105	77	57	31	17	8	4
1.7	204	188	170	151	133	101	76	57	32	18	9	5
1.8	177	166	152	137	123	96	74	56	32	19	10	5
1.9	155	147	137	125	113	91	71	55	33	20	10	6
2.0	138	131	123	114	105	86	69	54	33	20	11	6
2.1	123	118	112	104	97	81	66	53	33	21	12	7
2.2	111	107	102	96	89	76	63	51	33	21	12	7
2.3	100	97	93	88	83	71	60	50	33	21	12	7
2.4	91	88	85	81	77	67	57	48	33	22	13	8
2.5	83	81	78	75	71	63	54	46	32	22	13	8
2.6	76	75	72	70	66	59	52	44	32	22	13	8
2.7	70	69	67	65	62	56	49	43	31	22	14	9
2.8	65	64	62	60	58	53	47	41	30	22	14	9
2.9	60	59	58	56	54	50	44	39	29	21	14	9
3.0	56	55	54	53	51	47	42	38	29	21	14	9

Tabelle 1.27a Beiwerte i_z x 1000 fuer gleichmaessige Mantelreibung $\mu = 0.1$ (nur Druck) [1]

Table 1.27a Coefficients i_z x 1000 for uniform skin friction $\mu = 0.1$ (all values compression) [1]

m \ n	0.20	0.30	0.40	0.50	0.60	0.80	1.00	1.20	1.60	2.00	2.50	3.00
0.1	150	56	27	15	9	4	2	1	-	-	-	-
0.2	309	158	88	53	33	14	6	3	1	-	-	-
0.3	363	234	150	98	65	29	14	7	2	1	-	-
0.4	383	282	200	139	96	46	23	12	3	1	-	-
0.5	406	318	237	172	124	63	32	17	5	2	1	-
0.6	446	352	267	198	145	78	42	23	7	3	1	-
0.7	508	388	291	217	161	89	50	28	10	4	1	-
0.8	590	419	306	228	172	99	57	34	12	5	2	1
0.9	657	435	312	233	178	105	64	39	15	6	2	1
1.0	670	435	312	235	181	110	69	43	18	8	3	1
1.1	642	425	308	234	182	114	73	47	20	9	4	2
1.2	559	399	297	229	181	116	76	50	23	11	4	2
1.3	460	358	279	221	177	117	78	53	25	12	5	2
1.4	376	313	255	208	170	116	80	55	27	13	6	3
1.5	312	271	230	193	162	114	80	57	29	15	7	3
1.6	262	235	206	178	152	110	80	58	30	16	8	4
1.7	224	205	184	162	142	106	78	58	31	17	8	4
1.8	194	181	165	148	131	101	77	58	32	18	9	5
1.9	170	160	148	135	121	96	74	57	33	19	10	5
2.0	150	143	133	123	112	91	72	56	34	20	11	6
2.1	134	128	121	112	103	86	69	55	34	21	11	6
2.2	120	115	110	103	96	80	66	53	34	21	12	7
2.3	108	105	100	95	88	76	63	52	34	21	12	7
2.4	98	95	92	87	82	71	60	50	33	22	13	7
2.5	90	87	84	80	76	67	57	48	33	22	13	8
2.6	82	80	78	74	71	63	54	46	32	22	13	8
2.7	76	74	72	69	66	59	52	45	32	22	14	8
2.8	70	68	67	64	62	56	49	43	31	22	14	9
2.9	65	63	62	60	58	52	47	41	30	22	14	9
3.0	60	59	58	56	54	50	45	39	30	22	14	9

Tabbele 1.27b Beiwerte i_z x 1000 fuer gleichmaessige Mantelreibung $\mu = 0.3$ (nur Druck) [1]

Table 1.27b Coefficients i_z x 1000 for uniform skin friction $\mu = 0.3$ (all values compression) [1]

$$i_z = \sigma_z \cdot \frac{t^2}{P}$$

n / m	0.00	0.02	0.04	0.06	0.08	0.10	0.12	0.16
0.1		1509	1283	1012	766	571	426	245
0.2		816	782	731	668	600	532	408
0.3		595	584	567	545	518	489	426
0.4		515	510	501	490	476	460	423
0.5		506	502	496	488	478	466	438
0.6		556	552	545	536	525	512	481
0.7		684	677	665	649	630	608	560
0.8		980	957	923	879	831	780	680
0.9		1879	1727	1533	1340	1169	1024	870
1.0		6815	3405	2267	1698	1357	1129	842
1.1	1919	1860	1707	1514	1321	1150	1007	790
1.2	946	938	916	882	840	792	742	645
1.3	619	617	610	599	585	565	548	505
1.4	455	454	451	446	440	432	422	401
1.5	355	355	353	351	347	343	338	326
1.6	289	289	288	286	284	282	279	271
1.7	241	241	241	240	238	237	235	230
1.8	206	206	205	205	204	203	201	198
1.9	178	178	178	177	177	176	175	173
2.0	156	156	156	156	155	155	154	152
2.1	138	138	138	138	138	137	137	135
2.2	124	124	123	123	123	123	122	121
2.3	111	111	111	111	111	110	110	109
2.4	101	101	101	100	100	100	100	99
2.5	92	92	92	91	91	91	91	90
2.6	84	84	84	84	84	83	83	83
2.7	77	77	77	77	77	77	76	76
2.8	71	71	71	71	71	71	71	70
2.9	66	66	66	66	65	65	65	65
3.0	61	61	61	61	61	61	61	60

$$i_z = \sigma_z \cdot \frac{t^2}{P}$$

m \ n	0.00	0.02	0.04	0.06	0.08	0.10	0.12	0.16
0.1		1512	1285	1015	769	573	428	247
0.2		824	791	739	676	608	540	415
0.3		613	602	585	562	535	505	441
0.4		545	539	531	519	505	488	449
0.5		553	549	542	533	522	509	478
0.6		627	623	615	604	590	575	538
0.7		796	786	771	751	727	700	639
0.8		1167	1138	1094	1038	976	911	785
0.9		2281	2084	1835	1590	1374	1194	926
1.0		7918	3939	2612	1949	1550	1284	951
1.1	2341	2264	2068	1820	1574	1359	1180	913
1.2	1142	1133	1104	1060	1006	944	881	758
1.3	741	738	730	716	697	675	650	594
1.4	540	539	535	529	521	511	499	471
1.5	419	419	417	414	409	404	397	382
1.6	339	339	337	336	333	330	326	316
1.7	282	281	281	280	278	276	273	267
1.8	239	239	238	238	236	235	233	229
1.9	206	206	206	205	204	203	202	199
2.0	180	180	180	179	179	178	177	175
2.1	159	159	158	158	158	157	156	155
2.2	141	141	141	141	140	140	139	138
2.3	127	127	126	126	126	126	125	124
2.4	114	114	114	114	114	113	113	112
2.5	104	104	104	103	103	103	103	102
2.6	94	94	94	94	94	94	94	93
2.7	87	87	86	86	86	86	86	85
2.8	80	80	80	79	79	79	79	79
2.9	73	73	73	73	73	73	73	73
3.0	68	68	68	68	68	68	68	67

m \ n	0.20	0.30	0.40	0.50	0.60	0.80	1.00	1.20	1.60	2.00	2.50	3.00
0.1	152	58	29	16	10	4	2	1	-	-	-	-
0.2	316	164	92	56	35	15	6	3	-	-	-	-
0.3	377	246	158	103	68	30	13	6	1	-	-	-
0.4	407	301	213	147	101	47	22	10	2	-	-	-
0.5	442	345	255	184	130	64	31	15	4	1	-	-
0.6	497	388	290	212	153	78	40	21	6	2	-	-
0.7	576	430	316	231	169	90	48	26	8	2	1	-
0.8	673	465	332	243	179	99	56	31	10	3	1	-
0.9	744	479	337	247	185	106	62	36	13	5	1	-
1.0	750	478	337	249	189	112	67	41	15	6	2	1
1.1	732	473	336	250	191	116	72	45	18	7	3	1
1.2	649	451	329	249	193	120	76	49	21	9	3	1
1.3	537	409	313	243	191	122	79	52	53	10	4	2
1.4	440	360	289	231	186	123	82	55	25	12	5	2
1.5	364	312	261	216	178	121	83	57	27	13	6	3
1.6	305	271	235	200	169	119	84	59	29	15	6	3
1.7	260	236	210	183	158	115	83	60	31	16	7	3
1.8	224	207	188	167	147	110	82	60	32	17	8	4
1.9	195	183	168	152	136	105	80	60	33	18	9	4
2.0	172	163	151	139	125	100	77	59	34	19	10	5
2.1	153	146	137	127	116	94	75	58	34	20	10	5
2.2	136	131	124	116	107	89	72	57	35	21	11	6
2.3	123	119	113	106	99	83	69	55	35	21	12	6
2.4	111	108	103	98	91	78	65	54	35	22	12	7
2.5	101	98	95	90	85	74	62	52	34	22	13	7
2.6	93	90	87	83	79	69	59	50	34	22	13	8
2.7	85	83	80	77	73	65	57	48	33	22	13	8
2.8	78	77	74	72	68	61	54	46	33	23	14	8
2.9	72	71	69	67	64	58	51	44	32	22	14	9
3.0	67	66	64	62	60	54	49	43	31	22	14	9

Tabelle 1.27c — Beiwerte i_z x 1000 fuer gleichmaessige Mantelreibung $\mu = 0.5$ (nur Druck) [1]

Table 1.27c — Coefficients i_z x 1000 for uniform skin friction $\mu = 0.5$ (all values compression) [1]

m	i_r und and i_t (x 1000)		
	μ = 0.1	μ = 0.3	μ = 0.5
1.1	+325	+200	-16
1.2	+151	+89	-13
1.3	+95	+54	-10
1.4	+67	+37	-8
1.5	+51	+28	-7
1.6	+41	+22	-6
1.7	+33	+18	-5
1.8	+28	+15	-4
1.9	+24	+13	-3
2.0	+21	+11	-3
2.1	+18	+9	-3
2.2	+16	+8	-2
2.3	+15	+7	-2
2.4	+13	+7	-2
2.5	+12	+6	-2
2.6	+11	+6	-1
2.7	+10	+5	-1
2.8	+9	+5	-1
2.9	+8	+4	-1
3.0	+8	+4	-1

Tabelle 1.27d **Beiwerte i_r x 100 und i_t x 1000 fuer gleichmaessige Mantelreibung bei n = 0 (+ Zug, - Druck) [1]**

Table 1.27d **Coefficients i_r x 100 and i_t x 1000 for uniform skinfriction when n = 0 (+ tension, - compression) [1]**

(ii) Quelle (s. Seite 187)
Reference (Ref. PP. 187)

1.2.1.5.3 Dreieckfoermige Mantelreibung

(i) Formeln/Tabellen

1.2.1.5.3 Linearly varying skinfriction

(i) Formulae/Tables

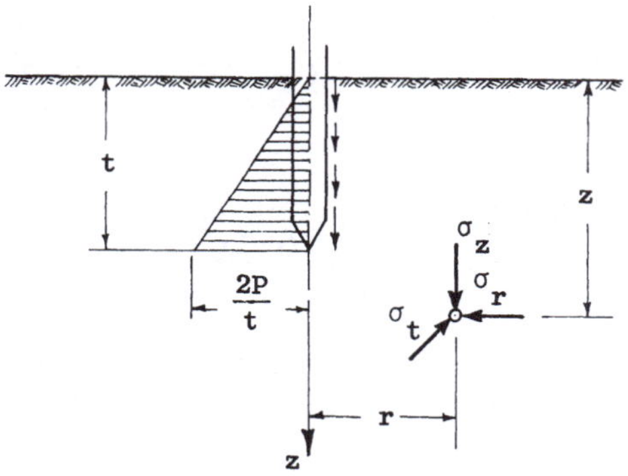

Mit den Bezeichungen:

With the designations:

$$r/t = n, \qquad z/t = m$$

$$A^2 = n^2+(m-1)^2, \; B^2 = n^2+(m+1)^2, \; F^2 = n^2+m^2$$

$$i_z = \sigma_z \cdot \frac{t^2}{P} = \frac{1}{4\pi(1-\mu)} \left[\frac{-2(2-\mu)}{A} + \frac{2(2-\mu)(4m+1)-2(1-2\mu)\left(\frac{m}{n}\right)^2(m+n)}{B} \right.$$

$$+ \frac{2(1-2\mu)\frac{m^3}{n^2} - 8(2-\mu)m}{F} + \frac{mn^2 + (m-1)^3}{A^3}$$

$$+ \frac{4\mu n^2 m + 4m^3 - 15n^2 m - 2(5+2\mu)\left(\frac{m}{n}\right)^2 (m+1)^3 + (m+1)^3}{B^3}$$

$$+ \frac{2(7-2\mu)mn^2 - 6m^3 + 2(5+2\mu)\left(\frac{m}{n}\right)^2 m^3}{F^3}$$

$$+\frac{6mn^2(n^2-m^2)+12\left(\frac{m}{n}\right)^2(m+1)^5}{B^5}-\left\{\frac{12\left(\frac{m}{n}\right)^2m^5+6mn^2(n^2-m^2)}{F^5}\right\}$$

$$-2(2-\mu)\ \ln\left(\frac{A+m-1}{F+m}\cdot\frac{B+m+1}{F+m}\right)\Bigg]$$

$$i_r=\sigma_r\cdot t^2/P=\frac{1}{4\pi(1-\mu)}\left[\frac{(1-2\mu)}{A}+\frac{(7-2\mu)-12m+12(1-\mu)\left(\frac{m}{n}\right)^2(m+1)}{B}\right.$$

$$+\frac{12m-12(1-\mu)\frac{m^3}{n^2}}{F}-\left\{\frac{(m-1)^3+mn^2}{A^3}\right\}-\left\{\frac{2(5+2\mu)\frac{m^5}{n^2}+4(5-\mu)mn^2}{F^3}\right\}$$

$$+\frac{3(m+1)^3-2m^3+(21-4\mu)mn^2+2(5+2\mu)\left(\frac{m}{n}\right)^2(m+1)^3}{B^3}$$

$$+\frac{6mn^2(m^2-n^2)-12\left(\frac{m}{n}\right)^2(m+1)^5}{B^5}-\left\{\frac{6mn^2(m^2-n^2)-12\frac{m^7}{n^2}}{F^5}\right\}$$

$$+(1-2\mu)\ln\left(\frac{A+m-1}{F+m}\right)+\left\{(1-2\mu)^2-6\right\}\ln\left(\frac{B+m+1}{F+m}\right)$$

$$\left.+2(1-\mu)(1-2\mu)\left\{\frac{m-1}{B+m+1}-\frac{m}{F+m}\right\}\right]$$

$$i_t=\sigma_t\cdot t^2/P=\frac{1}{4\pi(1-\mu)}\left[\frac{(1-2\mu)}{A}\left\{\frac{(1-2\mu)(3-4\mu)+6(1-2\mu)\left(\frac{m}{n}\right)^2(m+1)+6(2m-1)}{B}\right\}\right.$$

$$+\frac{6(1-2\mu)\frac{m^3}{n^2}+12m}{F}-(1-2\mu)\left\{\frac{2(m+1)^3+4mn^2-2\left(\frac{m}{n}\right)^2(m+1)^3}{B^3}\right\}$$

$$+\frac{2(m+1)^3+6mn^2-2m^3-6\left(\frac{m}{n}\right)^2(m+1)^3}{B^3}$$

$$+ \frac{\left(2m^3+4mn^2-2\frac{m^5}{n^2}\right)(1-2\mu)}{F^3} - \left\{\frac{6mn^2-6\frac{m^5}{n^2}}{F^3}\right\}$$

$$+ (1-2\mu)\ln\left(\frac{A+m-1}{F+m}\right) + \left\{(1-2\mu)^2-6\right\}\ln\left(\frac{B+m+1}{F+m}\right)$$

$$-2(1-\mu)(1-2\mu)\left\{\frac{m-1}{B+m+1} - \frac{m}{F+m}\right\} \Bigg]$$

Fig. 1.59 **Gleichungen fuer Spannungen infolge linear veraenderlicher Mantelreibung im Halbruam [1]**

Equations for stresses due to linear variation of skin friction in half-space [1]

$$i_z = \sigma_z \cdot \frac{t^2}{P}$$

m \ n	0.00	0.02	0.04	0.06	0.08	0.10	0.12	0.16
0.1		+872	+460	+264	+157	+95	+57	+20
0.2		+1262	+811	+563	+401	+288	+206	+102
0.3		+1437	+979	+718	+541	+411	+313	+175
0.4		+1498	+1037	+774	+592	+457	+352	+201
0.5		+1469	+1008	+744	+562	+427	+322	+171
0.6		+1340	+881	+620	+442	+312	+213	+75
0.7		+1057	+605	+354	+191	+77	4	105
0.8		+446	+21	187	300	359	385	385
0.9		1312	1509	1434	1276	1108	956	722
1.0		11064	5107	3180	2244	1699	1346	923
1.1	2682	2588	2347	2041	1738	1472	1250	924
1.2	1191	1180	1147	1096	1032	961	888	747
1.3	729	725	716	700	680	655	628	567
1.4	510	509	505	499	490	479	466	437
1.5	385	385	383	380	375	370	363	347
1.6	305	305	304	302	300	296	293	283
1.7	250	250	249	248	247	245	242	236
1.8	210	210	210	209	208	207	205	201
1.9	180	180	179	179	178	177	176	173
2.0	156	156	156	156	155	154	154	152
2.1	137	137	137	137	136	136	135	134
2.2	122	122	122	122	121	121	120	119
2.3	109	109	109	109	109	108	108	107
2.4	98	98	98	98	98	98	97	97
2.5	89	89	89	89	89	89	89	88
2.6	81	81	81	81	81	81	81	80
2.7	75	75	75	75	74	74	74	74
2.8	69	69	69	69	69	68	68	68
2.9	63	63	63	63	63	63	63	63
3.0	59	59	59	59	59	59	59	58

m \ n	0.20	0.30	0.40	0.50	0.60	0.80	1.00	1.20	1.60	2.00	2.50	3.00
0.1	+3	10	11	9	7	4	2	1	-	-	-	-
0.2	+43	16	29	28	23	14	7	4	1	-	-	-
0.3	+87	17	47	49	43	27	15	8	3	1	-	-
0.4	+101	26	66	71	63	40	23	13	4	2	-	-
0.5	+71	55	92	94	82	53	32	19	7	2	1	-
0.6	12	108	127	117	100	65	40	24	9	4	1	1
0.7	155	185	167	141	116	75	47	29	11	5	2	1
0.8	356	271	208	163	129	82	53	34	14	6	2	1
0.9	563	345	240	180	140	89	58	38	16	7	3	1
1.0	684	393	264	194	149	94	62	41	19	9	4	2
1.1	709	417	280	204	156	99	66	44	21	10	4	2
1.2	624	408	285	210	161	103	69	47	23	11	5	2
1.3	506	371	274	208	163	105	71	50	25	13	6	3
1.4	405	324	254	200	160	106	73	52	27	14	7	3
1.5	329	279	230	188	154	106	74	53	28	15	7	4
1.6	272	240	206	174	146	104	74	54	29	16	8	4
1.7	229	208	183	159	137	100	74	55	31	17	9	5
1.8	196	181	163	145	127	96	73	55	31	18	10	5
1.9	170	159	146	132	118	92	71	54	32	19	10	6
2.0	149	141	131	120	109	87	68	53	32	20	11	6
2.1	132	126	118	109	100	82	66	52	33	20	11	6
2.2	118	113	107	100	93	77	63	51	33	21	12	7
2.3	106	102	97	92	86	73	61	50	33	21	12	7
2.4	96	93	89	84	79	68	58	48	32	21	13	8
2.5	87	85	82	78	74	64	55	46	32	21	13	8
2.6	80	78	75	72	68	61	52	45	31	22	13	8
2.7	73	72	69	67	64	57	50	43	31	21	14	9
2.8	68	66	64	62	59	54	47	41	30	21	14	9
2.9	63	61	60	58	56	51	45	40	29	21	14	9
3.0	58	57	56	54	52	48	43	38	29	21	14	9

Tabelle 1.28a Beiwerte i_z x 1000 fuer linear veraenderliche Mantelreibung μ = 0.1 (+ Zug, andernfalls Druck) [1]

Table 1.28a Coefficients i_x x 1000 for linear variation of skin friction μ = 0.1 (+ tension, otherwise compression) [1]

$$i_z = \sigma_z \cdot \frac{t^2}{P}$$

m \ n	0.00	0.02	0.04	0.06	0.08	0.10	0.12	0.16
0.1		+1003	+529	+303	+180	+109	+66	+22
0.2		+1450	+532	+646	+459	+329	+236	+116
0.3		+1650	+1123	+823	+620	+470	+357	+198
0.4		+1719	+1189	+885	+677	+521	+401	+228
0.5		+1684	+1154	+850	+641	+486	+366	+193
0.6		+1533	+1006	+706	+503	+353	+240	+83
0.7		+1207	+688	+401	+214	+84	8	120
0.8		+504	+17	220	347	411	438	432
0.9		1516	1730	1632	1440	1239	1062	790
1.0		12130	5576	3459	2432	1835	1449	988
1.1	3074	2962	2674	2311	1954	1643	1386	1012
1.2	1360	1346	1307	1246	1170	1086	999	833
1.3	829	825	813	795	770	741	708	637
1.4	578	577	572	564	554	541	526	491
1.5	435	434	432	428	423	416	408	390
1.6	343	343	342	340	337	333	328	317
1.7	280	280	279	278	376	274	271	264
1.8	235	235	234	233	233	230	229	224
1.9	200	200	200	199	198	197	196	193
2.0	173	173	173	173	172	171	170	168
2.1	152	152	152	151	151	150	150	148
2.2	135	135	134	134	134	133	133	132
2.3	120	120	120	120	120	119	119	118
2.4	108	108	108	108	108	107	107	107
2.5	98	98	98	98	98	97	97	96
2.6	89	89	89	89	89	89	88	88
2.7	82	82	81	81	81	81	81	81
2.8	75	75	75	75	75	75	74	74
2.9	69	69	69	69	69	69	69	68
3.0	64	64	64	64	64	64	64	63

m \ n	0.20	0.30	0.40	0.50	0.60	0.80	1.00	1.20	1.60	2.00	2.50	3.00
0.1	+3	11	12	10	8	4	2	1	-	-	-	-
0.2	+49	19	33	32	27	15	8	4	1	-	-	-
0.3	+98	20	54	56	48	29	16	8	2	1	-	-
0.4	+113	31	75	80	70	44	25	13	4	1	-	-
0.5	+79	63	104	104	90	57	33	19	6	2	1	-
0.6	15	122	140	129	108	68	41	24	8	3	1	-
0.7	175	204	182	152	123	77	47	29	10	4	1	1
0.8	395	294	222	172	135	85	53	33	13	5	2	1
0.9	609	366	252	187	144	90	58	37	15	6	2	1
1.0	728	413	275	200	152	95	62	40	17	8	3	1
1.1	768	443	294	212	160	100	65	44	20	9	4	2
1.2	690	442	303	220	176	104	69	47	22	10	4	2
1.3	564	407	296	221	170	108	72	49	24	12	5	2
1.4	453	357	277	215	169	110	74	52	26	13	6	3
1.5	368	308	252	203	165	110	76	54	27	14	7	3
1.6	304	266	226	189	157	109	77	55	29	16	7	4
1.7	256	230	201	173	148	106	77	56	20	17	8	4
1.8	218	200	180	158	138	102	76	56	31	18	9	5
1.9	189	176	160	144	128	98	74	56	32	19	10	5
2.0	165	156	144	131	118	93	72	56	33	19	10	6
2.1	146	139	130	120	109	88	70	55	33	20	11	6
2.2	130	124	117	109	101	83	67	53	33	21	11	6
2.3	117	112	107	100	93	78	64	52	33	21	12	7
2.4	105	102	97	92	86	74	61	50	33	21	12	7
2.5	96	93	89	85	80	69	59	49	33	22	13	8
2.6	87	85	82	78	74	65	56	47	32	22	13	8
2.7	80	78	76	72	69	61	53	45	32	22	13	8
2.8	74	72	70	67	64	58	51	41	31	22	14	9
2.9	68	67	65	63	60	54	48	42	31	22	14	9
3.0	63	62	60	58	56	51	46	40	30	22	14	9

Tabelle 1.28c — Beiwerte i_z x 1000 fuer lineare veraenderliche Mantelreibung $\mu = 0.3$ (+ Zug, andernfalls Druck) [1]

Table 1.28c — Coefficients i_z x 1000 for linear variation of skin friction $\mu = 0.3$ (+ tension, otherwise compression) [1]

$$i_z = \sigma_z \cdot \frac{t^2}{p}$$

m \ n	0.00	0.02	0.04	0.06	0.08	0.10	0.12	0.16
0.1		+1238	+653	+374	+222	+134	+81	+26
0.2		+1789	+1148	+795	+565	+404	+289	+141
0.3		+2033	+1382	+1012	+761	+577	+436	+241
0.4		+2116	+1461	+1086	+829	+637	+489	+276
0.5		+2070	+1416	+1041	+784	+592	+444	+232
0.6		+1883	+1231	+862	+611	+428	+289	+97
0.7		+1477	+837	+484	+255	+97	14	148
0.8		+608	+9	280	431	505	532	517
0.9		1883	2130	1988	1735	1477	1251	914
1.0		14048	6420	3960	2770	2079	1634	1103
1.1	3780	3635	3263	2798	2344	1952	1631	1170
1.2	1664	1646	1595	1516	1419	1311	1200	989
1.3	1009	1004	989	965	933	895	853	761
1.4	700	698	692	682	668	652	663	588
1.5	524	523	520	515	508	500	490	466
1.6	412	411	410	407	403	398	392	378
1.7	335	335	333	332	329	326	323	314
1.8	279	279	278	277	276	274	271	265
1.9	237	237	237	236	235	233	232	228
2.0	205	204	204	204	203	202	201	198
2.1	179	179	178	178	177	177	176	174
2.2	158	158	157	157	157	156	155	154
2.3	140	140	140	140	140	139	139	137
2.4	126	126	126	125	125	125	124	123
2.5	114	114	113	113	113	113	112	112
2.6	103	103	103	103	103	102	102	102
2.7	94	94	94	94	94	93	93	93
2.8	86	86	86	86	86	86	86	85
2.9	79	79	79	79	79	79	79	78
3.0	73	73	73	73	73	73	73	72

n / m	0.20	0.30	0.40	0.50	0.60	0.80	1.00	1.20	1.60	2.00	2.50	3.00
0.1	+3	14	15	13	10	5	2	1	-	-	-	-
0.2	+58	25	42	39	32	18	9	4	1	-	-	-
0.3	+118	27	66	68	58	33	17	8	2	-	-	-
0.4	136	40	92	96	83	50	26	14	3	1	-	-
0.5	+93	78	125	123	105	64	35	19	5	1	-	-
0.6	20	146	165	149	123	75	43	24	7	2	-	-
0.7	212	239	209	171	136	83	49	28	9	3	1	-
0.8	466	335	247	188	145	88	53	32	11	4	1	-
0.9	693	404	272	199	152	93	57	35	13	5	1	-
1.0	805	448	294	211	159	97	61	39	15	6	2	1
1.1	875	490	318	225	168	102	65	42	17	7	2	1
1.2	809	503	335	238	177	108	69	45	20	9	3	1
1.3	669	471	334	244	184	113	73	48	22	10	4	1
1.4	539	418	317	241	186	117	76	51	24	11	5	2
1.5	438	362	291	231	183	119	79	54	26	13	5	2
1.6	361	312	262	216	177	119	81	56	28	14	6	3
1.7	303	270	231	199	167	117	82	58	30	15	7	3
1.8	258	235	209	182	156	113	82	59	31	17	8	4
1.9	222	206	187	166	145	109	81	59	32	18	9	4
2.0	194	182	167	151	135	104	79	59	33	19	9	5
2.1	171	162	150	138	124	99	77	59	34	20	10	5
2.2	152	145	136	126	115	93	74	58	34	20	11	6
2.3	136	130	123	115	106	88	71	56	35	21	11	6
2.4	122	118	112	105	98	83	68	55	35	22	12	7
2.5	111	107	102	97	91	78	65	53	35	22	12	7
2.6	101	98	94	89	84	73	62	52	34	22	13	7
2.7	92	90	87	83	78	69	59	50	34	23	13	8
2.8	84	83	80	77	73	65	56	48	33	23	14	8
2.9	78	76	74	71	68	61	54	46	33	23	14	8
3.0	72	71	69	66	64	58	51	44	32	23	14	9

Tabelle 1.28c Beiwerte i_z x 1000 fuer lineare veraenderliche Mantelreibung μ = 0.5 (+ Zug, andernfalls Druck) [1]

Tabel 1.28c Coefficients i_z x 1000 for linear variation of sking friction μ = 0.5 (+ tension, otherwise compression) [1]

m	i_r und/and i_t (x 1000)		
	p = 0.1	ν = 0.3	ν = 0.5
1.1	+536	+334	-20
1.2	+227	+137	-16
1.3	+132	+78	-13
1.4	+89	+51	-11
1.5	+65	+37	-9
1.6	+50	+28	-8
1.7	+40	+22	-6
1.8	+33	+18	-6
1.9	+28	+15	-5
2.0	+24	+12	-4
2.1	+20	+11	-4
2.2	+18	+9	-3
2.3	+16	+8	-3
2.4	+14	+7	-2
2.5	+13	+7	-2
2.6	+12	+6	-2
2.7	+11	+5	-2
2.8	+10	+5	-2
2.9	+9	+5	-1
3.0	+8	+4	-1

Tabelle 1.28d Beiwerte i_r x 1000 und i_t x 1000 fuer lineare veraenderliche Mantelreibung bei n = 0 (+Zug, -Druck) 1

Table 1.28d Coefficients i_r x 1000 and i_t x 1000 for linear vairation of skin firiction when n = 0 (+ tension, - compression)

(ii) Anwendungsbeispiel

Gegeben

Pfahl mit 0.50m Seitenlaenge des quadratischen Querschnitts und 10 m Tiefe traegt insgesamt 3000 kN, von denen 750 kN durch Spitzendruck getragen werden. Die uebrige Last wird getragen durch die Mantelreibung, deren verteilung aus der Skizze zu entnehmen ist.

(ii) Example

Given

A pile with a square section having a side dimension of 0.50 m and a 10 m embeded length carries a total load of 3000 kN. Of this, 750 kN is carried by point bearing and the remaining by shaft friction, which distribution is given in the sketch.

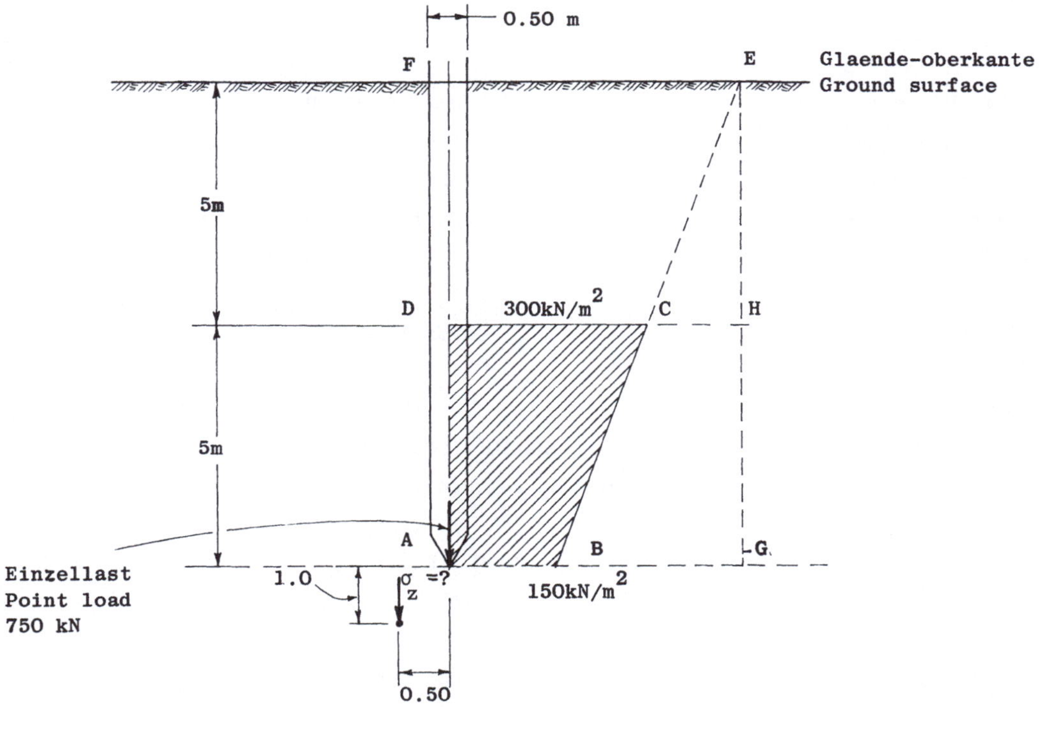

Gesucht

Die Spannung σ_z, 1.0 m unter der Pfahlspitze und 0.50 m in radialer Entfernung bei $\mu = 0.5$.

Required

The stress σ_z, at a depth of 1.0 m below the tip of the pile and at a radial distance of 0.50 m. Assume $\mu = 0.5$

Loesung | *Solution*

1) Spitzen-last | 1) Point load

$$n = r/t = \frac{0.50}{10} = 0.05$$

$$m = z/t = \frac{11}{10} = 1.10$$

Aus Tabelle 1.26c / From Table 1.26c } $i_z = 13.897$

$$\sigma_z = P \cdot i_z/t^2 = \frac{(750)(13.897)}{10^2} = 104.22 \text{ kN/m}^2$$

2) Mantelreibung: | Skin friction

Die Spannung infolge der Mantelreibung ABCD wird wie folget gerechnet: | The stress due to the skin friction ABCD is calculated as follows:

Anteil von ABCD = Anteil von AGEF-BGE-DHEF+CHE | Portion of ABCD = Portion of AGEF-BGE-DHEF+CHE

Fuer AGEF: | For AGEF:

$$n = 0.05$$

$$m = 1.10$$

$$P = 450(0.50)(4)(10) = 9000 \text{ kN}$$

$t = 10$ m; $i_z = 1.944$ (Tabelle 1.27C) (Table 1.27c)

$$\sigma_z = P \cdot \frac{i_z}{t^2} = \frac{(9000)(1.944)}{10^2} = 174.96 \text{ kN/m}^2$$

Fuer BGE: | For BGE:

$$n = 0.05$$

$$m = 1.10$$

$$P = \frac{1}{2}(300)(0.50)(4)(10) = 3000 \text{ kN}$$

$t = 10$ m; $i_z = 3.030$ (Tabelle 1.28c) (Table 1.28 c)

$$\sigma_z = \frac{p \cdot i_z}{t^2} = \frac{(3000)(3.030)}{10^2} = 90.9 \text{ kN/m}^2$$

Fuer DHEF

For DHEF

$$n = \frac{0.50}{5} = 0.10; \quad m = \frac{11}{5} = 2.2$$

$$P = (450)(0.50)(4)(5) = 4500 \text{ kN}$$

$t = 5$ m; $i_z = 0.140$ (Tabelle 1.27c)

(Table 1.27c)

$$\sigma_z = \frac{P \cdot i_z}{t^2} = \frac{(4500)(0.140)}{5^2} = 26.1 \text{ kN/m}^2$$

Fuer CHE

For CHE

$$n = 0.10; \quad m = 2.2$$

$$P = \frac{150}{2}(0.50)(4)(5) = 750 \text{ kN}$$

$t = 5$ m; $i_z = 0.156$ (Tabelle 1.28c)

(Table 1.28c)

$$\sigma_z = \frac{P \cdot i_z}{t^2} = \frac{(750)(0.156)}{5^2} = 4.7 \text{ kN/m}^2$$

Gesamete Spannungen

Total stresses

$$\sigma_z = 104.2 + 175.0 - 90.9 - 26.1 + 4.7$$

$$= 166.9 \text{ kN/m}^2$$ (druck)

(Compression)

(iii) Quelle (S. Seite 189)

Reference (Ref. PP. 189)

1.2.2 Waagerechte Lasten

1.2.2.1 Einzellast

(i) Formeln

1.2.2 Horizontal Loads

1.2.2.1 Point Load

(i) Formulae

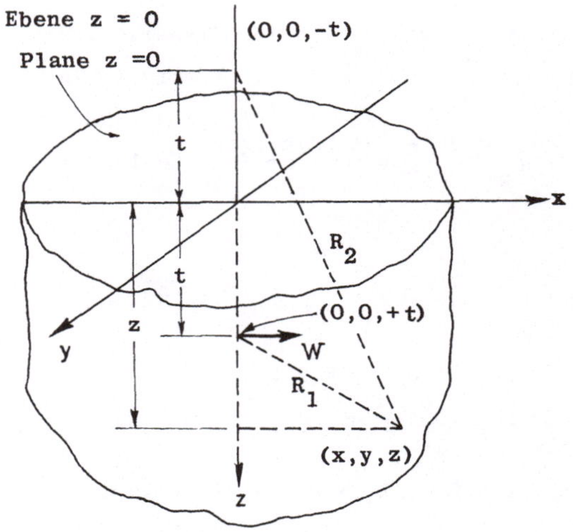

$$\sigma_z(x,z)_{y=0} = \frac{W\cdot x}{8\pi(1-\mu)}\left\{\frac{1-2\mu}{R_1^3} - \frac{1-2\mu}{R_2^3} - \frac{3(z-t)^2}{R_1^5}\right.$$

$$\left. - \frac{3(3-4\mu)(z+t)^2}{R_2^5} + \frac{6t}{R_2^5}\left[t+(1-2\mu)(z+t)+\frac{5z(z+t)^2}{R_2^2}\right]\right\}$$

Fig. 1.60a Gleichung fuer lotrechte Spannung infolge waagerechter Einzellast im Halbraum [1,2]

Equations for vertical stress due to horizontal point load in half-space [1,2]

$$\sigma_z(y=0) = i \cdot \frac{W}{t^2}$$

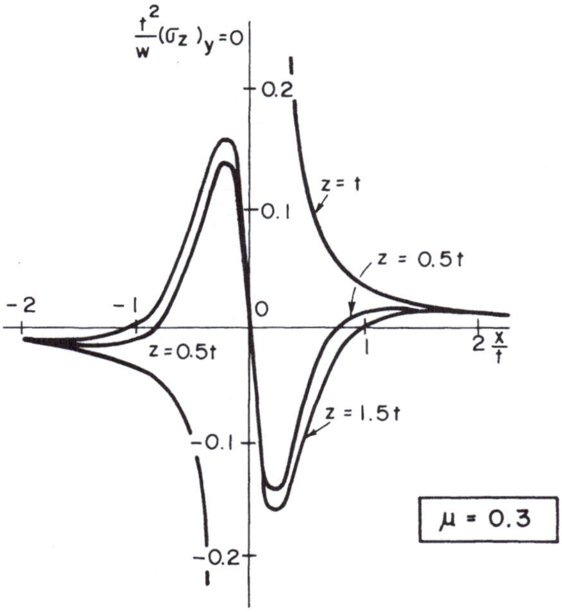

Fig. 1.60b Einflussbeiwert i zur Bestimmung der lotrechten Spannungen bei y = o infolge waagerchter Einzellast im Halbraum mit $\mu = 0.3$ [2]

Influence value i for determining the vertical stresses at y = 0 due to horizontal point load in a half space with $\mu = 0.3$ [2]

(ii) Quellen

References

[1] Harr (1966): Foundations of theoretical soil Mechanics. New York, PP 75.

[2] Mindlin (1936): Force at a point in the interior of a semi-infinite solid. Physic, Vol. 7, May, 1936.

1.2.2.2 Linienlast

(i) Formeln/Tafeln

1.2.2.2 Line Load

(i) Formulae/Charts

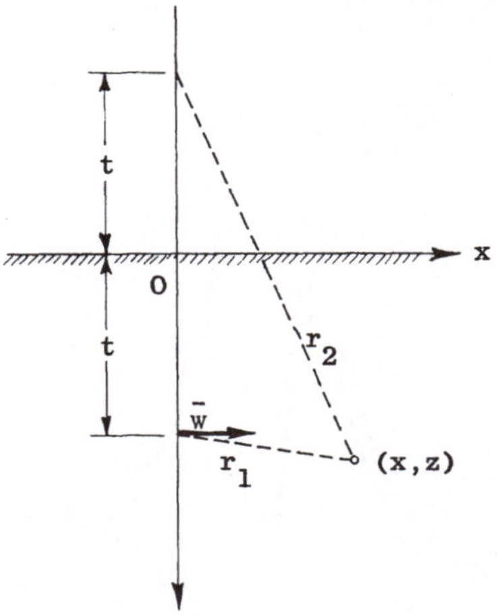

$$\sigma_z = \frac{\bar{w}x}{\pi}\left\{\frac{m+1}{2m}\left[\frac{(z-t)^2}{r_1^4} - \frac{t^2 - z^2 + 6tz}{r_2^4} + \frac{8tzx^2}{r_2^6}\right] - \frac{m-1}{4m}\left[\frac{1}{r_1^2} - \frac{1}{r_2^2} - \frac{4z(t+z)}{r_2^4}\right]\right\}$$

$$\tau_{xz} = \frac{\bar{w}}{\pi}\left\{\frac{m+1}{2m}\left[\frac{(z-t)x^2}{r_1^4} + \frac{(2tz + x^2)(t+z)}{r_2^4} - \frac{8tz(t+z)x^2}{r_2^6}\right] + \frac{m-1}{4m}\left[\frac{z-t}{r_1^2} + \frac{3z + t}{r_2^2} - \frac{4z(z+t)^2}{r_2^4}\right]\right\}$$

$$\sigma_x = \frac{\bar{w}x}{\pi}\left\{\frac{m+1}{2m}\left[\frac{x^2}{r_1^4} + \frac{x^2 + 8tz + 6t^2}{r_2^4} + \frac{8tz(t+z)^2}{r_2^6}\right] + \frac{m-1}{4m}\left[\frac{1}{r_1^2} + \frac{3}{r_2^2} - \frac{4z(t+z)}{r_2^4}\right]\right\}$$

Wobei $m = \frac{1-\mu}{\mu}$

where $m = \frac{1-\mu}{\mu}$

Fig. 1.61a Gleichungen zur Bestimmung verschiedener Spannungen infolge waagerechter Linienlast im Halbraum [1,2]

Equation for determining different stresses due to horizontal line load in half space [1,2]

$$\sigma_z = i \cdot \frac{\bar{w}}{t}$$

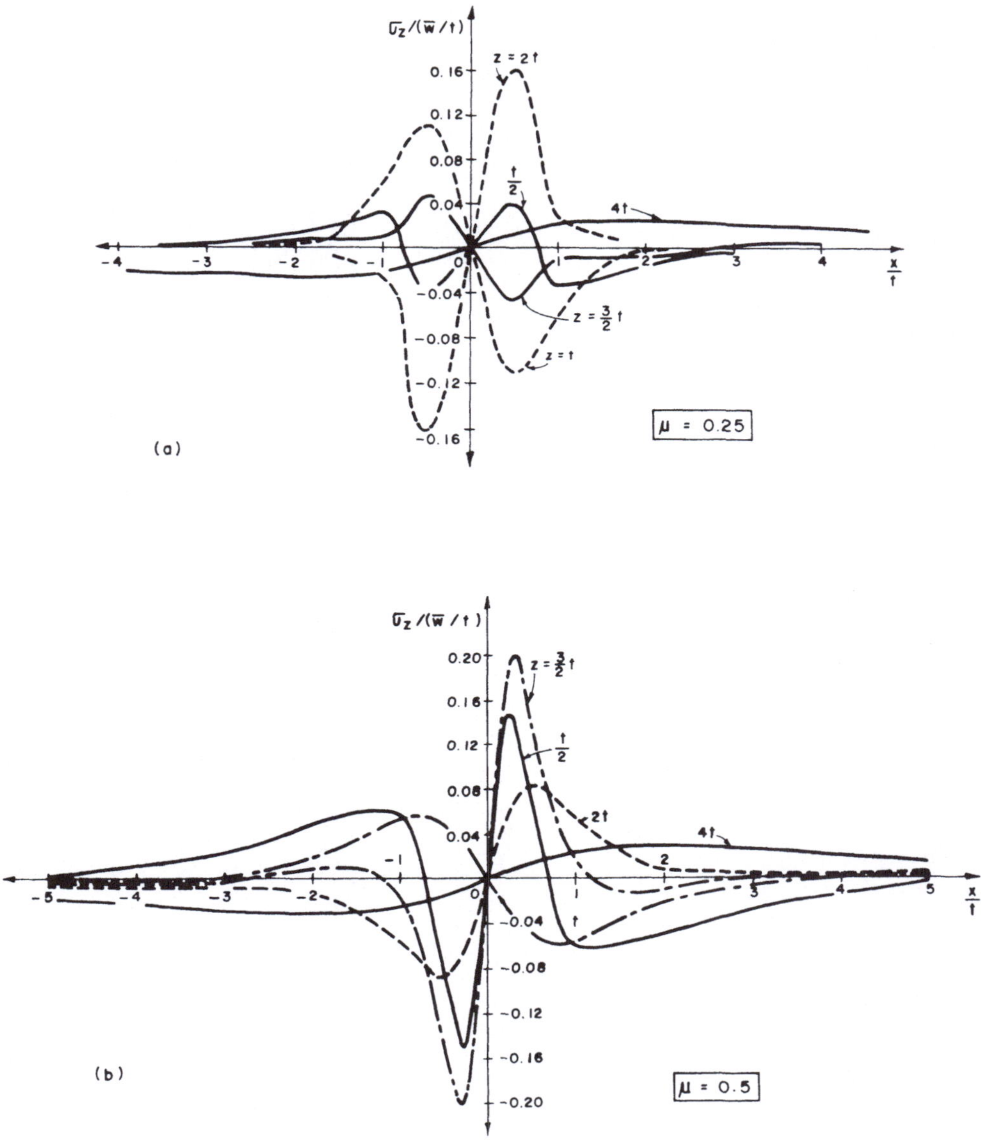

Fig. 1.61b Einflussbeiwert, i, zur Bestimmung der lotrechten Spannungen infolge waagerchter Linienlast im Halbraum [1]

Influence value i for determining the vertical stresses due to horizontal line load in half-space [1]

2
Two layers – Vertical surface loading
Zwei Schichten – Lotrechte Oberflaechenbelastung

2.1 EINZELLAST

(i) Formeln/Tabellen

2.1. POINT-LOAD

(i) Formulae/Tables

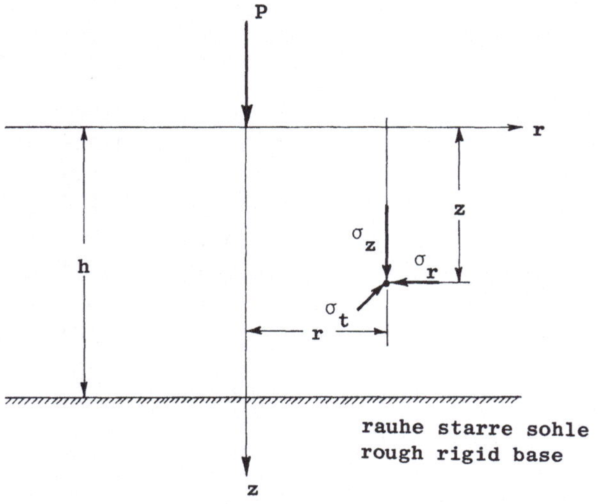

$$\sigma_z = \frac{P}{2\pi h^2} \cdot i_z \,, \qquad \sigma_r = \frac{P}{2\pi h^2} \cdot i_r$$

$$\sigma_t = \frac{P}{2\pi h^2} \cdot i_t \,, \qquad \tau_{rz} = \frac{P}{2\pi h^2} \cdot i_\tau$$

Fig. 1.62 Gleichungen zur Bestimmung von Spannungen infolge oberflaecher Einzellast auf zwei geschichteten Unterlagen [1,2]

Equations to determine the stresses due to surface point load acting on a two-layered system [1,2]

μ = 0	μ = 0.2
μ = 0.4	μ = 0.5

a) i_z

r/h \ z/h	1.0	0.9	0.8	0.7	0.6
0	4.422 4.358 4.454 4.695	4.993 5.028 5.259 5.603	5.779 5.853 6.113 6.457	7.005 7.089 7.330 7.631	9.017 9.098 9.297 9.542
0.1	4.308 4.243 4.329 4.556	4.844 4.874 5.091 5.417	5.571 5.640 5.887 6.215	6.680 6.761 6.900 7.279	8.448 8.525 8.717 8.951
0.2	3.989 3.919 3.980 4.171	4.431 4.548 4.626 4.905	5.005 5.060 5.270 5.554	5.826 5.894 6.093 6.345	7.022 7.089 7.257 7.463
0.3	3.523 3.447 3.472 3.612	3.842 3.840 3.963 4.175	4.225 4.259 4.416 4.638	4.713 4.763 4.917 5.118	5.316 5.369 5.502 5.666
0.4	2.982 2.902 2.890 2.974	3.178 3.157 3.221 3.359	3.386 3.399 3.496 3.649	3.601 3.630 3.734 3.876	3.782 3.817 3.910 4.027
0.5	2.432 2.352 2.306 2.341	2.528 2.490 2.500 2.571	2.606 2.597 2.639 2.726	2.642 2.651 2.705 2.791	2.595 2.613 2.665 2.738
0.6	1.923 1.844 1.775 1.770	1.947 1.898 1.865 1.882	1.943 1.918 1.912 1.945	1.886 1.878 1.888 1.927	1.747 1.750 1.767 1.803
0.7	1.481 1.407 1.323 1.291	1.460 1.405 1.343 1.321	1.412 1.376 1.336 1.329	1.319 1.293 1.276 1.279	1.165 1.155 1.143 1.152
0.8	1.114 1.048 0.959 0.911	1.070 1.015 0.936 0.891	1.006 0.964 0.902 0.869	0.910 0.879 0.833 0.813	0.770 0.752 0.720 0.710
0.9	0.821 0.764 0.671 0.623	0.768 0.716 0.632 0.576	0.702 0.659 0.587 0.540	0.615 0.581 0.523 0.490	0.504 0.480 0.436 0.415
1.0	0.593 0.545 0.466 0.412	0.541 0.494 0.412 0.355	0.481 0.439 0.365 0.314	0.408 0.374 0.312 0.272	0.324 0.297 0.248 0.223
1.25	0.238 0.213 0.163 0.123	0.201 0.172 0.116 0.072	0.164 0.135 0.080 0.037	0.127 0.099 0.050 0.015	0.090 0.066 0.025 0.000
1.50	0.074 0.067 0.045 0.023	0.055 0.042 0.016 -0.011	0.036 0.021 -0.007 -0.034	0.020 0.004 -0.022 -0.048	0.006 -0.010 -0.031 -0.062
1.75	0.007 0.010 0.009 -0.003	-0.003 -0.003 -0.007 -0.023	-0.011 -0.014 -0.020 -0.037	-0.018 -0.022 -0.028 -0.046	-0.021 -0.026 -0.031 -0.048
2.00	-0.017 -0.008 0.003 -0.005	-0.020 -0.014 -0.006 -0.016	-0.023 -0.019 -0.013 -0.024	-0.025 -0.022 -0.017 -0.030	-0.025 -0.024 -0.018 -0.032
2.50	-0.019 -0.008 0.007 0.004	-0.018 -0.009 -0.005 -0.012	-0.018 -0.009 0.003 -0.002	-0.016 -0.010 0.001 -0.005	-0.015 -0.009 0.001 -0.007
3.0	-0.010 -0.002 0.008 0.008	-0.009 -0.002 0.007 0.007	-0.008 -0.002 0.006 0.006	-0.007 -0.001 0.005 -0.003	-0.006 -0.001 0.005 0.002

z/h / /h	1.0		0.9		0.8		0.7		0.6	
3.5	-0.004 0.006	0.001 0.008	- 0.003 0.006	0.001 0.007	-0.002 0.005	0.001 0.006	-0.002 0.004	0.002 0.005	-0.001 0.004	0.002 0.004
4.0	-0.000 0.004	0.002 0.006	0.000 0.003	0.002 0.006	0.000 0.003	0.002 0.005	0.001 0.003	0.002 0.004	0.001 0.002	0.002 0.003
6.0	0.001 0.000	0.000 0.002	0.001 0.000	0.000 0.002	0.001 0.000	0.000 0.001	0.001 0.000	0.000 0.001	0.001 0.000	0.000 0.001
8.0	0.000 -0.000	-0.000 0.000	0.000 -0.000	-0.000 0.000	0.000 -0.000	-0.000 0.000	0.000 -0.000	-0.000 0.000	0.000 -0.000	-0.000 0.000
.0.0	0.000 -0.000	-0.000 -0.000	0.000 -0.000	-0.000 -0.000	0.000 -0.000	-0.000 -0.000	0.000 -0.000	-0.000 -0.000	0.000 -0.000	-0.000 -0.000

μ = 0	μ = 0.2
μ = 0.4	μ = 0.5

r/h \ z/h	0.5	0.4	0.3	0.2	0.1
0	12.50 12.57 12.73 12.91	19.10 19.15 19.25 19.39	33.55 33.59 33.64 33.75	75.11 75.13 75.15 75.26	300.0 300.0 300.0 300.0
0.1	11.37 11.44 11.58 11.76	16.45 16.51 16.60 16.73	25.82 25.86 25.92 26.01	43.04 43.06 43.08 43.17	53.07 53.08 53.07 53.17
0.2	8.738 8.797 8.926 9.080	11.05 11.10 11.18 11.29	13.49 13.52 13.57 13.64	13.36 13.38 13.40 13.44	5.400 5.415 5.404 5.438
0.3	5.970 6.018 6.121 6.241	6.426 6.466 6.536 6.610	6.067 6.098 6.135 6.167	4.028 4.050 4.058 4.053	0.980 0.994 0.981 0.947
0.4	3.827 3.862 3.936 4.019	3.554 3.585 3.634 3.679	2.741 2.766 2.791 2.795	1.419 1.437 1.440 1.403	0.280 0.293 0.279 0.205
0.5	2.397 2.418 2.461 2.512	1.977 1.998 2.026 2.049	1.323 1.341 1.355 1.344	0.594 0.609 0.605 0.561	0.112 0.124 0.108 0.032
0.6	1.499 1.507 1.523 1.549	1.132 1.143 1.153 1.162	0.690 0.701 0.702 0.692	0.288 0.298 0.290 0.257	0.057 0.067 0.051 -0.001
0.7	0.944 0.941 0.935 0.943	0.669 0.671 0.666 0.669	0.384 0.389 0.381 0.378	0.155 0.161 0.150 0.139	0.035 0.041 0.027 0.010
0.8	0.597 0.586 0.563 0.561	0.404 0.399 0.382 0.385	0.223 0.222 0.209 0.215	0.089 0.091 0.079 0.089	0.023 0.026 0.014 0.029
0.9	0.376 0.360 0.327 0.319	0.245 0.235 0.212 0.216	0.131 0.126 0.109 0.124	0.052 0.051 0.039 0.064	0.015 0.016 0.007 0.042
1.0	0.233 0.214 0.177 0.166	0.146 0.133 0.107 0.111	0.075 0.068 0.050 0.069	0.029 0.025 0.015 0.047	0.010 0.008 0.002 0.046
1.25	0.056 0.036 0.007 -0.007	0.028 0.013 -0.007 -0.007	0.010 -0.001 -0.012 0.000	0.001 -0.007 -0.010 0.014	-0.001 -0.006 -0.003 0.029
1.50	-0.006 -0.018 -0.034 -0.048	-0.012 -0.022 -0.031 -0.039	-0.013 -0.021 -0.023 -0.025	-0.010 -0.017 -0.013 -0.010	-0.007 -0.012 -0.004 0.003
1.75	-0.023 -0.028 -0.030 -0.046	-0.028 -0.026 -0.025 -0.039	-0.018 -0.022 -0.017 -0.030	-0.014 -0.017 -0.009 -0.021	-0.010 -0.012 -0.002 -0.013
2.00	-0.023 -0.023 -0.017 -0.033	-0.020 -0.020 -0.014 -0.030	-0.017 -0.017 -0.009 -0.026	-0.013 -0.013 -0.005 -0.021	-0.010 -0.010 0.001 -0.018
2.50	-0.013 -0.009 0.000 -0.008	-0.011 -0.007 0.000 -0.009	-0.009 -0.006 0.001 -0.010	-0.007 -0.005 0.001 -0.010	-0.006 -0.004 0.002 -0.011
3.0	-0.005 -0.001 0.004 -0.001	-0.004 -0.001 0.003 -0.001	-0.003 -0.000 0.003 -0.002	-0.002 0.000 0.003 -0.003	-0.001 0.001 0.002 -0.003

r/h \ z/h	0.5	0.4	0.3	0.2	0.1
3.5	-0.000 0.002 0.003 0.003	0.000 0.002 0.003 0.002	0.001 0.002 0.002 0.001	0.001 0.002 0.002 0.001	0.001 0.002 0.002 0.000
4.0	0.001 0.002 0.002 0.003	0.002 0.002 0.001 0.002	0.002 0.002 0.001 0.002	0.002 0.002 0.001 0.002	0.002 0.002 0.001 0.002
6.0	0.001 0.000 -0.000 0.001	0.001 0.000 -0.000 0.001	0.001 0.000 -0.000 0.001	0.001 0.000 -0.000 0.001	0.001 0.000 -0.000 0.001
8.0	0.000 -0.000 -0.000 0.000	0.000 -0.000 -0.000 0.000	0.000 -0.000 -0.000 0.000	0.000 -0.000 -0.000 0.000	0.000 -0.000 -0.000 0.000
10.0	0.000 -0.000 -0.000 -0.000	0.000 -0.000 -0.000 -0.000	0.000 -0.000 -0.000 -0.000	0.000 -0.000 -0.000 -0.000	0.000 -0.000 -0.000 -0.000

μ = 0	μ = 0.2
μ = 0.4	μ = 0.5

b) i_t

z/h → r/h ↓	1.0	0.9	0.8	0.7	0.6
0	0 1.089 2.970 4,695	-0.444 0.407 1.785 2.981	-0.795 -0.065 1.024 1.902	-1.150 -0.452 0.492 1.184	-1.598 -0.845 0.073 0.672
0.1	0 1.060 2.886 4.557	-0.431 0.400 1.742 2.905	-0.769 -0.057 1.004 1.859	-1.104 -0.429 0.487 1.159	-1.513 -0.797 0.082 0.658
0.2	0 0.979 2.653 4.171	-0.394 0.378 1.621 2.690	-0.697 -0.039 0.945 1.736	-0.70 -0.368 0.470 1.088	-1.293 -0.672 0.103 0.618
0.3	0 0.861 2.315 3.613	-0.341 0.344 1.440 2.373	-0.595 -0.014 0.855 1.552	-0.810 -0.287 0.440 0.480	-1.014 -0.516 0.123 0.557
0.4	0 0.725 1.926 2.975	-0.281 0.302 1.227 2.001	-0.482 0.009 0.745 1.330	-0.633 -0.207 0.397 0.849	-0.743 -0.367 0.134 0.482
0.5	0 0.587 1.538 2.341	-0.222 0.257 1.006 1.618	-0.373 0.025 0.625 1.096	-0.472 -0.138 0.345 0.707	-0.577 -0.247 0.131 0.401
0.6	0 0.460 1.183 1.770	-0.169 0.211 0.796 1.259	-0.279 0.035 0.507 0.871	-0.339 -0.086 0.288 0.568	-0.346 -0.159 0.118 0.321
0.7	0 0.351 0.882 1.291	-0.124 0.169 0.610 0.946	-0.201 0.038 0.398 0.668	-0.235 -0.050 0.231 0.440	-0.223 -0.100 0.099 0.247
0.8	0 0.261 0.640 0.912	-0.089 0.131 0.454 0.688	-0.141 0.036 0.303 0.496	-0.158 -0.027 0.179 0.330	-0.137 -0.061 0.077 0.182
0.9	0 0.190 0.452 0.623	-0.062 0.099 0.329 0.485	-0.096 0.031 0.223 0.356	-0.103 -0.014 0.133 0.238	-0.080 -0.037 0.056 0.128
1.0	0 0.136 0.311 0.413	-0.042 0.073 0.232 0.331	-0.064 0.025 0.160 0.247	-0.064 -0.007 0.096 0.165	-0.042 -0.023 0.038 0.085
1.25	0 0.053 0.108 0.124	-0.014 0.030 0.086 0.107	-0.018 0.011 0.061 0.083	-0.013 -0.002 0.035 0.053	0.003 -0.008 0.009 0.019
1.5	0 0.016 0.030 0.023	-0.002 0.009 0.086 0.023	-0.001 0.002 0.018 0.017	0.004 -0.003 0.009 0.008	0.014 -0.005 -0.002 -0.005
1.75	0 0.002 0.006 -0.003	0.001 0.002 0.006 -0.001	0.004 -0.002 0.005 -0.002	0.008 -0.003 0.002 -0.005	0.014 -0.004 -0.002 -0.009
2.0	0 -0.002 0.002 -0.004	0.002 -0.002 0.003 -0.003	0.005 -0.003 0.003 -0.003	0.008 -0.003 0.002 -0.004	0.011 -0.003 0.001 -0.005
2.5	0 -0.002 0.005 0.005	0.001 -0.002 0.005 0.005	0.003 -0.001 0.006 0.004	0.004 -0.001 0.006 0.004	0.005 -0.001 0.006 0.004
3.0	0 -0.000 0.006 0.009	0.001 -0.000 0.006 0.008	0.001 -0.000 0.006 0.008	0.002 -0.000 0.006 0.008	0.002 0.000 0.007 0.008

z/h r/h	1.0	0.9	0.8	0.7	0.6
3.5	0 0.001 0.004 0.008	0.000 0.001 0.004 0.008	0.001 0.001 0.005 0.008	0.001 0.001 0.005 0.008	0.001 0.001 0.005 0.008
4.0	0 0.001 0.003 0.007	0.000 0.001 0.003 0.007	0.000 0.001 0.003 0.007	0.000 0.001 0.003 0.007	0.000 0.001 0.004 0.007
6.0	-0.000 0.000 0.000 0.002	-0.000 0.000 0.000 0.002	-0.000 0.000 0.000 0.002	-0.000 0.000 0.000 0.002	-0.000 0.000 0.000 0.002
8.0	-0.000 -0.000 -0.000 0.001	-0.000 -0.000 -0.000 0.001	-0.000 -0.000 0.000 0.001	-0.000 -0.000 0.000 0.001	-0.000 -0.000 0.000 0.001
10.0	-0.000 -0.000 -0.000 0.000	-0.000 -0.000 -0.000 0.000	-0.000 -0.000 -0.000 0.000	-0.000 -0.000 -0.000 0.000	-0.000 -0.000 -0.000 0.000

μ = 0	μ = 0.2
μ = 0.4	μ = 0.5

r/h \ z/h	0.5	0.4	0.3	0.2	0.1
0	-2.273 -1.347 -0.318 0.274	-3.460 -2.143 -0.775 -0.067	-5.960 -3.725 -1.487 -0.393	-12.99 -8.033 -3.121 -0.733	-50.61 -30.71 -10.91 -1.101
0.1	-2.101 -1.245 -0.287 0.267	-3.054 -1.900 -0.695 -0.069	-4.758 -3.003 -1.245 -0.391	-7.822 -4.929 -2.082 -0.729	-6.674 -4.342 -2.115 -1.100
0.2	-1.681 -0.996 -0.214 0.248	-2.161 -1.365 -0.518 -0.074	-2.600 -1.705 -0.870 -0.385	-2.002 -1.420 -0.903 -0.716	4.276 2.237 0.099 -1.089
0.3	-1.200 -0.712 -0.134 0.218	-1.299 -0.847 -0.348 -0.080	-1.064 -0.779 -0.489 -0.373	0.206 -0.097 -0.439 -0.688	3.849 1.996 0.052 -1.056
0.4	-0.788 -0.471 -0.072 0.183	-0.691 -0.482 -0.228 -0.086	-0.279 -0.303 -0.319 -0.353	0.757 0.246 -0.298 -0.644	2.738 1.349 -0.120 -0.991
0.5	-0.486 -0.296 -0.033 0.145	-0.324 -0.261 -0.156 -0.090	0.062 -0.090 -0.234 -0.326	0.788 0.279 -0.256 -0.586	1.913 0.877 -0.226 -0.896
0.6	-0.283 -0.181 -0.014 0.108	-0.119 -0.136 -0.116 -0.092	0.190 -0.006 -0.189 -0.294	0.681 0.231 -0.238 -0.517	1.353 0.566 -0.273 -0.780
0.7	-0.154 -0.108 -0.007 0.074	-0.011 -0.070 -0.093 -0.091	0.221 0.023 -0.162 -0.259	0.554 0.171 -0.221 0.443	0.972 0.363 -0.283 -0.658
0.8	-0.073 -0.064 -0.007 0.045	-0.042 -0.035 -0.079 -0.087	0.213 0.027 -0.142 -0.223	0.440 0.119 -0.202 -0.371	0.708 0.231 -0.269 -0.540
0.9	-0.025 -0.039 -0.010 0.023	0.065 -0.018 -0.069 -0.082	0.190 0.023 -0.124 -0.189	0.346 0.079 -0.180 -0.305	0.522 0.144 -0.244 -0.434
1.0	0.003 -0.024 -0.014 0.005	0.072 -0.010 -0.061 -0.075	0.162 0.016 -0.107 -0.158	0.271 0.050 -0.156 -0.247	0.388 0.087 -0.212 -0.343
1.25	0.028 -0.010 -0.017 -0.017	0.062 -0.006 -0.043 -0.056	0.103 0.002 -0.070 -0.096	0.148 0.011 -0.099 -0.138	0.192 0.018 -0.132 -0.181
1.5	0.087 -0.006 -0.014 -0.020	0.044 -0.005 -0.026 -0.037	0.063 -0.004 -0.039 -0.054	0.081 -0.002 -0.054 -0.072	0.099 -0.002 -0.070 -0.089
1.75	0.021 -0.005 -0.007 -0.014	0.029 -0.005 -0.012 -0.021	0.038 -0.005 -0.018 -0.027	0.046 -0.005 -0.024 -0.034	0.052 -0.006 -0.031 -0.040
2.0	0.015 -0.004 -0.001 -0.007	0.019 -0.004 -0.003 -0.009	0.023 -0.004 -0.005 -0.011	0.026 -0.004 -0.007 -0.012	0.028 -0.005 -0.009 -0.013
2.5	0.007 -0.001 0.006 0.005	0.008 -0.001 0.006 0.005	0.008 -0.001 0.006 0.005	0.009 -0.001 0.007 0.006	0.009 -0.001 0.007 0.007
3.0	0.003 0.000 0.007 0.008	0.003 0.000 0.007 0.009	0.003 0.000 0.007 0.009	0.003 0.000 0.008 0.010	0.003 0.001 0.008 0.011

z/h r/h	0.5	0.4	0.3	0.2	0.1
3.5	0.001 0.001 0.005 0.008	0.001 0.001 0.006 0.009	0.001 0.001 0.006 0.009	0.001 0.001 0.006 0.009	0.001 0.001 0.006 0.010
4.0	0.000 0.001 0.004 0.007	0.000 0.001 0.004 0.007	0.000 0.001 0.004 0.007	0.000 0.001 0.004 0.008	0.000 0.001 0.004 0.008
6.0	-0.000 0.000 0.000 0.002	-0.000 0.000 0.000 0.002	-0.000 0.000 0.000 0.002	-0.000 0.000 0.000 0.002	-0.000 0.000 0.000 0.002
8.0	-0.000 -0.000 0.000 0.001	-0.000 -0.000 0.000 0.001	-0.000 -0.000 0.000 0.001	-0.000 -0.000 0.000 0.001	-0.000 -0.000 0.000 0.001
10.0	-0.000 -0.000 -0.000 0.000	-0.000 -0.000 -0.000 0.000	-0.000 -0.000 -0.000 0.000	-0.000 -0.000 -0.000 0.000	-0.000 -0.000 -0.000 0.000

μ = 0	μ = 0.2
μ = 0.4	μ = 0.5

c) i_r

r/h \ z/h	1.0		0.9		0.8		0.7		0.6	
0	0	1.090	-0.444	0.407	-0.795	-0.065	-1.150	-0.452	-1.598	-0.845
	2.969	4.695	1.785	2.981	1.024	1.902	0.492	1.184	0.073	0.672
0.1	0	1.061	-0.405	0.431	-0.718	0.002	-1.014	-0.327	-1.350	-0.613
	2.886	4.556	1.777	2.947	1.071	1.937	0.601	1.288	0.288	0.884
0.2	0	0.980	-0.301	0.488	-0.516	0.172	-0.673	-0.156	-0.767	-0.677
	2.653	4.171	1.753	2.846	1.192	2.018	0.875	1.537	0.792	1.366
0.3	0	0.863	0.167	0.552	-0.263	0.376	-0.275	0.338	-0.161	0.484
	2.315	3.612	1.695	2.672	1.320	2.081	1.172	1.793	1.283	1.824
0.4	0	0.726	-0.038	0.595	-0.030	0.547	0.056	0.615	0.264	0.845
	1.926	2.974	1.540	2.492	1.397	2.077	1.377	1.946	1.573	2.071
0.5	0	0.589	0.062	0.604	0.140	0.649	0.269	0.768	0.475	0.990
	1.537	2.341	1.441	2.133	1.411	1.983	1.449	1.957	1.641	2.092
0.6	0	0.462	0.125	0.576	0.238	0.678	0.370	0.808	0.529	0.980
	1.183	1.770	1.260	1.812	1.314	1.811	1.404	1.847	1.554	1.953
0.7	0	0.352	0.155	0.522	0.278	0.650	0.391	0.769	0.494	0.885
	0.882	1.291	1.066	1.492	1.182	1.589	1.282	1.660	1.384	1.732
0.8	0	0.263	0.159	0.452	0.276	0.586	0.364	0.686	0.421	0.756
	0.639	0.911	0.876	1.195	1.022	1.349	1.121	1.437	1.186	1.485
0.9	0	0.192	0.148	0.378	0.250	0.504	0.315	0.586	0.339	0.624
	0.451	0.623	0.701	0.934	0.856	1.113	0.949	1.210	0.990	1.244
1.0	0	0.137	0.128	0.307	0.213	0.420	0.258	0.485	0.261	0.503
	0.311	0.412	0.549	0.715	0.700	0.899	0.784	0.997	0.812	1.026
1.25	0	0.054	0.075	0.282	0.119	0.266	0.133	0.291	0.115	0.263
	0.108	0.123	0.278	0.344	0.391	0.492	0.453	0.578	0.470	0.608
1.5	0	0.017	0.036	0.080	0.054	0.121	0.055	0.140	0.036	0.136
	0.030	0.023	0.131	0.156	0.204	0.258	0.247	0.313	2.263	0.356
1.75	0	0.008	0.015	0.053	0.019	0.067	0.041	0.074	-0.001	0.069
	0.006	-0.003	0.061	0.104	0.104	0.139	0.133	0.203	0.147	0.227
2.0	0	0.021	0.004	0.011	0.003	0.021	-0.003	0.025	-0.015	0.025
	0.002	-0.005	0.031	0.039	0.055	0.078	0.073	0.110	0.085	0.134
2.5	0	-0.002	-0.003	0.000	-0.006	0.002	-0.010	0.003	-0.015	0.003
	0.005	0.004	0.013	0.019	0.021	0.035	0.029	0.049	0.036	0.063

r/h \ z/h	1.0		0.9		0.8		0.7		0.6	
3.0	0	-0.001	-0.003	-0.001	-0.005	-0.000	-0.007	-0.000	-0.010	-0.000
	0.006	0.008	0.009	0.015	0.013	0.023	0.016	0.037	0.020	0.038
3.5	0	-0.000	-0.002	-0.000	-0.003	-0.000	-0.004	-0.000	-0.005	-0.000
	0.004	0.008	0.006	0.012	0.008	0.016	0.010	0.021	0.008	0.025
4.0	0	0.000	-0.001	0.000	-0.002	0.000	-0.002	0.000	-0.003	0.000
	0.003	0.006	0.004	0.009	0.005	0.012	0.007	0.015	0.008	0.017
6.0	0.000	0.001	0.000	0.001	0.000	0.001	0.000	0.001	0.000	0.001
	0.001	0.001	0.001	0.002	0.001	0.003	0.001	0.004	0.002	0.004
8.0	0.000	0.000	0.000	0.000	0.000	0.000	0.000	0.000	0.000	0.000
	0.000	0.000	0.000	0.000	0.000	0.000	0.000	0.000	0.000	0.000
10.0	0.000	0.000	0.000	0.000	0.000	0.000	0.000	0.000	0.000	0.000
	0.000	0.000	0.000	0.000	0.000	0.000	0.000	0.000	0.000	0.000

μ = 0	μ = 0.2
μ = 0.4	μ = 0.5

r/h \ z/h	0.5	0.4	0.3	0.2	0.1
0	-2.273 -1.347 -0.318 0.274	-3.460 -2.143 -0.775 -0.067	-5.960 -3.725 -1.487 -0.393	-12.99 -8.033 -3.121 -0.733	-50.61 -30.71 -10.92 -1.101
0.1	-1.784 -0.878 0.125 0.711	-2.308 -1.054 0.257 0.947	-2.686 -0.627 1.442 2.464	-0.311 3.865 8.002 10.02	23.14 34.75 46.27 51.95
0.2	-0.729 0.112 1.049 1.606	0.273 0.843 2.014 2.641	1.325 3.011 4.709 5.556	5.468 8.351 11.20 12.58	7.069 12.50 17.83 20.42
0.3	0.173 0.926 1.774 2.289	0.937 1.884 2.888 3.437	2.277 3.586 4.912 5.581	3.474 5.412 7.322 8.242	0.401 3.347 6.214 7.561
0.4	0.636 1.296 2.048 2.515	1.212 2.000 2.845 3.321	1.776 2.785 3.814 4.344	1.509 2.855 4.180 4.817	-1.197 0.614 2.359 3.146
0.5	0.750 1.320 1.980 2.397	1.043 1.695 2.405 2.815	1.099 1.885 2.696 3.122	0.438 1.413 2.377 2.841	-1.473 -0.259 0.905 1.412
0.6	0.683 1.173 1.747 2.115	0.765 1.306 1.906 2.259	0.589 1.213 1.866 2.216	-0.077 0.657 1.390 1.737	-1.395 -0.528 0.306 0.743
0.7	0.553 0.972 1.469 1.791	0.514 0.966 1.475 1.780	0.261 0.766 1.303 1.597	-0.307 0.264 0.844 1.134	-1.228 -0.578 0.056 0.547
0.8	0.418 0.775 1.204 1.484	0.321 0.701 1.136 1.400	0.065 0.479 0.929 1.182	-0.394 0.063 0.537 0.785	-1.050 -0.544 -0.037 0.412
0.9	0.302 0.605 0.973 1.216	0.185 0.505 0.878 1.110	-0.048 0.297 0.779 0.903	-0.409 -0.036 0.362 0.585	-0.885 -0.479 -0.058 0.320
1.0	0.209 0.466 0.781 0.991	0.091 0.363 0.684 0.888	-0.108 0.182 0.511 0.712	-0.390 -0.080 0.262 0.467	-0.740 -0.292 -0.045 0.256
1.25	0.064 0.226 0.445 0.593	-0.024 0.176 0.381 0.535	-0.146 0.087 0.282 0.444	-0.296 -0.089 0.158 0.331	-0.463 -0.181 0.023 0.173
1.5	0.001 0.112 0.254 0.355	-0.054 0.067 0.224 0.342	-0.124 0.008 0.179 0.310	-0.204 -0.062 0.125 0.269	-0.285 -0.134 0.073 0.135
1.75	-0.023 0.051 0.148 0.234	-0.055 0.031 0.140 0.240	-0.093 -0.002 0.124 0.238	-0.135 -0.038 0.107 0.216	-0.175 -0.069 0.092 0.102
2.0	-0.028 0.019 0.091 0.152	-0.046 0.023 0.092 0.163	-0.066 -0.002 0.091 0.171	-0.086 -0.018 0.089 0.176	-0.104 -0.031 0.090 0.082
2.5	-0.021 0.002 0.042 0.076	-0.026 -0.002 0.047 0.087	-0.042 -0.001 0.052 0.098	-0.036 -0.003 0.057 0.107	-0.039 -0.003 0.062 0.041

r/h \ z/h	0.5	0.4	0.3	0.2	0.1
3.0	-0.012 -0.000 0.024 0.045	-0.014 0.000 0.027 0.053	-0.015 0.001 0.031 0.059	-0.015 0.001 0.034 0.065	-0.015 0.002 0.037 0.019
3.5	-0.006 0.000 0.014 0.029	-0.007 0.000 0.016 0.034	-0.007 0.001 0.018 0.037	-0.007 0.001 0.020 0.040	-0.006 0.002 0.021 0.008
4.0	-0.003 0.001 0.009 0.020	-0.003 0.001 0.010 0.022	-0.003 0.001 0.011 0.024	-0.003 0.001 0.011 0.026	-0.003 0.001 0.012 0.028
6.0	0.000 0.001 0.002 0.004	0.000 0.001 0.002 0.005	0.000 0.001 0.002 0.005	0.000 0.001 0.002 0.005	0.000 0.001 0.002 0.005
8.0	0.000 0.000 0.000 0.001	0.000 0.000 0.000 0.001	0.000 0.000 0.000 0.001	0.000 0.000 0.000 0.001	0.000 0.000 0.000 0.001
10.0	0.000 0.000 0.000 0.000	0.000 0.000 0.000 0.000	0.000 0.000 0.000 0.000	0.000 0.000 0.000 0.000	0.000 0.000 0.000 0.000

μ = 0	μ = 0.2
μ = 0.4	μ = 0.5

d) i_τ

r/h \ z/h	1.0		0.9		0.8		0.7		0.6	
0	0	0	0	0	0	0	0	0	0	0
	0	0	0	0	0	0	0	0	0	0
0.1	0.258	0.330	0.312	0.342	0.462	0.472	0.731	0.731	1.204	1.199
	0.436	0.530	0.375	0.395	0.470	0.459	0.713	0.690	1.175	1.148
0.2	0.473	0.607	0.564	0.623	0.814	0.835	1.246	1.248	1.957	1.949
	0.806	0.980	0.687	0.728	0.834	0.815	1.216	1.173	1.904	1.853
0.3	0.616	0.797	0.719	0.804	0.999	1.032	1.456	1.462	2.138	2.129
	1.064	1.294	0.896	0.957	1.037	1.017	1.423	1.367	2.069	2.001
0.4	0.678	0.889	0.774	0.880	1.025	1.071	1.408	1.420	1.916	1.909
	1.195	1.456	0.995	1.074	1.085	1.071	1.381	1.322	1.843	1.766
0.5	0.671	0.896	0.747	0.868	0.939	0.996	1.211	1.231	1.527	1.524
	1.214	1.481	1.001	1.094	1.024	1.020	1.198	1.144	1.459	1.380
0.6	0.614	0.839	0.668	0.798	0.796	0.864	0.963	0.992	1.128	1.131
	1.149	1.406	0.941	1.044	0.904	0.915	0.969	0.997	1.073	1.002
0.7	0.529	0.746	0.564	0.697	0.638	0.713	0.725	0.761	0.791	0.801
	1.033	1.268	0.843	0.952	0.765	0.791	0.752	0.726	0.753	0.696
0.8	0.435	0.636	0.455	0.586	0.490	0.570	0.523	0.566	0.532	0.550
	0.893	1.102	0.730	0.841	0.631	0.671	0.570	0.562	0.514	0.476
0.9	0.344	0.526	0.353	0.480	0.363	0.446	0.365	0.413	0.345	0.368
	0.750	0.932	0.616	0.725	0.512	0.565	0.428	0.438	0.346	0.328
1.0	0.262	0.424	0.265	0.384	0.261	0.343	0.246	0.297	0.214	0.242
	0.616	0.773	0.510	0.616	0.412	0.474	0.322	0.348	0.232	0.234
1.25	0.112	0.229	0.111	0.206	0.098	0.170	0.075	0.127	0.045	0.079
	0.352	0.463	0.301	0.394	0.235	0.309	0.164	0.219	0.091	0.132
1.5	0.032	0.114	0.032	0.103	0.023	0.082	0.009	0.054	-0.009	0.024
	0.191	0.274	0.170	0.251	0.135	0.209	0.091	0.158	0.045	0.105
1.75	-0.004	0.054	-0.002	0.050	-0.006	0.039	-0.012	0.024	-0.020	0.007
	0.102	0,169	0.096	0.164	0.079	0.146	0.055	0.120	0.030	0.091
2.0	-0.016	0.025	-0.013	0.024	-0.013	0.019	-0.014	0.012	-0.016	0.004
	0.056	0.111	0.055	0.113	0.048	0.106	0.036	0.094	0.022	0.078
2.5	-0.014	0.007	-0.011	0.007	-0.008	0.007	-0.006	0.006	-0.003	0.004
	0.020	0.059	0.020	0.061	0.019	0.060	0.016	0.057	0.012	0.053

r/h \ z/h	1.0	0.9	0.8	0.7	0.6
3.0	-0.007 0.004 0.009 0.037	-0.004 0.005 0.008 0.038	-0.001 0.005 0.007 0.037	0.002 0.005 0.006 0.036	0.005 0.005 0.004 0.033
3.5	-0.002 0.004 0.003 0.025	0.001 0.004 0.002 0.025	0.003 0.004 0.001 0.024	0.006 0.004 0.000 0.023	0.008 0.004 -0.001 0.021
4.0	0.001 0.004 -0.000 0.017	0.003 0.004 -0.001 0.017	0.005 0.004 -0.002 0.016	0.007 0.003 -0.003 0.015	0.009 0.003 -0.004 0.014
6.0	0.001 0.002 -0.004 0.004	0.003 0.001 -0.004 0.004	0.004 0.001 -0.005 0.004	0.005 0.001 -0.005 0.003	0.006 0.001 -0.005 0.003
8.0	0.001 0.001 -0.003 0.001	0.001 0.001 -0.004 0.001	0.002 0.000 -0.004 0.001	0.002 0.000 -0.004 0.001	0.003 0.000 -0.004 0.001
10.0	0.000 0.000 -0.002 0.001	0.001 0.000 -0.003 0.000	0.001 0.000 -0.003 0.000	0.001 0.000 0.003 0.000	0.002 0.000 -0.003 0.000

μ = 0	μ = 0.2
μ = 0.4	μ = 0.5

z/h \ z/h	0.5		0.4		0.3		0.2		0.1	
0	0	0	0	0	0	0	0	0	0	0
	0	0	0	0	0	0	0	0	0	0
0.1	2.093	2.086	3.957	3.949	8.480	8.472	21.42	21.42	53.01	53.00
	2.060	2.035	3.923	3.905	8.448	8.440	21.40	21.40	52.99	53.01
0.2	3.154	3.141	5.230	5.215	8.750	8.734	13.18	13.16	10.69	10.68
	3.091	3.043	5.165	5.128	8.689	8.668	13.12	13.13	10.65	10.67
0.3	3.116	3.099	4.416	4.395	5.735	5.713	5.791	5.771	2.781	2.765
	3.032	2.965	4.327	4.270	5.649	5.610	5.717	5.702	2.727	2.739
0.4	2.517	2.499	3.080	3.056	3.262	3.237	2.540	2.516	0.928	0.909
	2.421	2.340	2.976	2.901	3.161	3.101	2.450	2.142	0.862	0.851
0.5	1.820	1.804	1.966	1.943	1.785	1.758	1.164	1.138	0.347	0.326
	1.723	1.636	1.856	1.771	1.675	1.599	1.065	1.007	0.273	0.241
0.6	1.232	1.220	1.199	1.178	0.962	0.937	0.541	0.515	0.124	0.102
	1.142	1.057	1.091	1.004	0.851	0.769	0.439	0.372	0.045	0.007
0.7	0.797	0.790	0.706	0.689	0.505	0.482	0.294	0.215	0.027	0.005
	0.720	0.646	0.607	0.527	0.391	0.322	0.139	0.075	-0.052	-0.087
0.8	0.495	0.495	0.398	0.386	0.249	0.229	0.086	0.064	-0.016	-0.038
	0.435	0.378	0.313	0.249	0.152	0.089	-0.008	-0.059	-0.094	-0.118
0.9	0.294	0.299	0.290	0.202	0.104	0.088	0.008	-0.012	-0.035	-0.056
	0.252	0.216	0.140	0.096	0.020	-0.023	-0.078	-0.110	-0.109	-0.119
1.0	0.163	0.173	0.095	0.093	0.023	0.012	-0.031	-0.048	-0.041	-0.061
	0.139	0.124	0.042	0.020	-0.047	-0.068	-0.107	-0.119	-0.110	-0.105
1.25	0.011	0.031	-0.022	-0.015	-0.046	-0.049	-0.052	-0.063	-0.032	-0.049
	0.022	0.053	-0.038	-0.013	-0.082	-0.058	-0.102	-0.075	-0.088	-0.057
1.5	-0.026	-0.005	-0.038	-0.029	-0.044	-0.043	-0.036	-0.045	-0.014	-0.031
	0.002	0.056	-0.034	0.015	-0.059	-0.012	-0.068	-0.024	-0.060	-0.020
1.75	-0.026	-0.008	-0.029	-0.020	-0.026	-0.027	-0.017	-0.026	0.001	-0.017
	0.004	0.063	-0.017	0.038	-0.032	0.019	-0.040	0.006	-0.040	0.001
2.0	-0.017	-0.004	-0.015	-0.010	-0.010	-0.013	-0.002	-0.012	0.011	-0.008
	0.008	0.062	-0.005	0.046	-0.016	0.032	-0.023	0.020	-0.028	0.009
2.5	-0.000	0.003	0.004	0.001	0.009	0.000	0.015	-0.000	0.021	-0.000
	0.006	0.046	0.001	0.039	-0.006	0.030	-0.012	0.021	-0.019	0.011

z/h \ r/h	0.5		0.4		0.3		0.2		0.1	
3.0	0.008	0.004	0.012	0.004	0.015	0.003	0.019	0.002	0.023	0.001
	0.001	0.030	-0.003	0.026	-0.007	0.021	-0.011	0.015	-0.016	0.008
3.5	0.011	0.004	0.013	0.003	0.017	0.003	0.019	0.002	0.022	0.001
	-0.003	0.019	-0.006	0.016	-0.008	0.013	-0.011	0.009	-0.015	0.005
4.0	0.011	0.003	0.014	0.003	0.016	0.002	0.018	0.002	0.020	0.001
	-0.006	0.012	-0.007	0.010	-0.009	0.008	-0.011	0.006	-0.013	0.003
6.0	0.007	0.001	0.008	0.001	0.009	0.001	0.010	0.000	0.011	0.000
	-0.006	0.003	-0.006	0.002	-0.007	0.002	-0.007	0.001	-0.008	0.001
8.0	0.004	0.000	0.004	0.000	0.005	0.000	0.005	0.000	0.006	0.000
	-0.004	0.001	-0.004	0.001	-0.004	0.000	-0.005	0.000	-0.005	0.000
10.0	0.002	0.000	0.002	0.000	0.003	0.000	0.003	0.000	0.003	0.000
	-0.003	0.000	-0.003	0.000	-0.003	0.000	-0.003	0.000	-0.003	0.000

Tabelle 1.29 Einflussbeiwerte fuer die Gleichungen in Fig. 1.6 [2]

Table 1.29 Influence values for the equations in Fig. 1. [2]

(ii) Voraussetzungen:

Elastisch isotroper homogene Schicht, die auf einer rauhen starren Unterlage sitzt. Die waagerechten Dehnungen zwischen den zwei Schichten sind gleich Null.

(ii) Assumptions:

Elastic isotropic homogeneous layer resting on a rough rigid base. The horizontal displacements between the two layers are zero.

(iii) Quellen

References

[1] Burmister (1956): Stress and displacement characteristics of a two-layer rigid base soil system, influence diagrams and practical applications. Proc. Highw. Res. Bd 35, PP 773.

[2] Poulos (1967): Stresses and displacements in an elastic layer underlain by a rough, rigid base. Geotechnique XVII, PP 378.

2.2 LINIENLAST

(i) Formeln/Tafeln

2.2 LINE LOAD

(i) Formulae/Charts

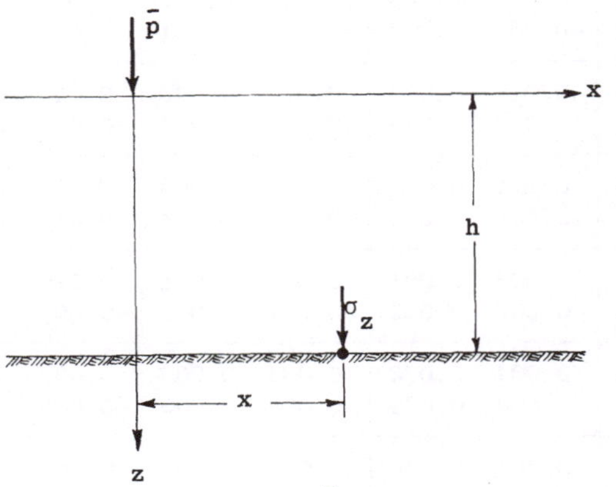

$$\sigma_z = \frac{\bar{p}}{\pi h} \cdot i$$

Fig. 1.63a Gleichung zur Bestimmung der lotrechten Spannung an der Schichtgrenze infolge Linienlast an der Oberflaeche bei $\mu = 0$ [2]

Equation for determining the vertical stress at the base of the layer due to line load at the surface for $\mu = 0$ [2]

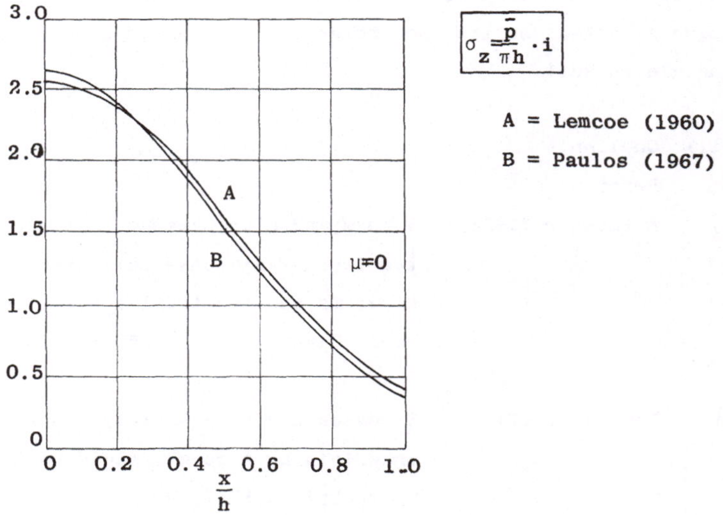

Fig. 1.63b Einflussbeiwert i fuer die Gleichung in Fig. 1.63a

Influence value i for the equation in Fig. 1.63a

(ii) Voraussetzung:

Wie bei Einzellast (2.1)

(ii) Assumption:

As in Point Load (2.1)

(iii) Quellen

References

[1] Lemcoe (1960): Stresses in Layered elastic solids. J. Eng'g Mech. Div. Am. Soci. Civ. Eng'rs 86, 4.

[2] Paulos (1967): Stresses and displacements in an elastic layer underlain by a rough, rigid base. Geotechnique XVII, PP 378.

2.3 STREIFENLAST

(i) Formeln/Tafeln

2.3 STRIP LOADING

(i) Formulae/Charts

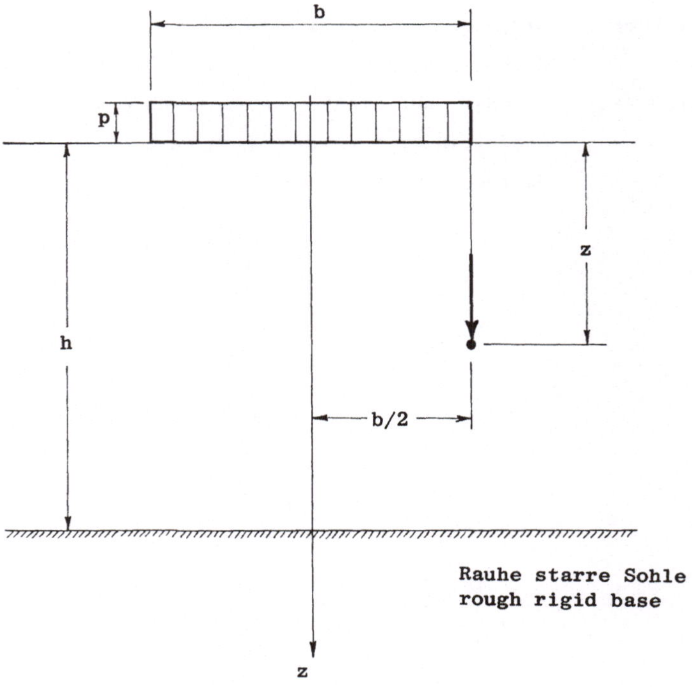

$$\sigma_z = \frac{p}{\pi} \cdot i_z$$

$$\tau = \frac{p}{\pi} \cdot i_\tau$$

Fig. 1.64a Spannungen unter dem Randeiner unendlich langen Streifenlast [1]

Stresses under the edge of a strip loading with an infinite length [1]

$$\sigma_z = \frac{p}{\pi} \cdot i_z$$

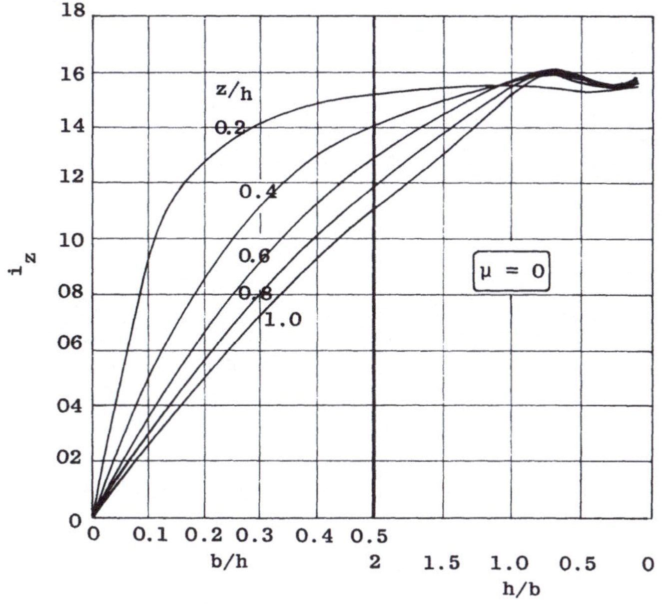

(a)

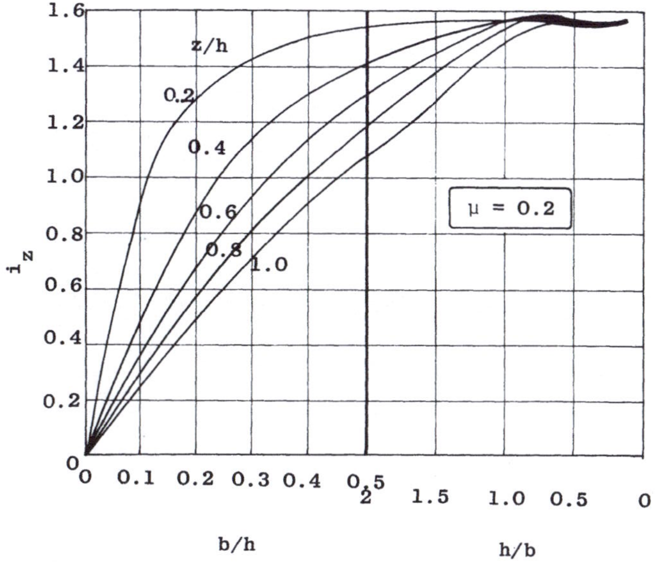

(b)

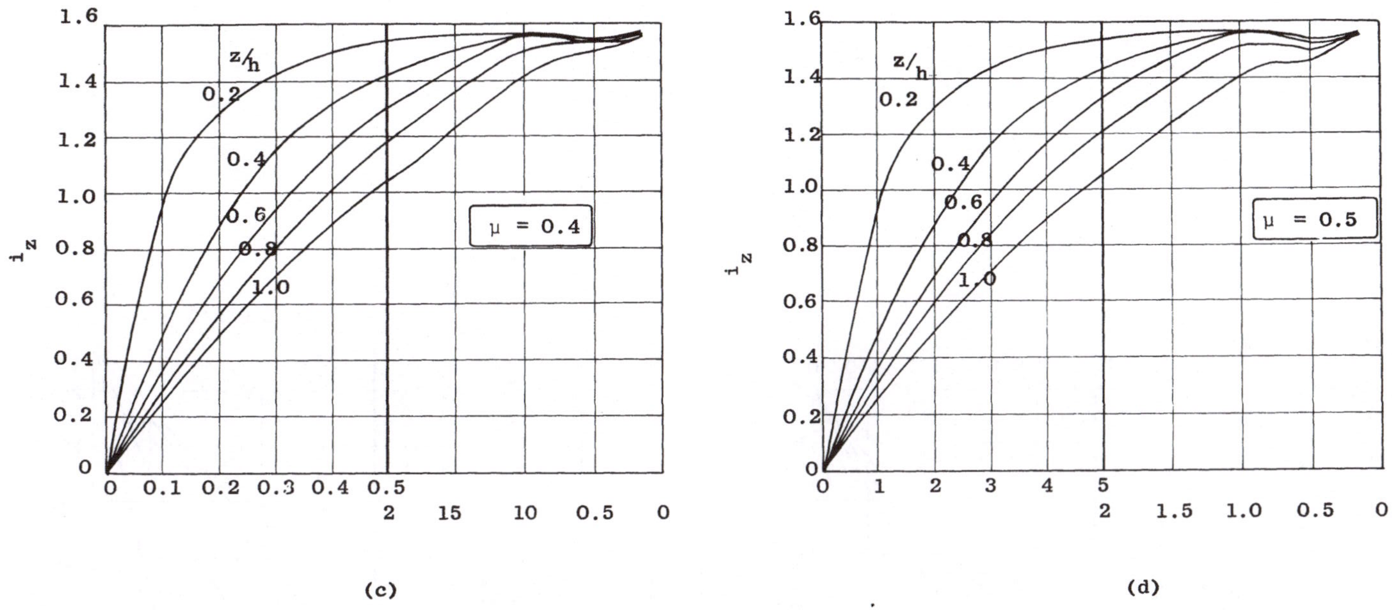

Fig. 1.64b Einflussbeiwerte i_z [1]
Influence values i_z [1]

$$\tau = \frac{p}{\pi} \cdot i_\tau$$

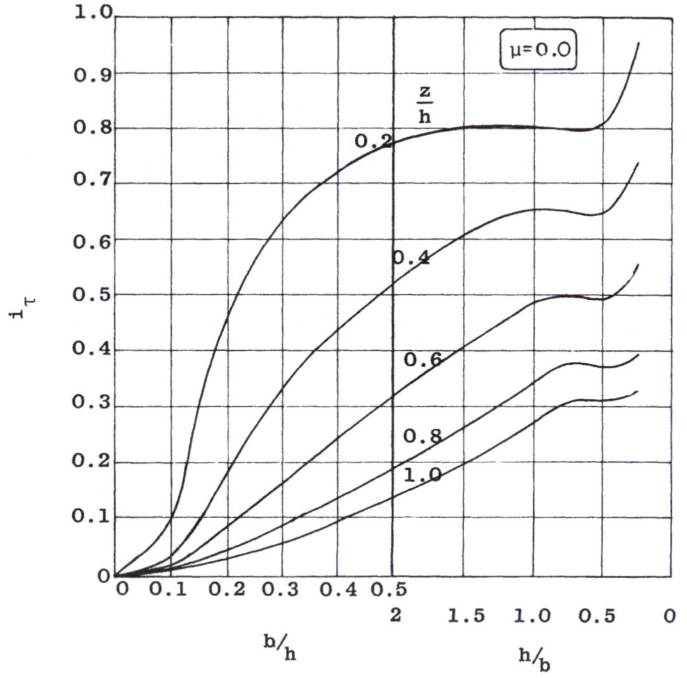

(a)

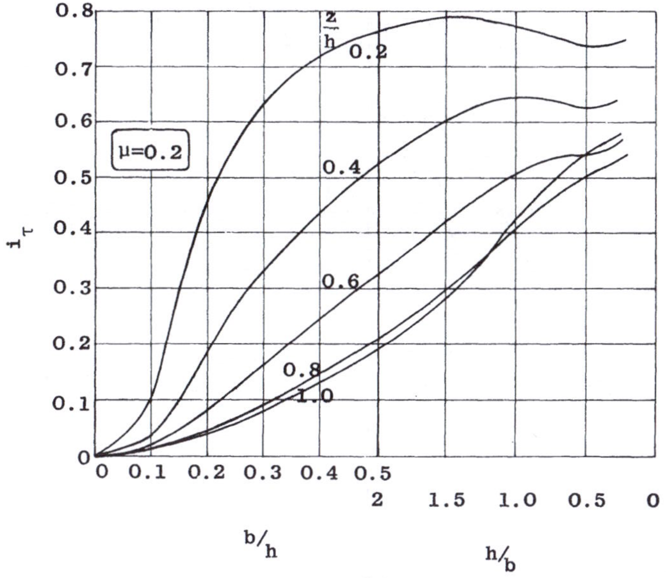

(b)

Fig. 1.64c Einflussbeiwerte i_τ [1]
Influence values i_τ [1]

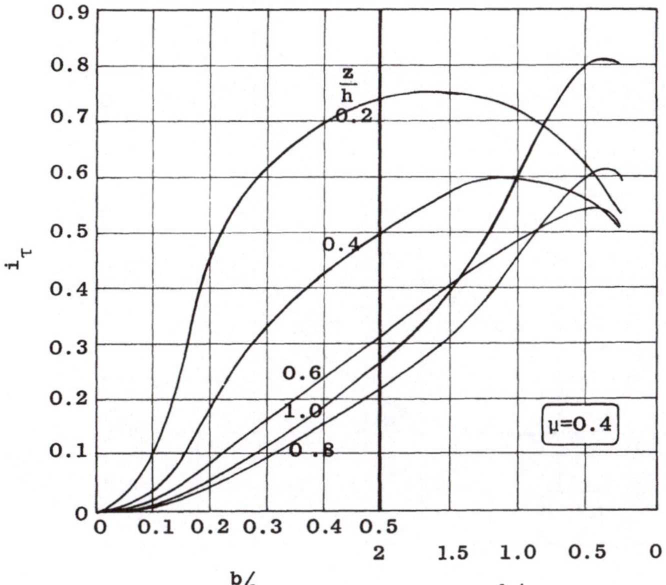

(c)

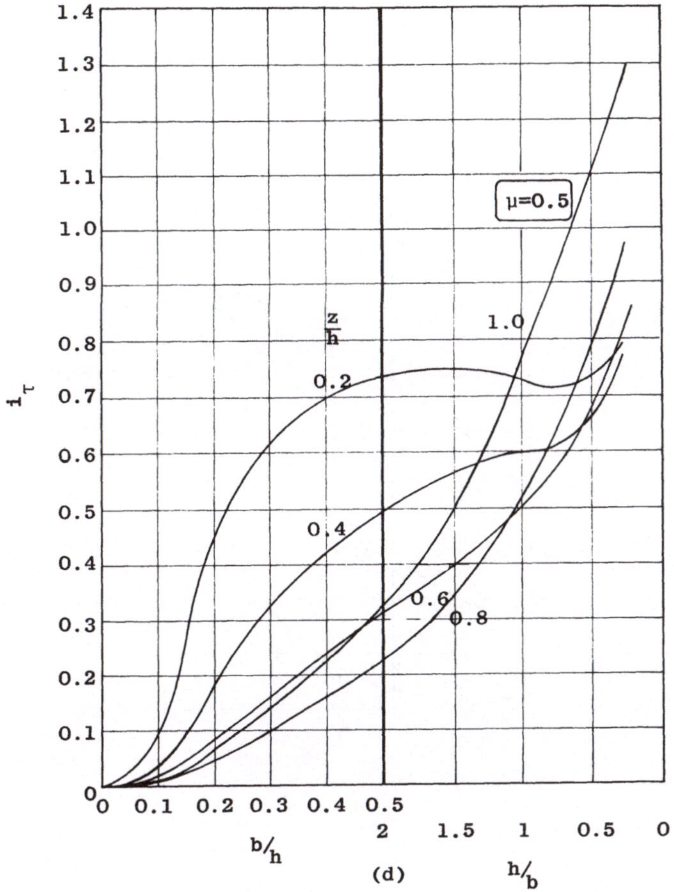

(d)

(ii) Quelle
Reference
Paulos (1967): Stesses and displacements in an elastic layer underlain by a rough, rigid base. Geotechnique XVII, P. 378.

2.4 GLEICHFOERMIGE RECHTECK-LASTEN

(i) Formeln/Tafeln

2.4 UNIFORMLY DISTRIBUTED RECTANGULAR LOADS

(i) Formulae/Charts

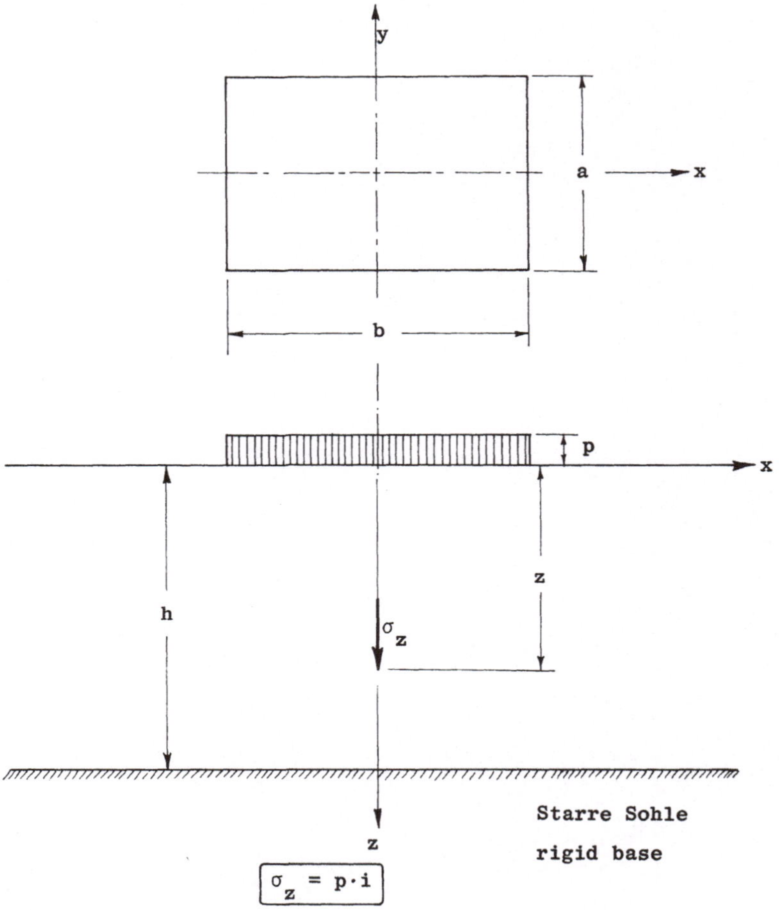

Fig. 1.65a Gleichung fuer lotrechte Spannung unter dem Mittelpunkt der Rechtecklast [1]

Equation for vertical stress under the centre of the rectangular load [1]

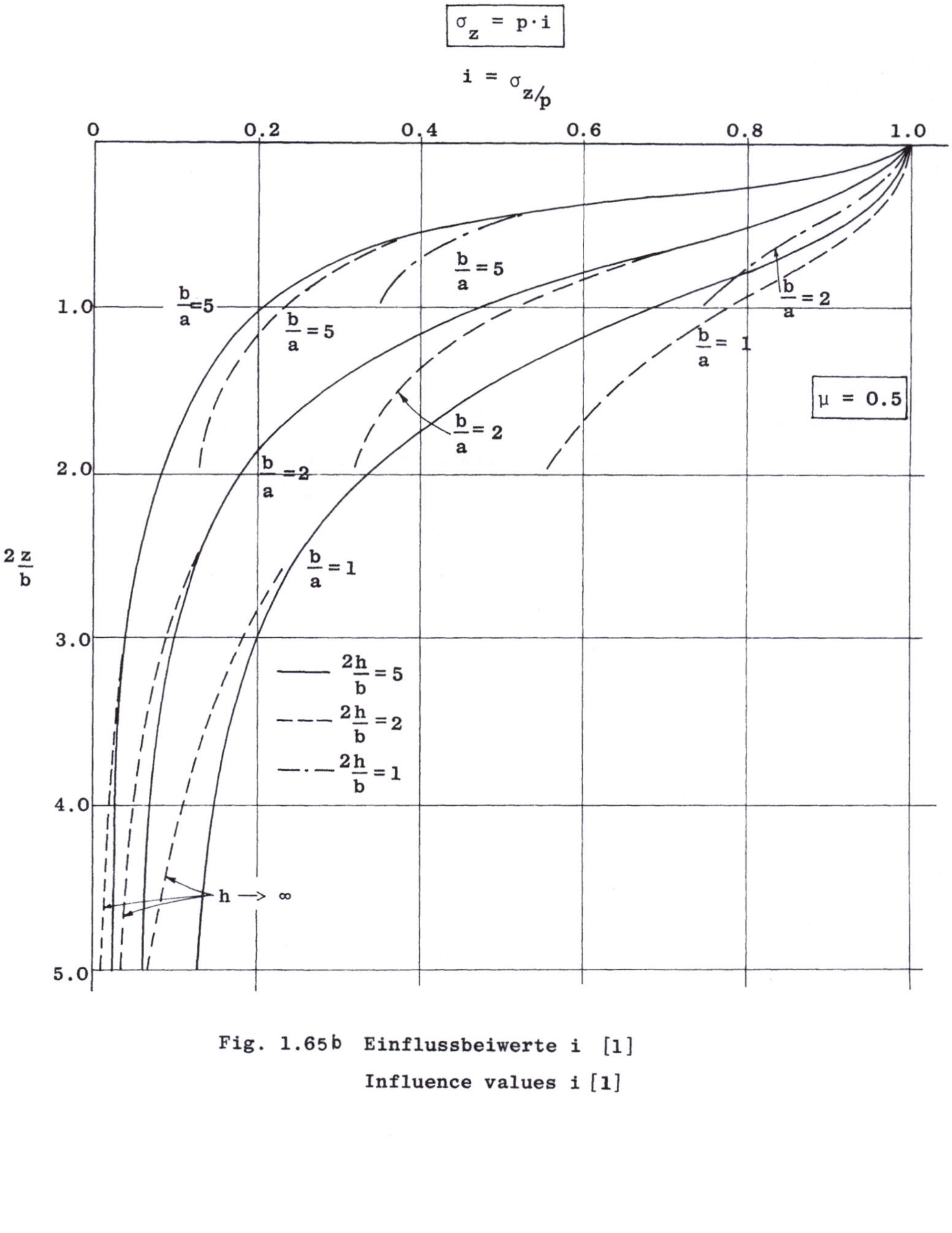

Fig. 1.65b Einflussbeiwerte i [1]
Influence values i [1]

(ii) Voraussetzung:

Elastisch isotrope homogenen Schicht, die auf einer reibungsfreien starren Unterlage sitzt. Die Spannung zwischen den zwei Schichten ist gleich null.

(ii) Assumption:

Elastic isotropic homogeneous layer which rests on a frictionless rigid base. The stress between the two layers is zero.

(iii) Quelle

Reference

[1] Sovine (1961): Stresses and sisplacements in a limited layers of uniform thickness, resting on a rigid base, and subjected to a uniformly distributed flexible load of rectangular shape. Proc. 5. ICSMFE, Paris, ud. 1, PP. 823.

2.5 AXIALSYMMETRISCHE LASTEN

2.5 AXIAL SYMMETRIC LOADS

(i) Formeln/Tafeln

(i) Formulae/Charts

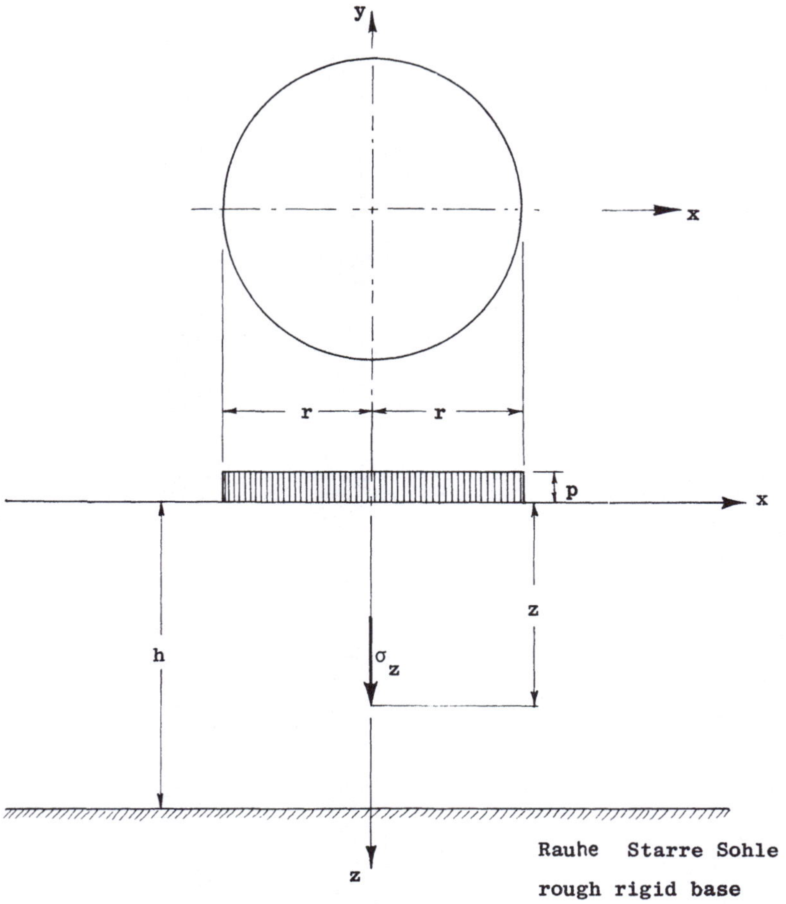

$$\sigma_z = p \cdot i_z$$

Fig. 1.66a Gleichung fuer die lotrechte Spannung unter dem Mittelpunkt einer gleichmaessig belasteten Kreisflaeche [1]

Equation for vertical stress under the centre of uniformly loaded circular area [1]

$\sigma_z = p \cdot i_z$

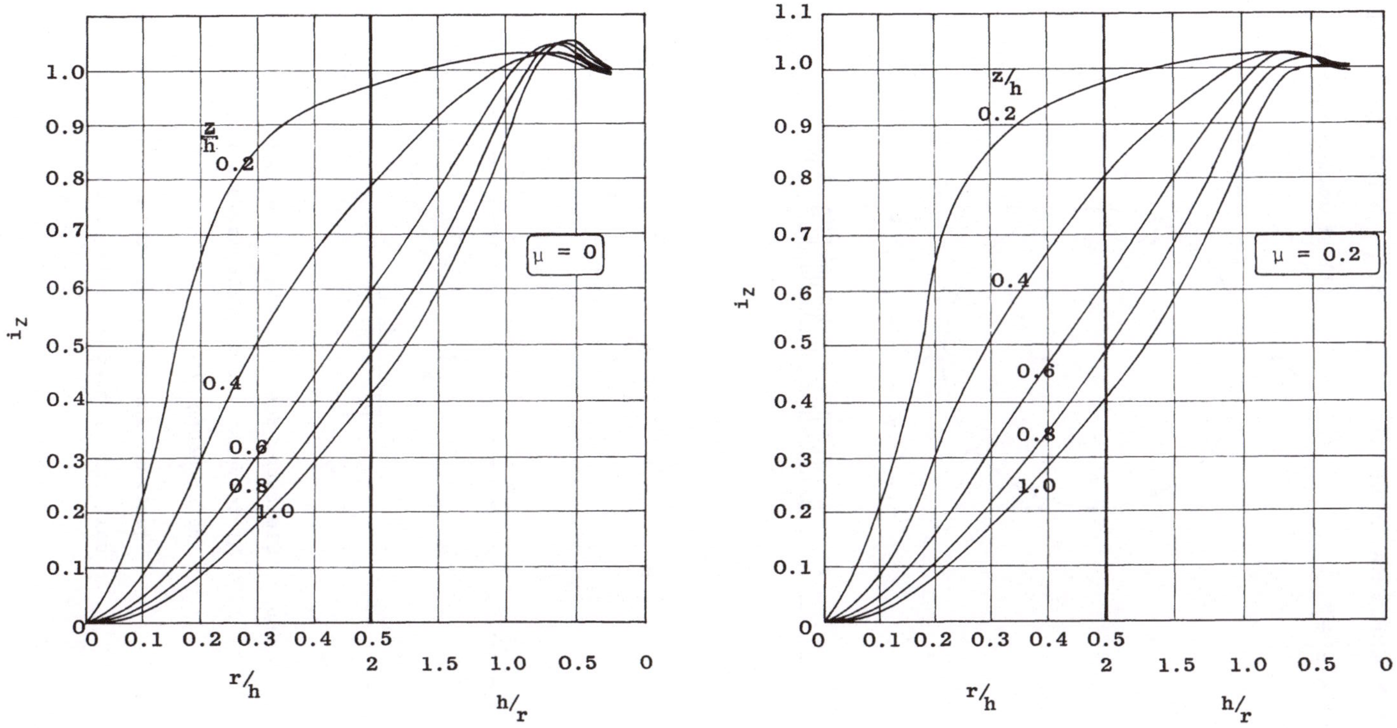

(a) (b)

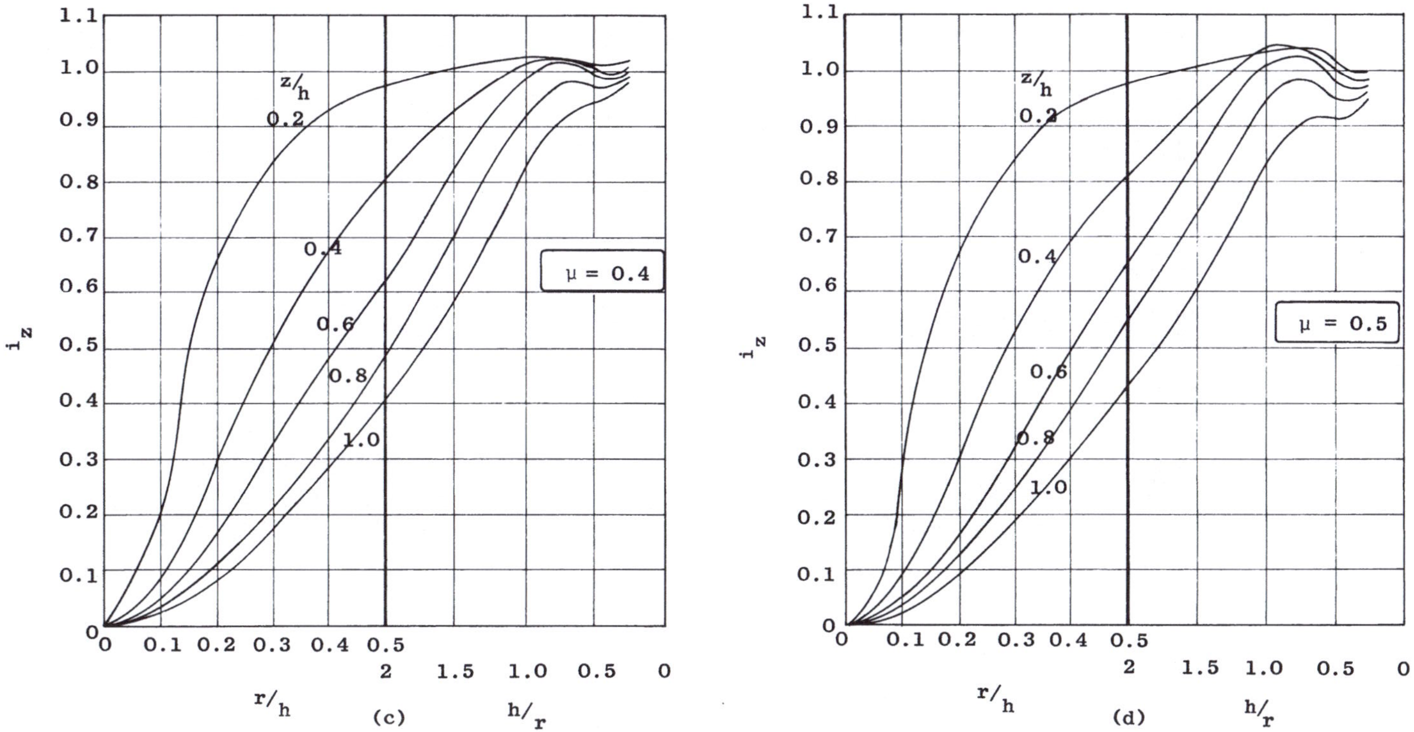

Fig. 1.66b Einflussbeiwerte i [1]
Influence value i [1]

(ii) Vorausetzung: Wie in Abschn. 1.1.2.1

(ii) Assumption: As in Chap. 1.1.2.1.

(iii) Quelle
Reference

[1] Paulos (1967): Stresses and displacements in an elastic layer underlain by a rough rigid base. Geotechnique XVII, S. 378.

2.6 BELIEBIGE LASTEN MIT BELIEBIGEM GRUNDRISS

(i) Formeln/Tafeln

2.6 LOADING WITH ANY SHAPE

(i) Formulae/Charts

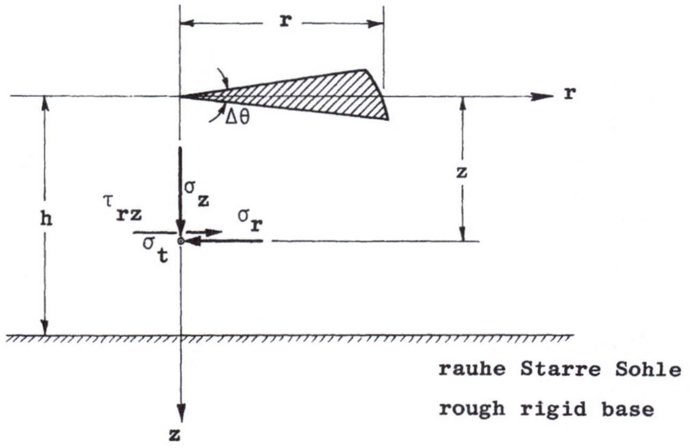

$$\sigma_z = \frac{\Delta\theta}{2\pi} \cdot p \cdot i_z$$

$$\sigma_r = \frac{\Delta\theta}{2\pi} \cdot p \cdot i_r$$

$$\sigma_t = \frac{\Delta\theta}{2\pi} \cdot p \cdot i_t$$

$$\tau_{rz} = \frac{\Delta\theta}{2\pi} \cdot p \cdot i_\tau$$

Fig. 1.67a Gleichungen zur Ermittlung von Spannungen unter beliebigen Flaechenlasten nach Sektormethode [1]

Equations for determining stresses under any spread loading according to the Sector Method [1]

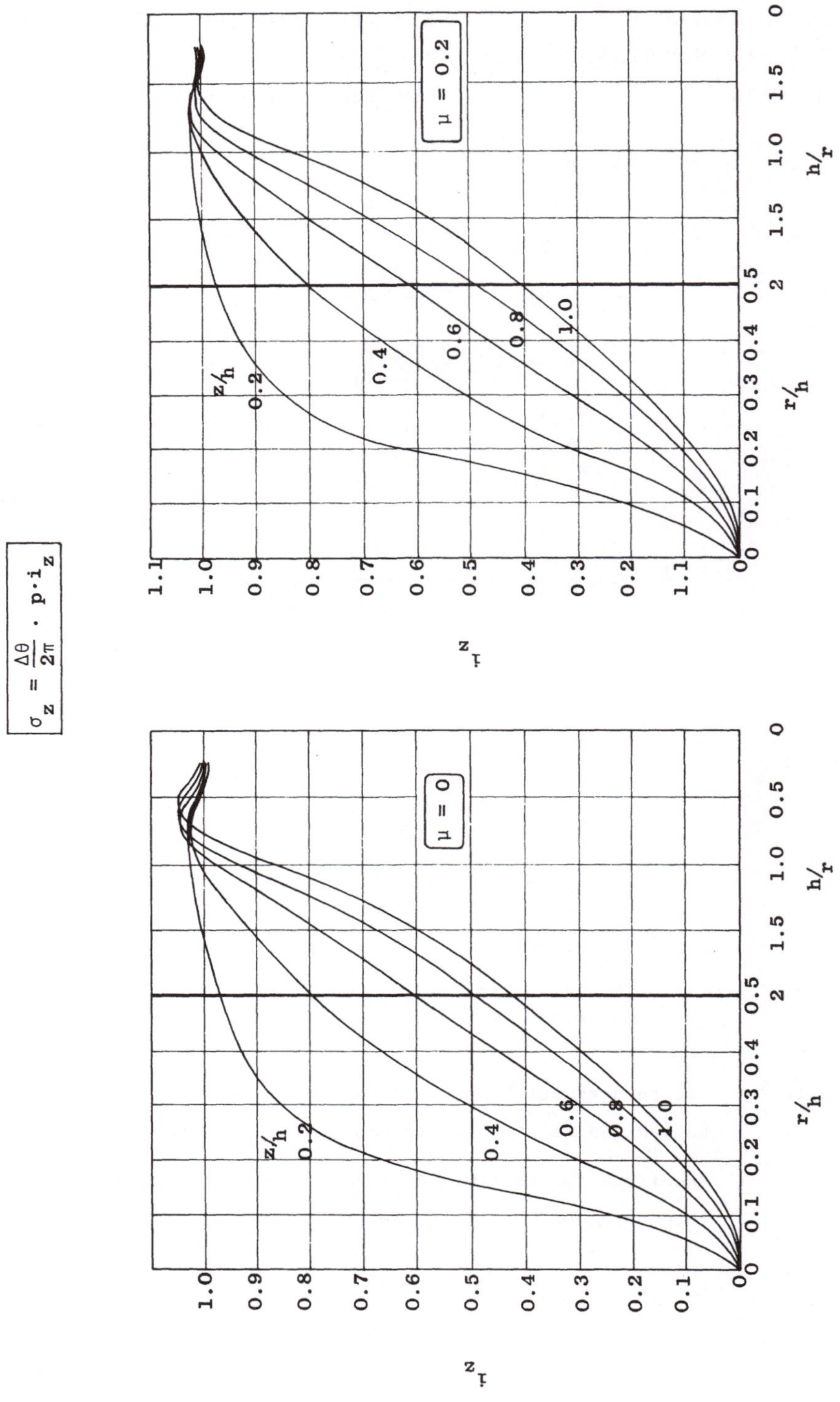

σz = Δθ/2π · p·iz
μ = 0
μ = 0.2
iz
r/h
h/r
z/h
0.2
0.4
0.6
0.8
1.0
(a)
(b)

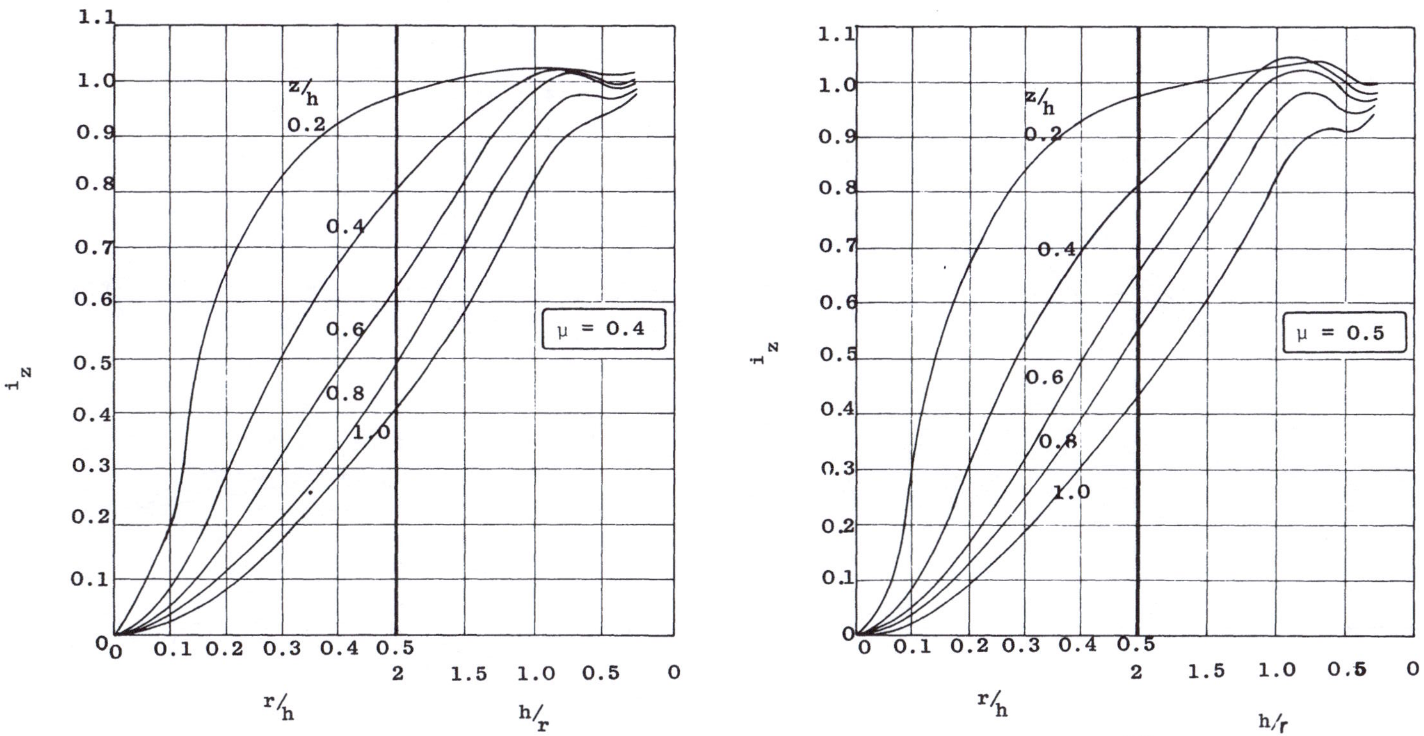

(c) (d)

Fig. 1.67b Einflussbeiwerte i_z [1]
Influence values i_z [1]

$$\sigma_r = \frac{\Delta\theta}{2\pi} \; p \cdot i_r$$

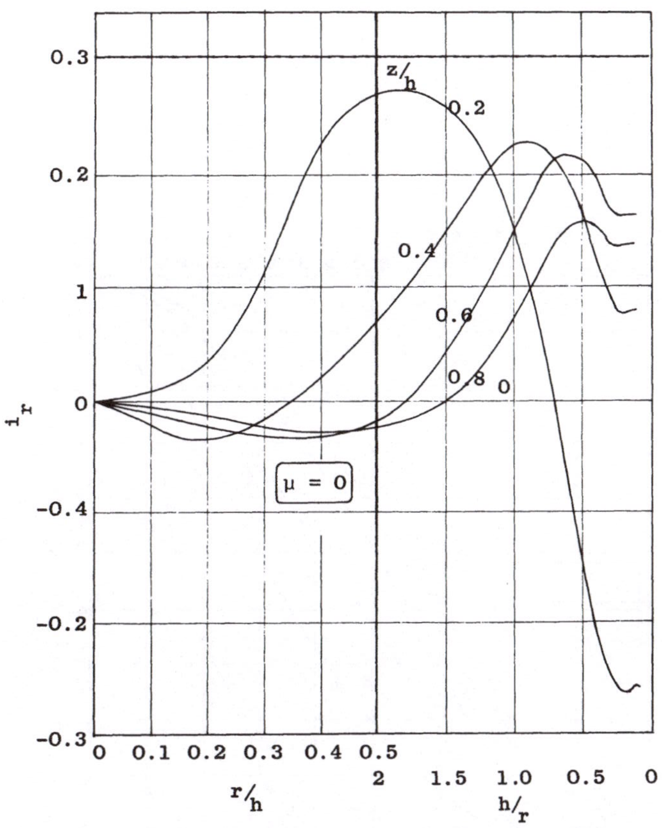

(a)

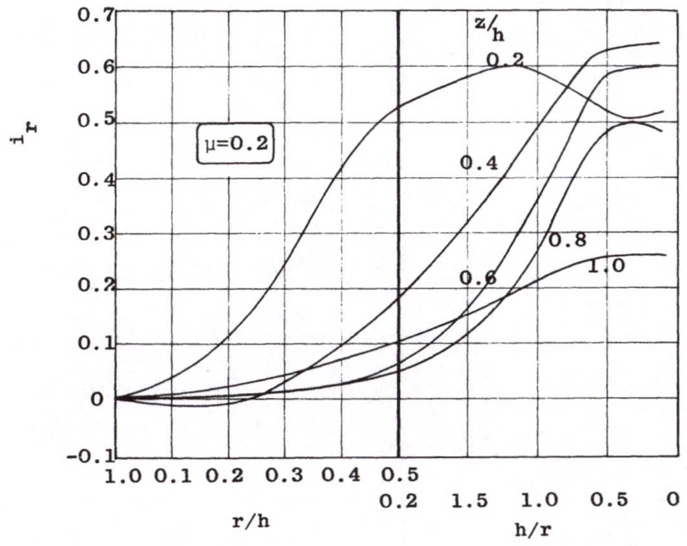

(b)

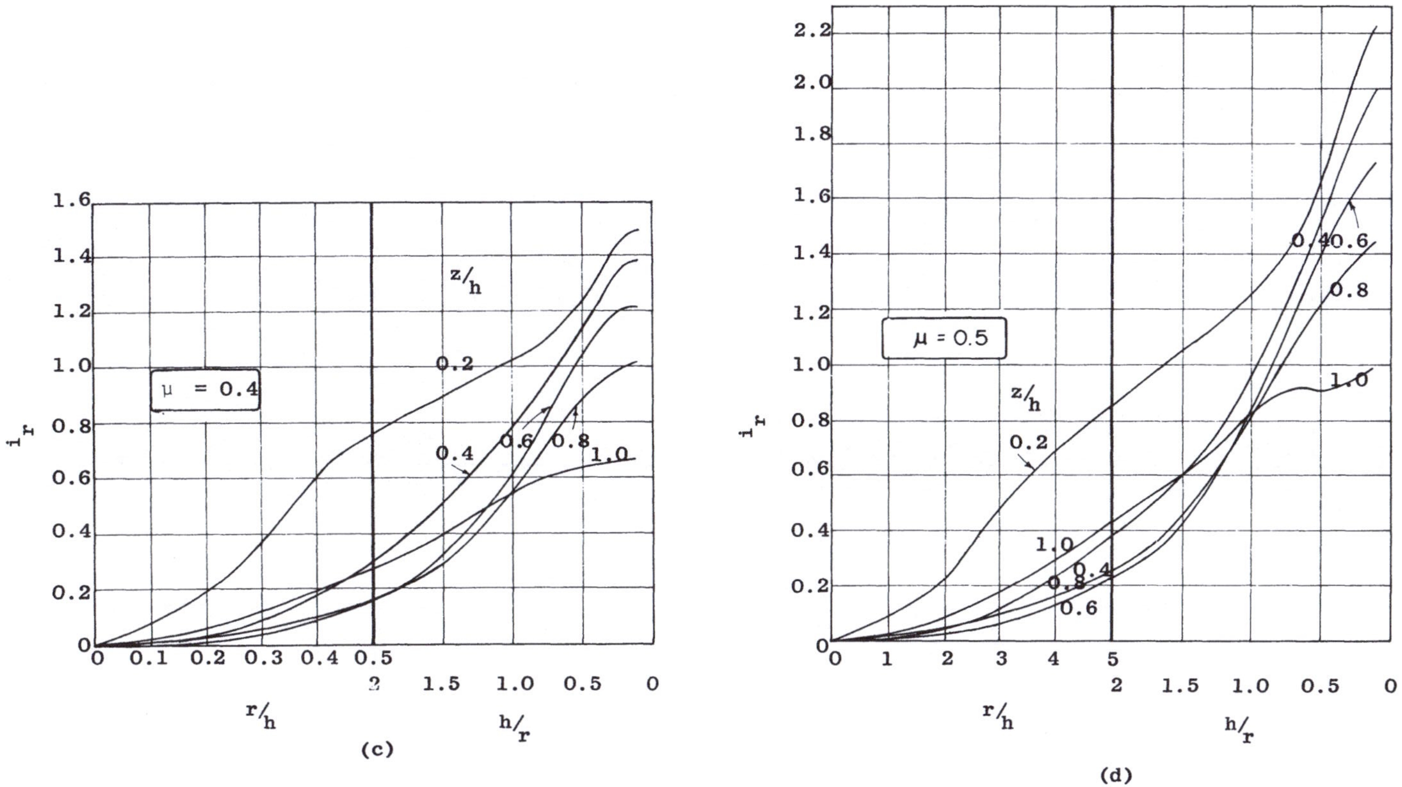

Fig. 1.67c Einflussbeiwerte i_r [1]
Influence values i_r [1]

$$\sigma_t = \frac{\Delta\theta}{2\pi} \cdot p \cdot i_t$$

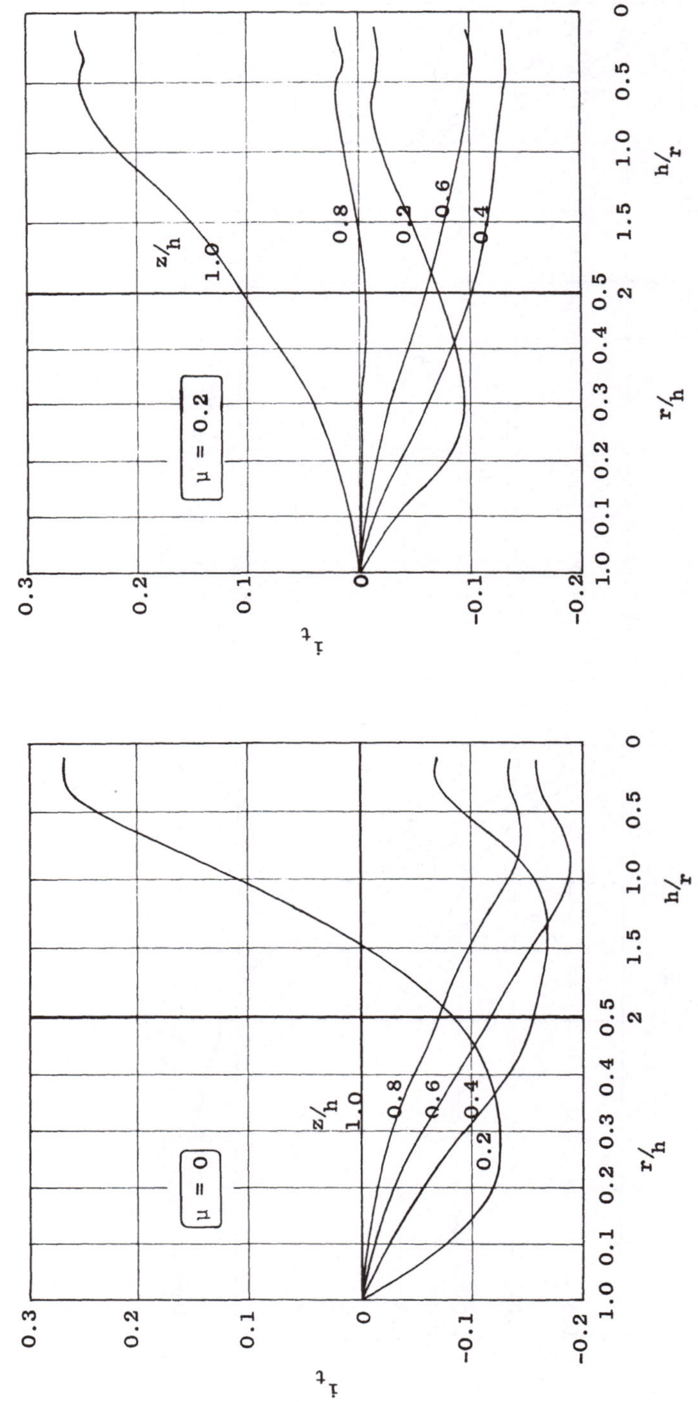

(a) (b)

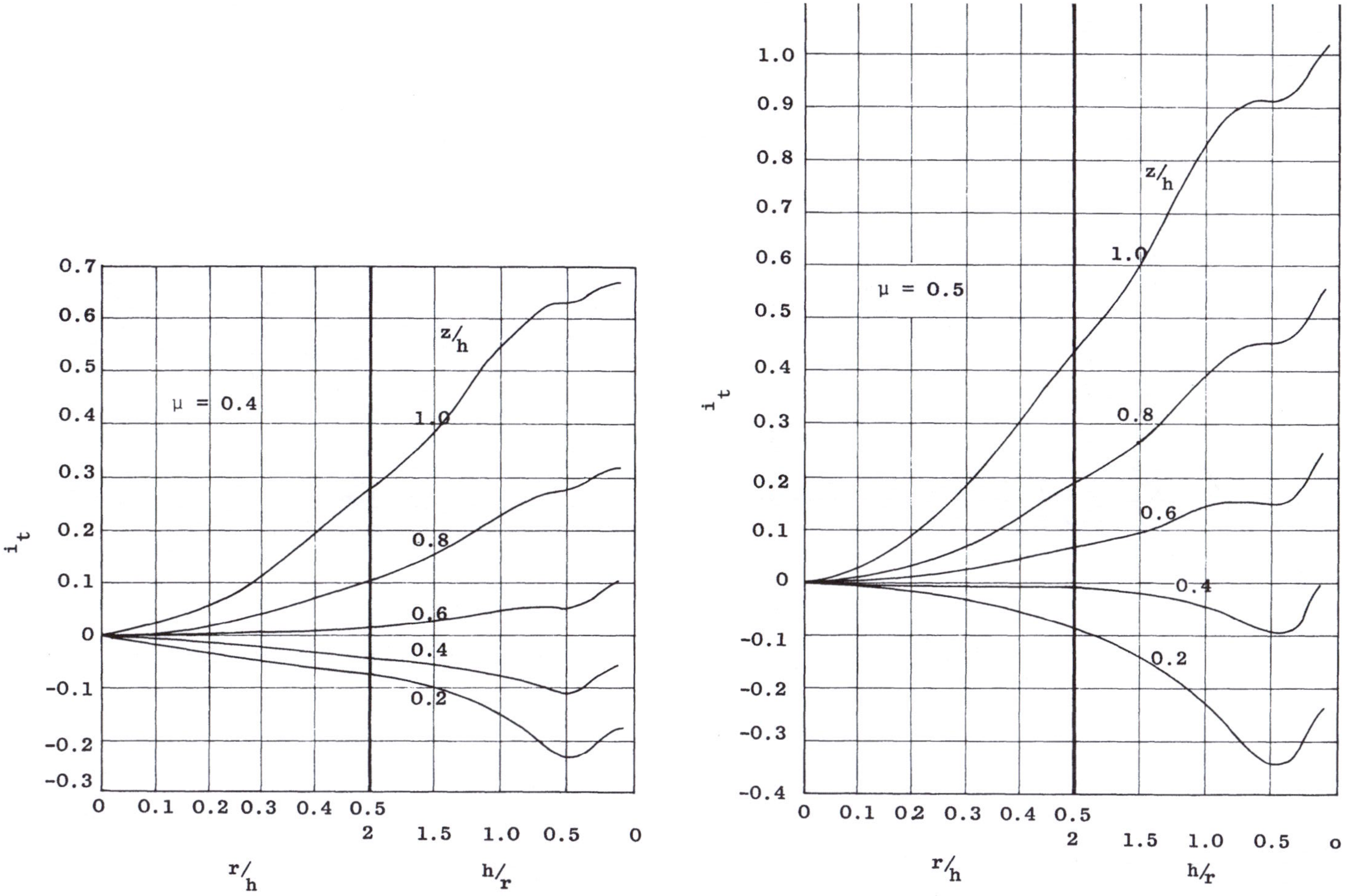

Fig. 1.67d Einflussbeiwerte i_t [1]

Influence values i_t [1]

$$\tau_{rz} = \frac{\Delta\theta}{2\pi} \cdot p \cdot i_{\tau}$$

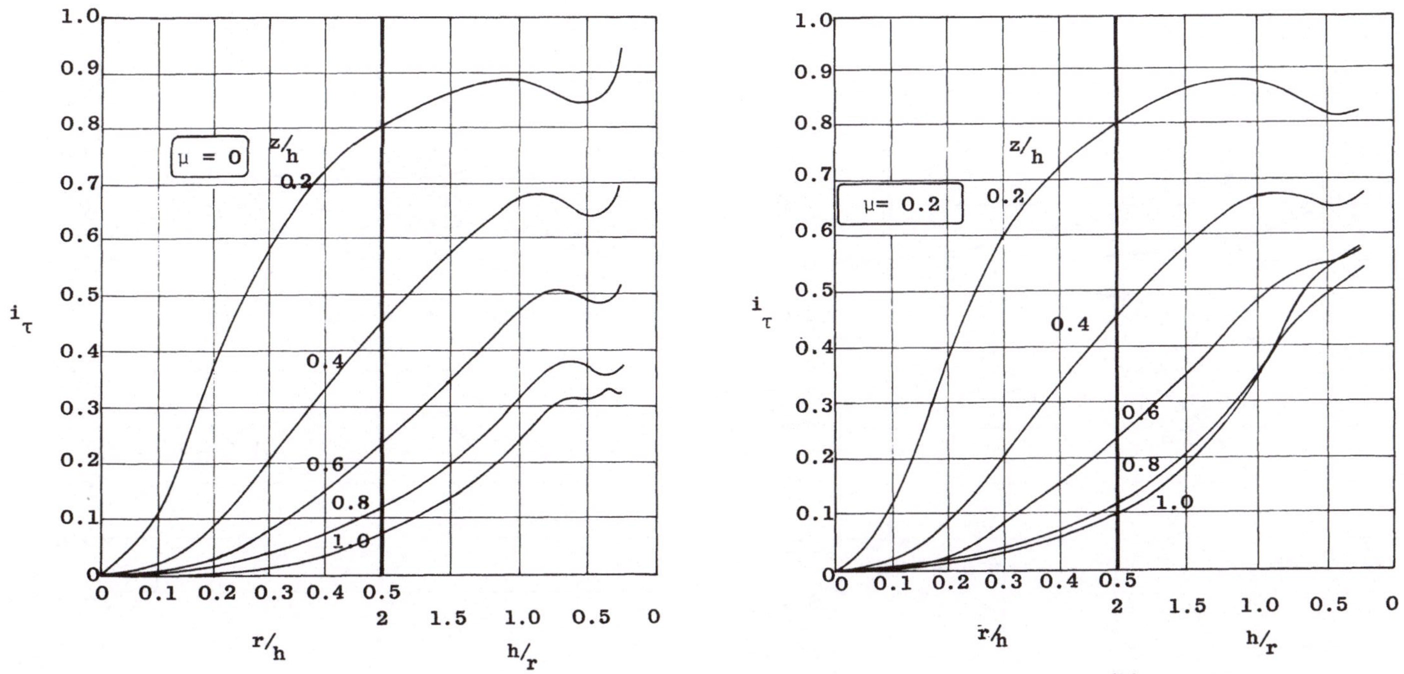

(a) (b)

Fig. 1.67e Einflussbeiwerte i_τ [1]
Influence values i_τ [1]

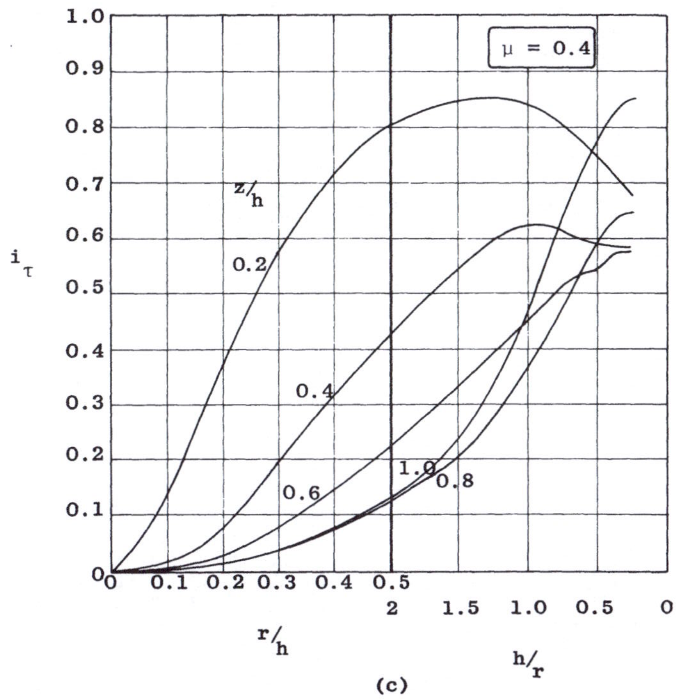

(c)

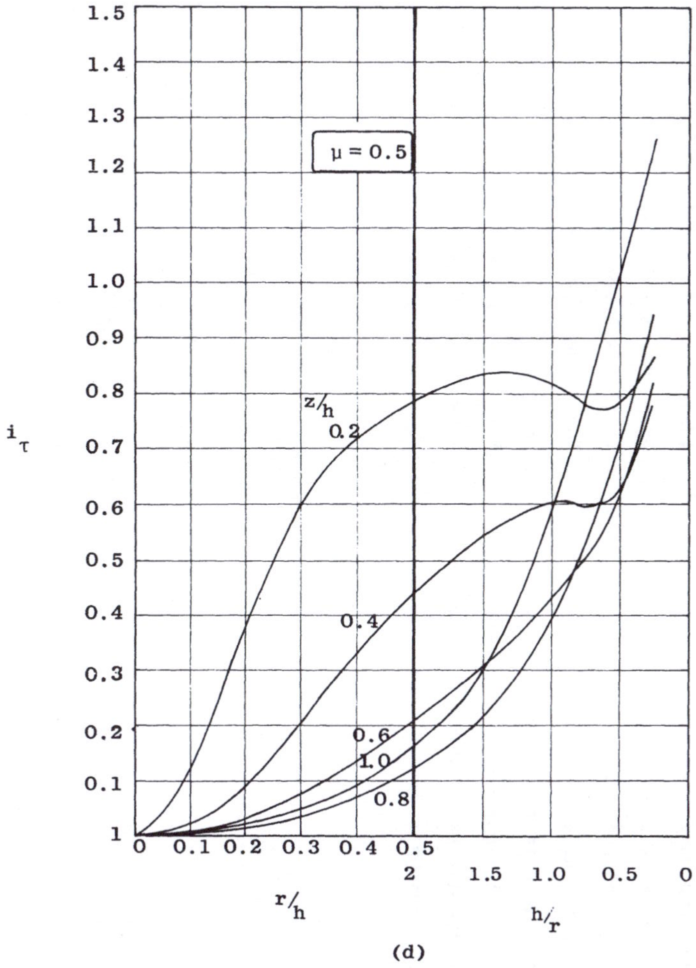

(d)

(ii) Voraussetzungen:

Wie bei Einzellast (2.1)

(iii) Erlaeuterung:

Mit den Einflussbeiwerten fuer die Spannungen infolge einer Sektorbelastung, koennen die Spannungen unter einer beliebigen Lastflaechenform gerechnet werden (Fig. 1.67f).

Assumptions:

As in Point Load (2.1)

(iii) Explanatory note:

The stresses under any given loaded area may be claculated using the influence values for stresses due to a sector loading (Fig. 1.67f).

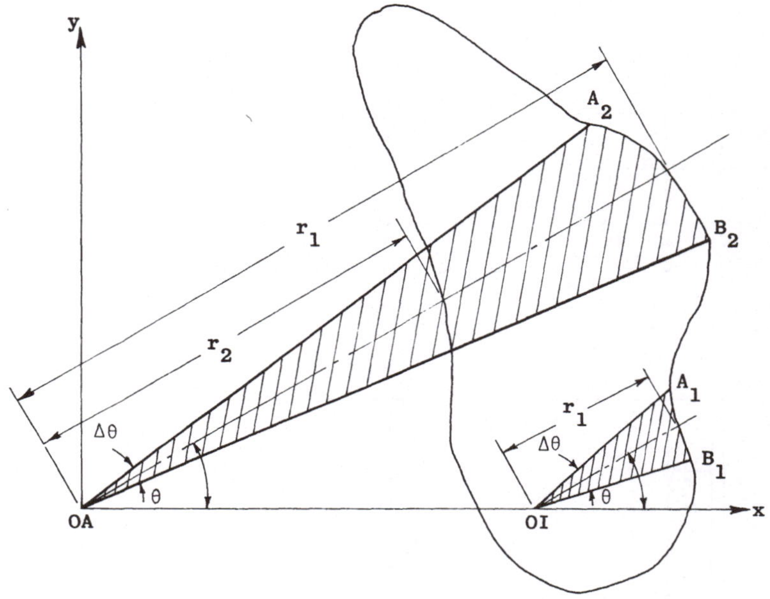

OA, OI sind Punkte, wo die Spannungen gerechnet werden sollen.

OA, OI are points where the stresses are to be calculated.

Fig. 1.67f Sketormethode [1]
Sector method [1]

Man teilt die vorhandene Lastflaeche in einer beliebigen Anzahl von Sektoren (je groesser die Anzahl der Sektoren, desto besser ist die Genauigkeit der Spannungsberechnung). Fuer die Berechnung der Spannungen summiert man die Einflussbeiwerte fuer die vershiedenen Sektoren in folgender Form fuer die Spannungen unter dem Punkt OA.

One divides the given loaded area in sectors (the bigger the number of subdivisions the better would be the accuracy of calculation). In order to calculate stresses under the point OA one sums the influence values for the different sectors as follows:

$$\sigma_z = \frac{p}{2\pi}\sum \Delta\theta(i_{z1}-i_{z2})$$

$$\sigma_x = \frac{p}{2\pi}\sum \left\{(i_{r1}-i_{r2})\Delta\theta \cdot \cos^2\theta + (i_{t1}-i_{t2})\Delta\theta \cdot \sin^2\theta\right\}$$

$$\sigma_y = \frac{p}{2\pi}\sum \left\{(i_{r1}-i_{r2})\cdot\Delta\theta\cdot\sin^2\theta + (i_{t1}-i_{t2})\Delta\theta \cdot \cos^2\theta\right\}$$

wobei i_1 und i_2 die Einflussbeiwerte fuer die Radien r_1 und r_2 darstellen. Analog koennen die Spannungen im Punkt O_I ermittlet werden.

where i_1 and i_2 are the influence value for the raii r_1 and r_2. In the same fashion the stresses under Point O_I may be determined.

(iii) Anwendungsbeispiel [1]

Gegeben

Gleichmaessig belastete Kreisflaeche mit Radius r auf einer Schicht mit der Staerke h = r.

(iii) Example [1]

Given

Uniformly loaded circular area with radius r on a layer with a thickness h = r.

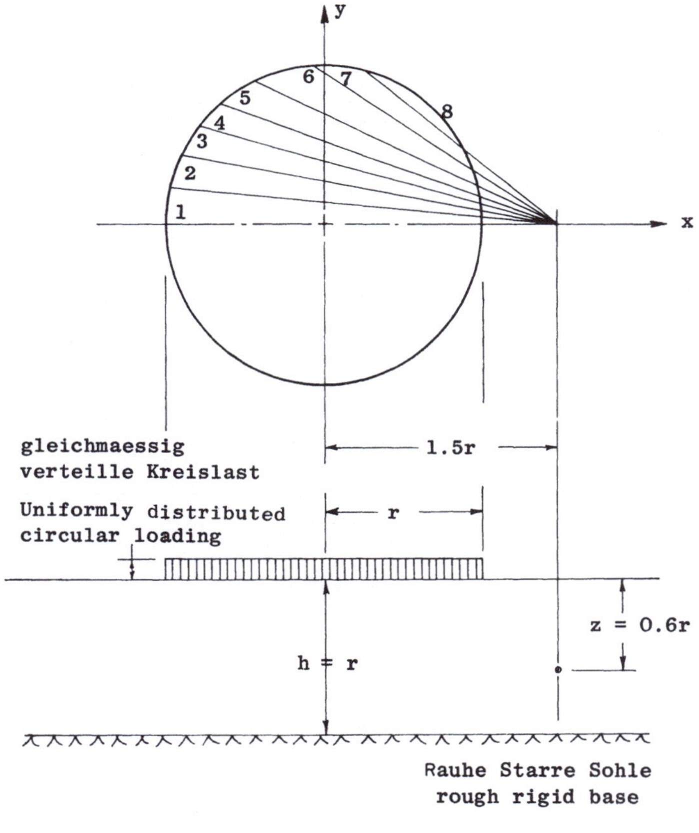

Gesucht

Die Spannungen σ_z, σ_x, σ_y im Punkt mit den Ordinaten $x = 1.5r$; $y = 0$; $z = 0.6r$ fuer $\mu = 0.5$.

Required

The stresses σ_z, σ_x, σ_y at a point with the coordinates $x = 1.5r$; $y = 0$; $z = 0.6r$ for $\mu = 0.5$.

Loesung

Solution

Sektor Sector No.	$\frac{r_1}{h}$	$\frac{r_2}{h}$	$\Delta\theta^o$	θ^o	$\sin^2\theta$	$\cos^2\theta$	i_{z1}	i_{z2}	i_{t1}	i_{t2}	i_{r1}	i_{r2}
1	2.49	0.50	10	5.0	0.0076	0.992	0.967	0.645	0.15	0.07	1.48	0.23
2	2.42	0.52	5	12.5	0.0468	0.953	0.969	0.673	0.15	0.07	1.47	0.25
3	2.33	0.53	5	17.5	0.0904	0.910	0.971	0.687	0.15	0.07	1.46	0.26
4	2.21	0.56	5	22.5	0.1475	0.854	0.975	0.724	0.15	0.08	1.43	0.29
5	2.05	0.61	5	27.5	0.213	0.787	0.981	0.781	0.15	0.09	1.39	0.35
6	1.86	0.68	5	32.5	0.289	0.711	0.992	0.850	0.15	0.10	1.33	0.44
7	1.63	0.77	4	37.0	0.371	0.630	1.009	0.915	0.16	0.12	1.25	0.54
8	1.40	0.90	3	40.5	0.578	0.578	0.026	0.980	0.16	0.14	1.12	0.69

Sektor Sector No.	$(i_{z1}-i_{z2})\Delta\theta$	$(i_{t1}-i_{t2})\times \Delta\theta\ \sin^2\theta$	$(i_{r1}-i_{r2})\times \Delta\theta\ \cos^2\theta$	$(i_{t1}-i_{t2})\times \Delta\theta\ \cos^2\theta$	$(i_{r1}-i_{r2})\times \Delta\theta\ \sin^2\theta$
1	3.22	0.01	12.40	0.79	0.10
2	1.48	0.02	5.82	0.38	0.28
3	1.41	0.04	5.46	0.36	0.54
4	1.26	0.05	4.86	0.30	0.84
5	1.00	0.06	3.99	0.23	1.11
6	0.71	0.07	3.17	0.18	1.29
7	0.38	0.06	1.79	0.10	1.05
8	0.14	0.03	0.74	0.03	0.54
$\Sigma =$	9.60	0.34	38.23	2.37	5.75

$$\sigma_z = \frac{9.60p}{180} = \underline{0.053p}$$

$$\sigma_x = \frac{(0.34 + 38.23)p}{180} = \underline{0.214p}$$

$$\sigma_y = \frac{(2.37 + 5.57)p}{180} \quad \underline{0.045p}$$

(iv) **Quelle**

Reference

[1] Paulos (1967): Stresses and displacements in an elastic layer underlain by a rough, rigid base. Geotechnique XVII, PP. 378

2.7 GRENZFLAECHENSPANNUNGEN

2.7.1 Rechteckige Gleichlast

(i) Formeln/Tabelle

2.7 STRESSES AT INTERFACE

2.7.1 Uniformly Distributed rectangular load

(i) Formulae/Table

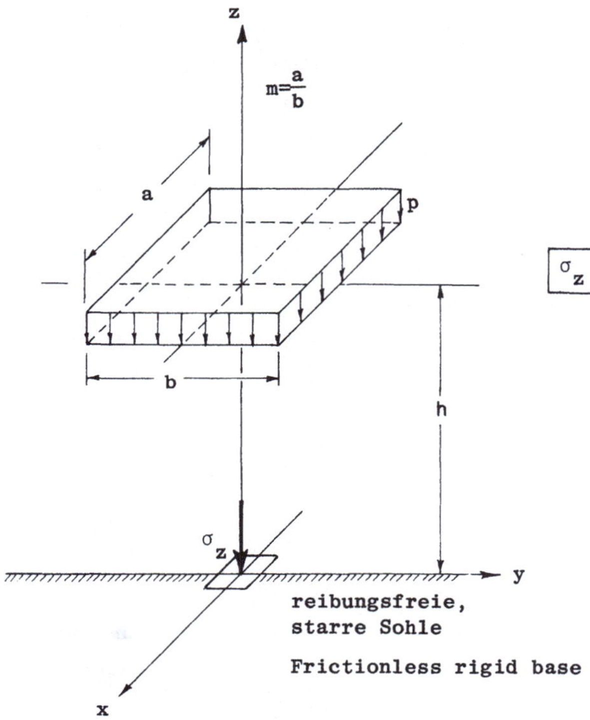

$\sigma_z = p \cdot i$

Fig. 1.68 Lotrechte Spannung, σ_z, in der Fuge zwischen elastischer und unnachgiebiger Schicht unter dem Mittelpunkt der Spannungsflaeche [1]

Vertical stress, σ_z, at the interface between elastic and rigid layers for a point located under the centre of the loaded area [1]

$\sigma_z = p \cdot i$

$\frac{2h}{b}$	Rechteck Rectangle						Unendlicher Sfreifen Infinite Strip
	m = 1	m = 2	m = 3	m = 5	m = 7	m = 10	m = ∞
0	0	0	0	0	0	0	0
0.2	0.096	0.098	0.098	0.099	0.099	0.099	0.100
0.5	0.226	0.231	0.233	0.236	0.237	0.238	0.239
1.0	0.403	0.427	0.435	0.441	0.444	0.446	0.452
2.0	0.609	0.698	0.727	0.748	0.757	0.764	0.784
3	0.711	0.856	0.910	0.952	0.964	0.982	1.018
5	0.800	1.010	1.119	1.201	1.238	1.256	1.323
7	0.842	1.094	1.223	1.346	1.402	1.442	1.532
10	0.873	1.155	1.309	1.475	1.556	1.619	1.758
∞	0.946	1.300	1.527	1.826	2.028	2.246	∞

Tabelle 1.30 Einflussbeiwert, i [1]

Table 1.30 Influence value i [1]

(ii) Voraussetzung: Beliebige Poissonnummer

(ii) Assumption: For any Poisson's number

(iii) Quelle

Reference

[1] Harr (1966): Foundations of theoretical soil mechanics, New York, PP. 97.

2.7.2 Kreisfoermige Gleichlast

2.7.2 Uniformly Distributed Circular Loading

2.7.2.1 Spannungen zwischen elastischer und unnachgiebiger Schicht

2.7.2.1 Stresses between elastic and rigid layers

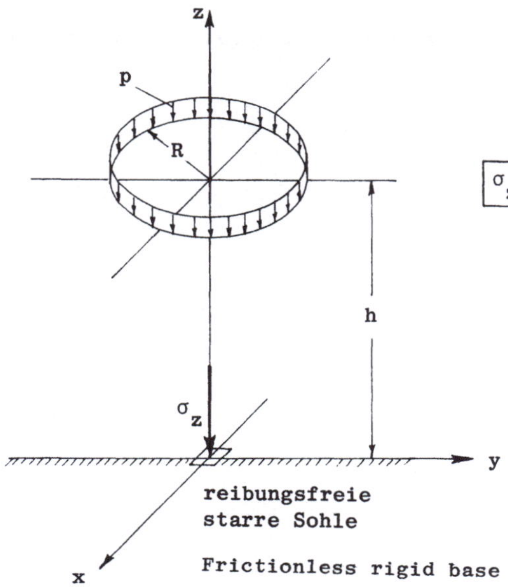

$\sigma_z = p \cdot i$

Einflussbeiwerte $i = f(\frac{h}{R})$
Influence values

h/R	i
0	0.000
0.2	0.096
0.5	0.225
1	0.396
2	0.578
3	0.661
5	0.740
7	0.776
10	0.818
∞	0.840

Fig. 1.69a **Lotrechte Spannung, σ_z, in der Fuge zwischen elastischer und unnachgiebiger Schicht unter dem Mittlepunkt der Spannungsflaeche [1]**

Vertical stress, σ_z, at the interface between elastic and rigid layers for a point located under the centre of the loaded area [1]

(ii) Voraussetzung:

Wie im Absch 1.1.2.7.1 (ii)

(ii) Assumption:

As in Chapter 1.1.2.7.1 (ii)

(iii) Quelle

Reference

[1] Harr (1966): Foundations of theoretical soil Mechanics New York, PP 97.

2.7.2.2 Spannungen Zwischen Zwei elastischen Schichten

(i) Formeln/Tafeln

2.7.2.2 Stresses between two elastic layers

(i) Formulae/Charts

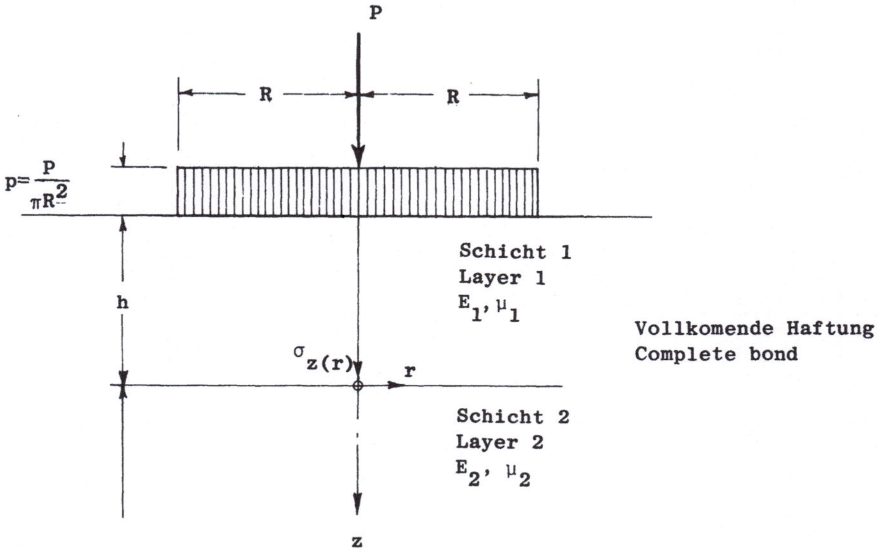

$\chi = E_1/E_2$

$\mu_1 = \mu_2 = 0.3; \quad \sigma_{z(R)} = p \cdot \bar{i}$

Fig. 1.69b Spannung, σ_z, in der Schichtfuge Zwischen Zwei elastischen Schichten [1]

Stress, σ_z, in the interface between two elastic layers [1]

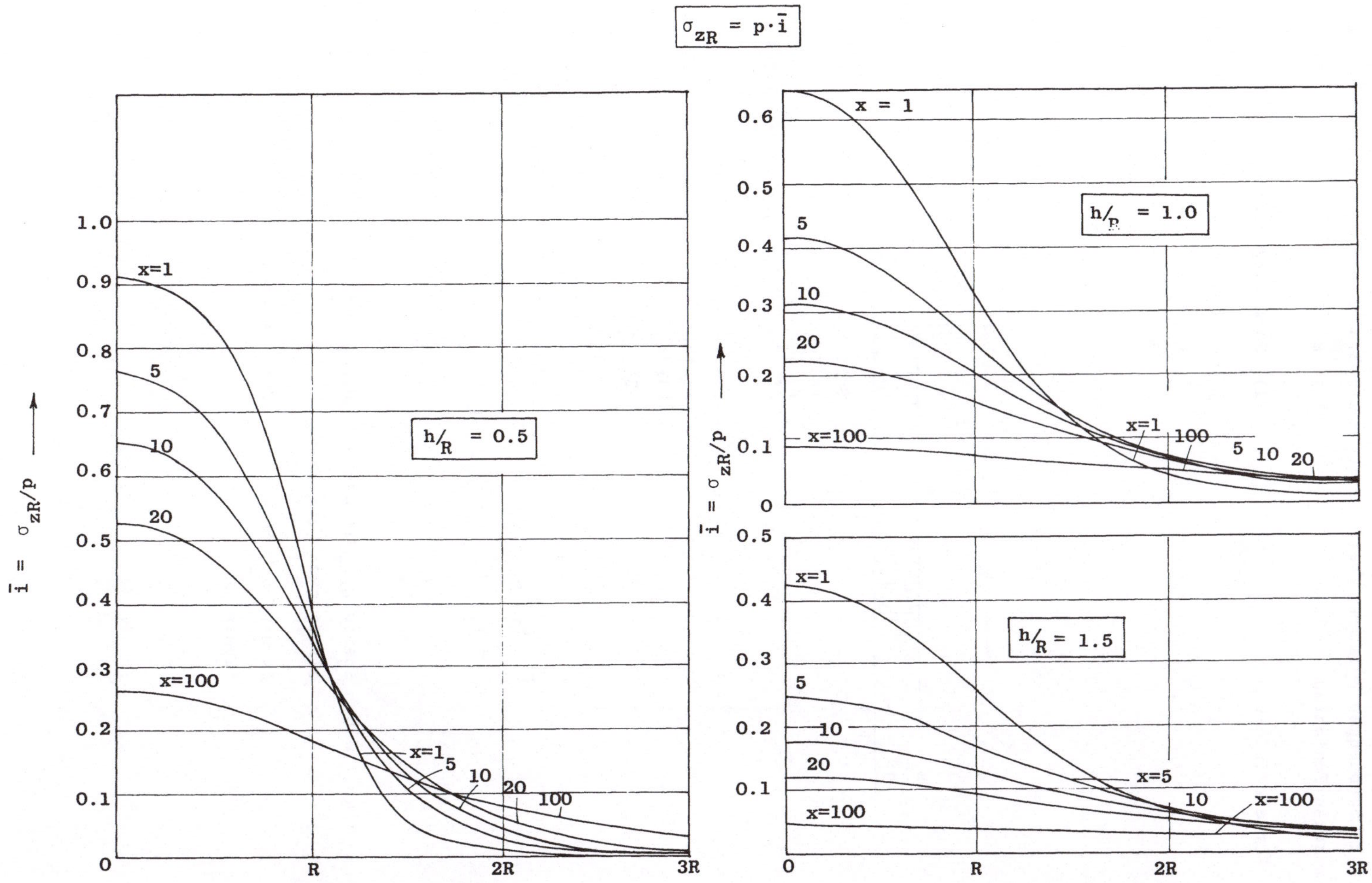

a) Abstand von der Belastungsachse
Distance from load axis

b) Abstand von der Belastungsache
Distance from load axis

Fig. 1.69c Einflussbeiwerte i [1]
Influence values
(a,b) beliebige Stelle
any point
(c,d) R = 0 (mittelpunkt)
(centre)

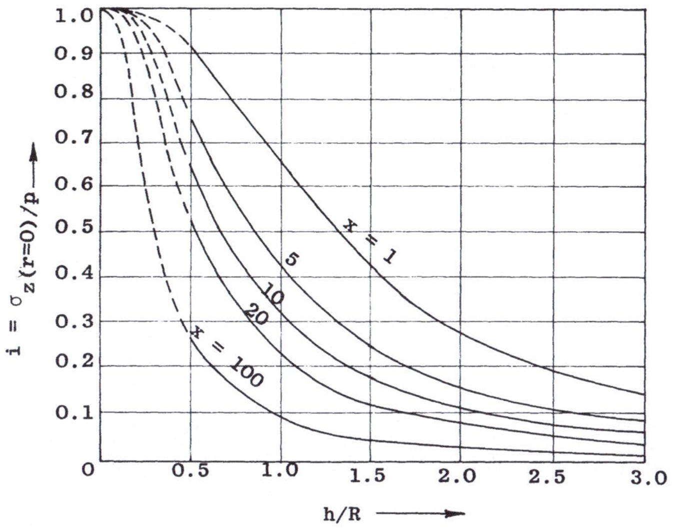

(c)

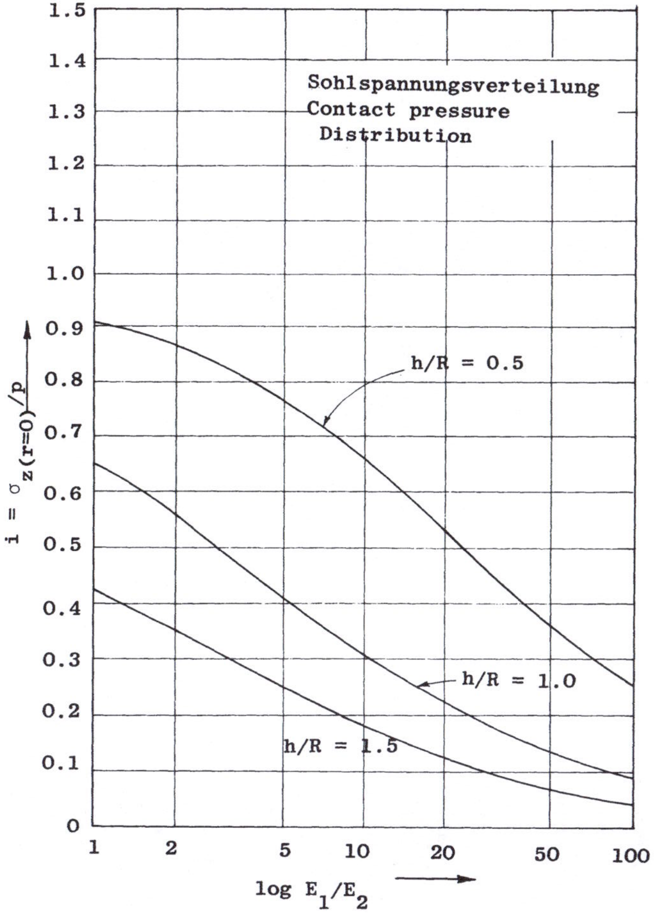

(d)

(ii) Voraussetzung	(ii) Assumption
Vollkomende Haftung zwischen den Schichten	Complete bond between the layers

(iii) Quelle

Reference

[1] Jelinek/Pauke (1970): Berechnung der Spannungsverteilung in einem Zweischichtensystem Bautechnik 47, S. 48.

2.7.2.3

Spannungen in der Fuge zwischen zwei elastischen Schichten unter dem Mittelpunkt der Spannungsflaeche

2.7.2.3

Stress at the interface between two elastic layers for a point located under the centre of the loaded area

(i) Formeln/Tabelle

(i) Formulae/Tables

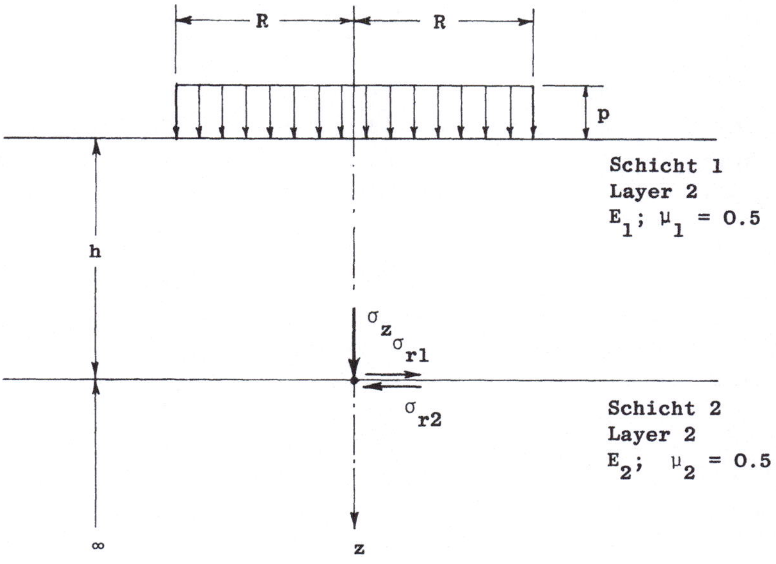

$$\sigma_z = p \cdot i_z$$

$$\sigma_{r1} = p \cdot i_{r1}$$

$$\sigma_{r2} = p \cdot i_{r2}$$

Fig. 1.69d Gleichungen zur Berechnung der Spannungen σ_t und σ_r in der Fuge zwischen Zwei elastischen Schichter unter dem Mittelpunkt der Spannungsflaeche [1]

Equation for determining the stresses σ_z and σ_r at the interface between two elastic layers for a point located below the centre of the loaded area [1]

Rauhe Grenzflaeche Rough Interface

h/R	$E_2/E_1 = 0.05$			$E_2/E_1 = 0.1$		
	σ_z	σ_{r2}	σ_{r1}	σ_z	σ_{r2}	σ_{r1}
1.0	+0.2045	+0.0857	-2.173	+0.2916	+0.105	-1.597
1.111	0.1744	0.0695	-1.924	0.2523	0.0847	-1.424
1.25	0.1451	0.0544	-1.664	0.2129	0.0667	-1.249
1.429	0.1170	0.0414	-1.394	0.1741	0.0506	-1.061
1.667	0.0904	0.0303	-1.117	0.1364	0.0366	-0.862
2.0	0.0659	0.0206	-0.8424	0.1006	0.0148	-0.657
2.5	0.0441	0.0128	-0.5809	0.0681	0.0155	-0.458
3.333	0.0258	0.0071	-0.3481	0.0401	0.0085	-0.276
5.0	0.0188	0.0031	-0.1623	0.0185	0.0037	-0.130
10.0	0.0031	0.0008	-0.0418	0.0047	0.0009	-0.030

h/R	$E_2/E_1 = 0.3$			$E_2/E_1 = 0.5$			$E_2/E_1 = 1$	
	σ_z	σ_{r2}	σ_{r1}	σ_z	σ_{r2}	σ_{r1}	σ_z	σ_{r1}
1.0	+0.4629	+0.1227	-0.6710	+0.5469	+0.1230	-0.3010	-0.646	+0.116
1.111	0.4108	0.0985	-0.6300	0.4905	0.0979	-0.2949	0.589	0.097
1.25	0.3554	0.0764	-0.5743	0.4298	0.0747	-0.2802	0.524	0.067
1.429	0.2976	0.0567	-0.5053	0.3638	0.0542	-0.2554	0.450	0.046
1.667	0.2384	0.0398	-0.4240	0.2948	0.0368	-0.2212	0.369	0.028
2.0	0.1796	0.0258	-0.3327	0.2240	0.0230	-0.1780	0.284	0.016
2.5	0.1236	0.0154	0.2370	0.1557	0.0129	-0.1298	0.200	0.008
3.333	0.0739	0.0081	-0.1453	0.0936	0.0062	-0.0812	0.121	0.003
5.0	0.0344	0.0033	-0.0693	0.0439	0.0024	-0.0392	0.057	0.001
10.0	0.0088	0.0008	-0.0180	0.0112	0.0006	-0.0100	0.015	0.000

h/R	E_2/E_1 = 0.0005			$E_2/_1$ = 0.0025		
	σ_z	σ_{r2}	σ_{r1}	σ_z	σ_{r2}	σ_{r1}
0.5	+0.0437	+0.0355	-16.40	+0.1147	+0.0833	-12.45
1.0	0.0121	0.0092	-5.718	0.0341	0.0222	-4.706
1.111	0.0099	0.0075	-4.834	0.0282	0.0181	-4.010
1.25	0.0079	0.0059	-3.991	0.0227	0.0143	-3.337
1.429	0.0061	0.0045	-3.195	0.0178	0.0110	-2.694
1.667	0.0046	0.0034	-2.456	0.0133	0.0080	-2.087
2.0	0.0032	0.0023	-1.782	0.0094	0.0056	-1.526
2.5	0.0021	0.0015	-1.188	0.0062	0.0036	-1.023
3.333	0.0012	0.0009	-0.6928	0.0035	0.0020	-0.6000
5.0	0.0005	0.0003	-0.3164	0.0016	0.0009	-0.2754
10.0	0.0001	0.0001	-0.0805	0.0004	0.0002	-0.0703

h/R	E_2/E_1 = 0.005			E_2/E_1 = 0.01		
	σ_z	σ_{r2}	σ_{r1}	σ_z	σ_{r2}	σ_{r1}
0.5	+0.1696	+0.1184	-10.08	+0.0809	+0.0449	-3.52
1.0	0.0529	0.0323	-4.069	0.0676	0.0365	-3.04
1.111	0.0438	0.0261	-3.493	0.0552	0.0289	-2.57
1.25	0.0355	0.0208	-2.929	0.0436	0.0222	-2.10
1.429	0.0278	0.0158	-2.380	0.0330	0.0163	-1.64
1.667	0.0213	0.0116	-1.850	0.0237	0.0113	-1.22
2.0	0.0150	0.0080	-1.358	0.0156	0.0072	-0.82
2.5	0.0098	0.0054	-0.9133	0.0090	0.0040	-0.49
3.333	0.0056	0.0028	-0.5343	0.0041	0.0018	-0.23
5.0	0.0023	0.0011	-0.5343	0.0010	0.0004	-0.06
10.0	0.0006	0.0002	-0.0627			

Reibungslose Grenzflaeche　　　　Frictionless Interface

h/R	$E_2/E_1 = 0.1$		$E_2/E_1 = 0.3$		$E_2/E_1 = 0.5$	
	$\sigma_z = \sigma_{r2}$	σ_{r1}	$\sigma_z = \sigma_{r2}$	σ_{r1}	$\sigma_z = \sigma_{r2}$	σ_{r1}
1.0	0.3050	-1.8624	0.5030	-1.1068	0.5979	-0.8122
1.111	0.2633	-1.6689	0.4444	-1.0222	0.5345	-0.7630
1.25	0.2221	-1.4598	0.3833	-0.9202	0.4662	-0.6980
1.429	0.1813	-1.2352	0.3198	-0.7993	0.3933	-0.6159
1.667	0.1420	-1.0006	0.2554	-0.6640	0.3172	-0.5181
2.0	0.1046	-0.7610	0.1916	-0.5156	0.2403	-0.4074
2.5	0.0710	-0.5290	0.1314	-0.3650	0.1663	-0.2913
3.333	0.0416	-0.3184	0.0782	-0.2228	0.0998	-0.1793
5.0	0.0149	-0.0563	0.0363	-0.1056	0.0467	-0.0873
10.0	0.0049	-0.0381	0.0092	-0.0271	0.0119	-0.0220

Tabelle 1.31　Einflussbeiwerte der Spannungen im Zweischichtensystem [1]

Table 1.31　Influence value of the stresses in a two layer system [1]

(ii) Quelle

Reference

[1] Kezdi (1970): Handbuch der Bodenmechanik 2, S. 200.

2.7.3 Kreisfoermige Hohlparabolische Last

2.7.3 Circular Loading of a Concave Parabolic Intensity

(i) Formeln/Tafeln

(i) Formulae/Charts

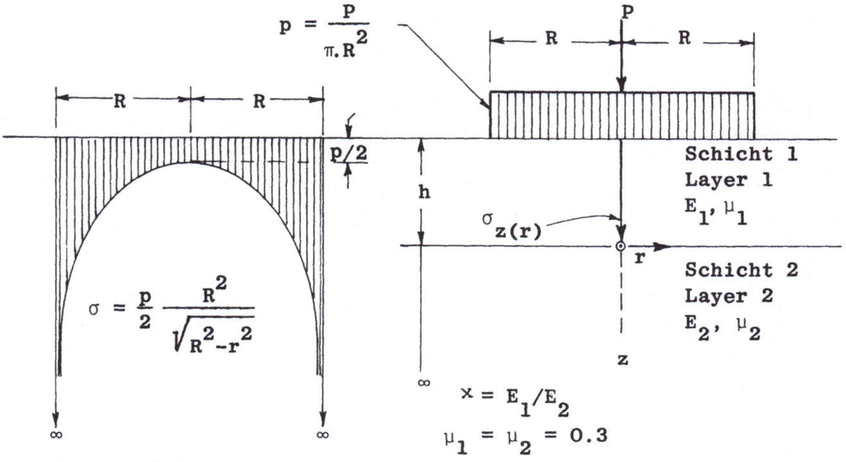

Sohlspannungsverteilung

Contact pressure distribution

$$\boxed{\sigma_{z(r)} = p \cdot i}$$

Fig. 1.70 a **Spannung, σ_z, in der Schichtfuge zwischen Zwei elastischen Schichten [1]**

Stress, σ_z, in the interface between two elastic layers [1]

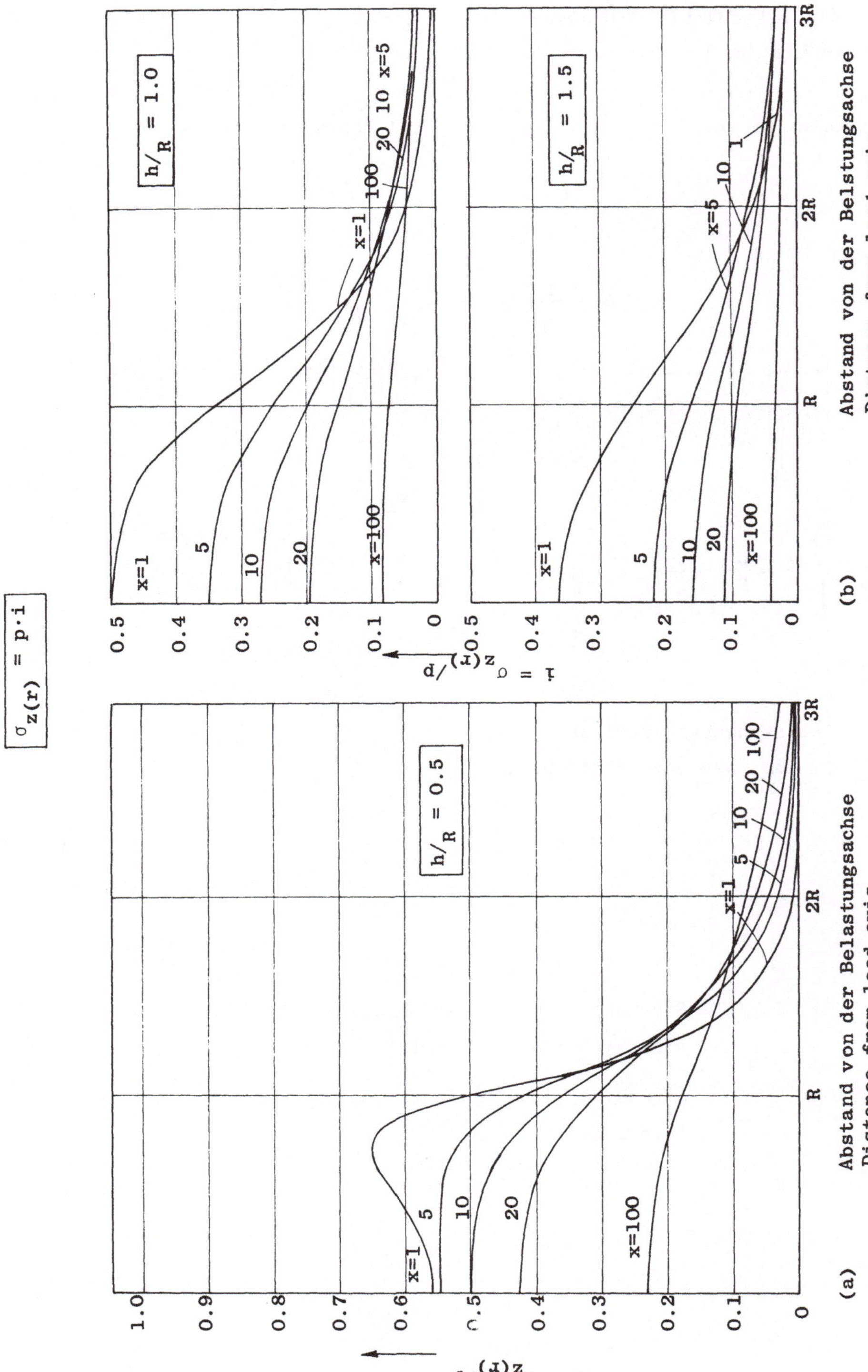

$\sigma_{z(r)} = p \cdot i$
$i = \sigma_{z(r)}/p$
$h/_R = 0.5$
$h/_R = 1.0$
$h/_R = 1.5$
x=1
5
10
20
x=100
x=5
100
R
2R
3R
(a)
(b)
Abstand von der Belastungsachse
Distance from load axis

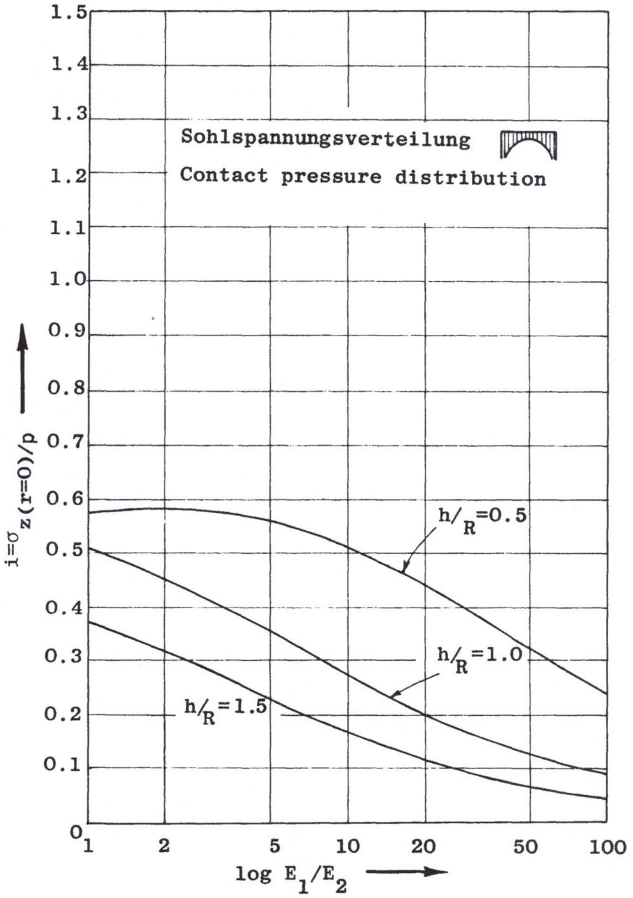

(c)

Fig. 1.70b Einflussbeiwerte i [1]

Influence values i [1]

(a,b) beliebige Stelle
any position

(c) r = 0 (Mittelpunkt)
(Centre)

(ii) Voraussetzung: Wie im Absch. 1.1.2.7.2

(ii) Assumption: as in Chap. 1.1.2.7.2

(iii) Quelle
Reference

[1] Jelinek/Pauke (1970): Berechnung der Spannungsverteilung in einem Zweischichtensystem. Bautechnik 47, S. 48.

2.7.4 Kreisfoermige Vollparabolische Last

2.7.4 Circular Loading of a Convex Parablic Intensity

(i) Formeln/Tafeln

(i) Formulae/Charts

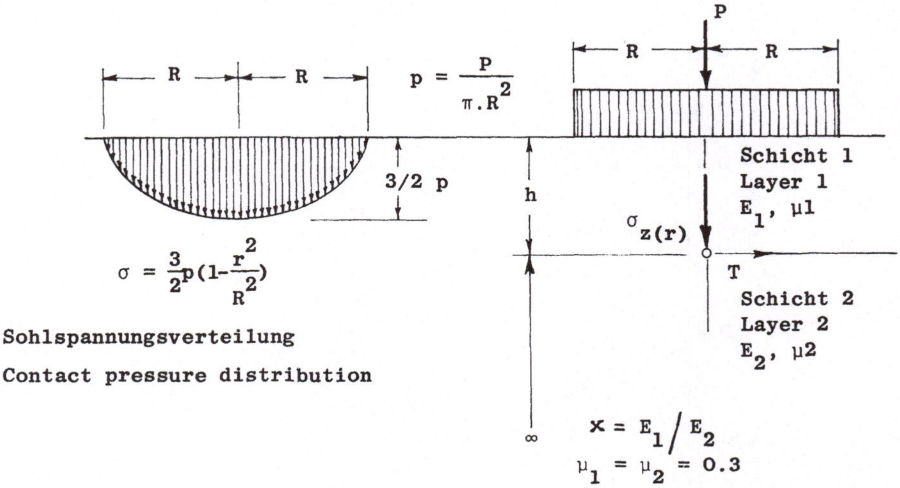

$$\boxed{\sigma_{z(r)} = p \cdot i}$$

Fig. 1.71 a **Spannung, σ_z, in der Schichtfuge Zwischen Zwei elastischen Schichten [1]**

Stress, σ_z, in the interface between two elastic layers [1]

$\sigma_{z(r)} = p \cdot i$

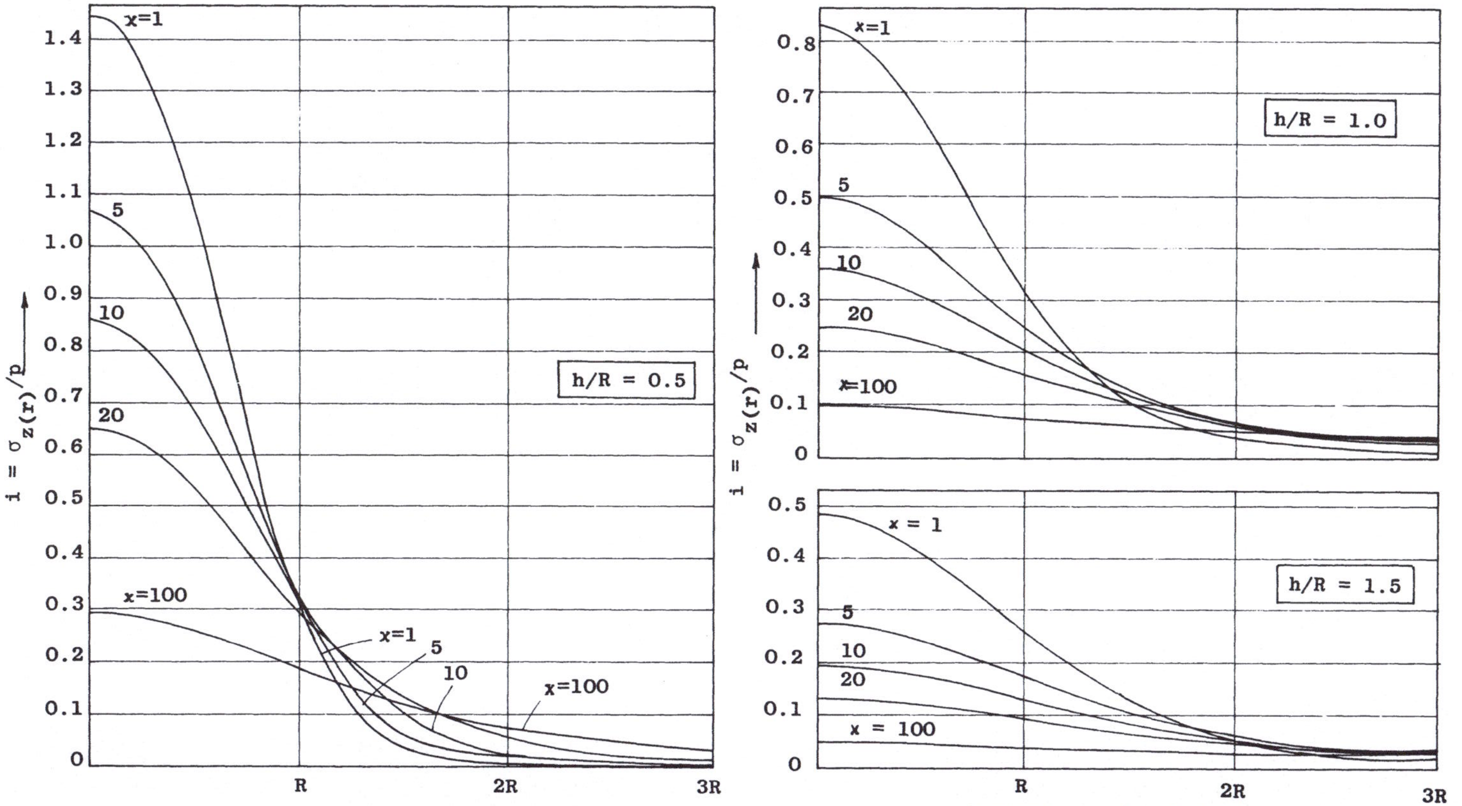

Abstand von der Belastungsachse
Distance from load axis

(a)

Abstand von der Belastungsachse
Distance from load axis

(b)

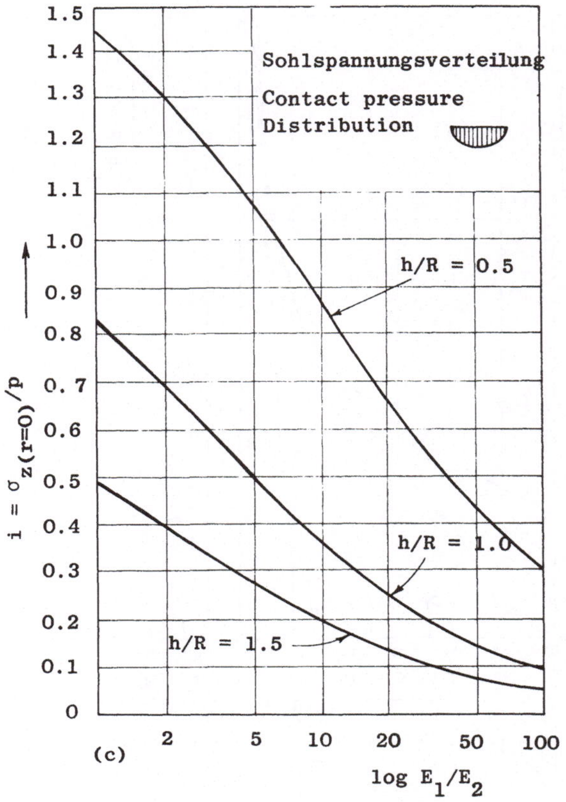

Fig. 1.71b Einflussbeiwerte i

Influence values i

(a,b) beliebige stelle

(c) r = 0 (mittelpunkt)
(centre)

(ii) Voraussetzung: wie im Absch. 1.1.2.7.2

(ii) Assumption: as in Chap. 1.1.2.7.2

(iii) Quelle : (S. Seite 273)
Reference (Ref. PP. 273)

2.7.5 Einzellast

(i) Formeln/Tafeln

2.7.5 Point Load

(i) Formulae/Charts

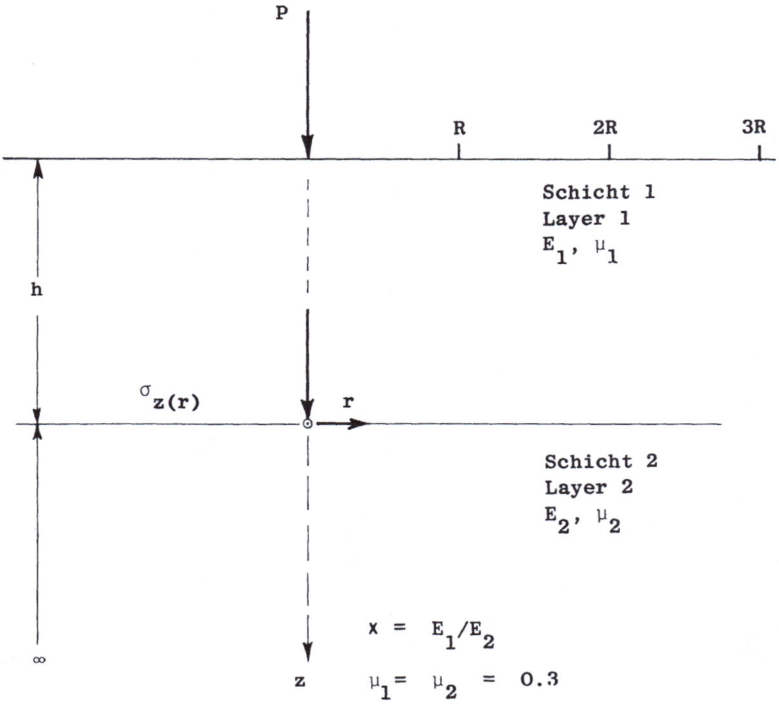

$$\boxed{\sigma_{z(r)} = p \cdot i}$$

Fig. 1.72a Spannung, σ_z, in der Schichtfuge zwischen zwei elastischen Schichten [1]

Stress, σ_z, in the interface between two elastic layers [1]

$\sigma_{z(r)} = p \cdot i$

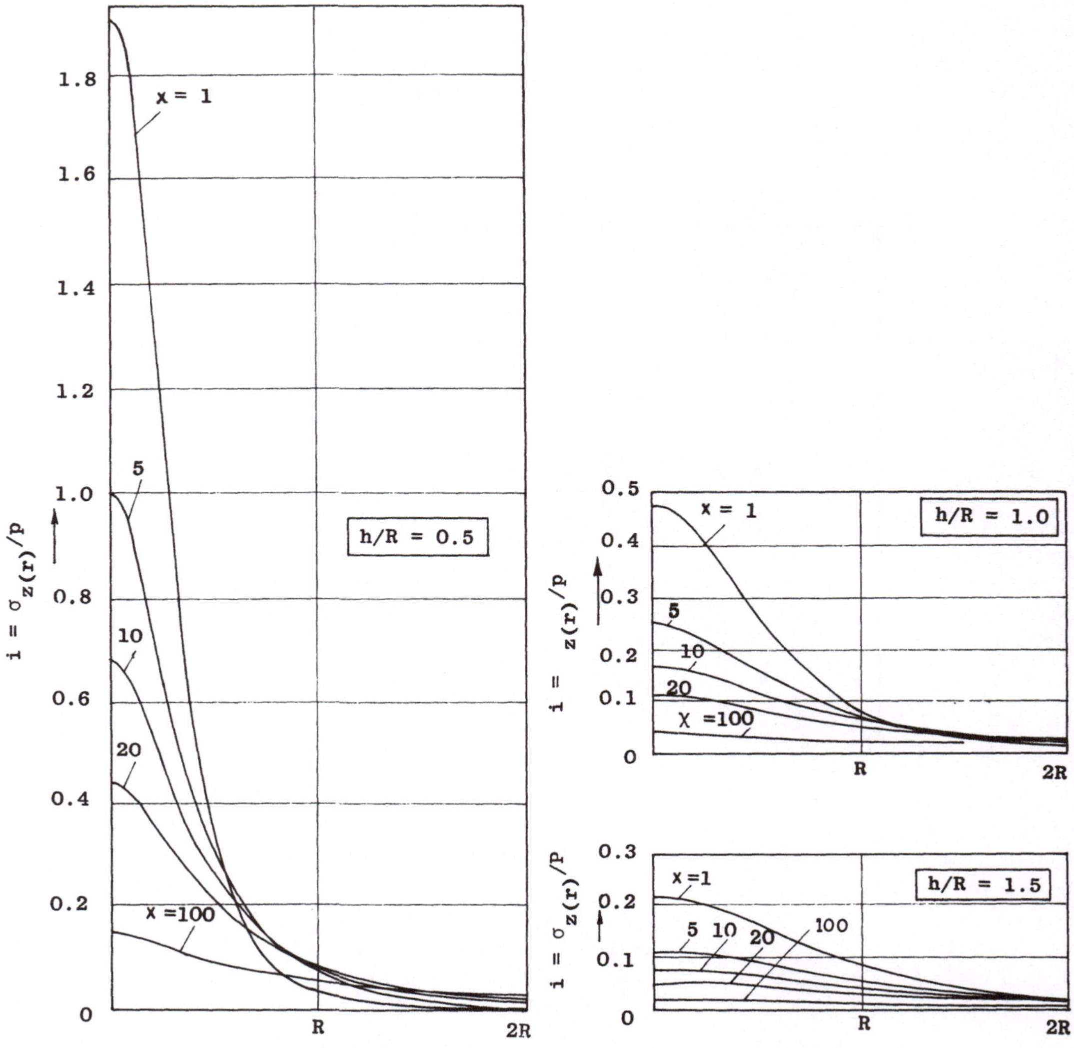

Abstand von der Belastungsachse
Distance from load axis

Abstand von der Belastungsachse
Distance from load axis

Fig. 1.72b Einflussbeiwerte i
Influence values i

(ii) Voraussetzung: Wie im Absch. 1.1.2.7.2

(ii) Assumption: As in Chap. 1.1.2.7.2

(iii) Quelle: (S. Seite 273)
Reference: (Ref. PP. 273)

2.7.6 Axialsymmetrische Horizontallast

(i) Formeln/Tafeln

2.7.6 Axially Symmetrical Horizontal Loading

(i) Formulae/Charts

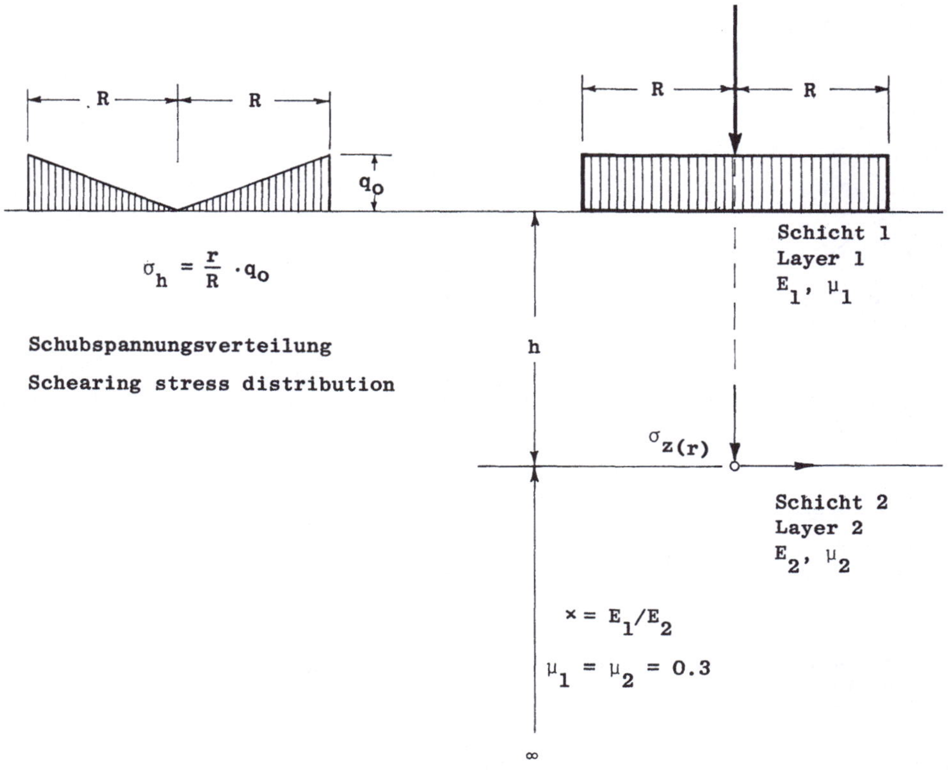

$$\boxed{\sigma_{z(r)} = q_o \cdot i}$$

Fig. 1.73a Spannung, σ_z, in der Schichtfuge zwishen zwei elastischen Schichten [1]

Stress, σ_z, in the interface between two elastic layers [1]

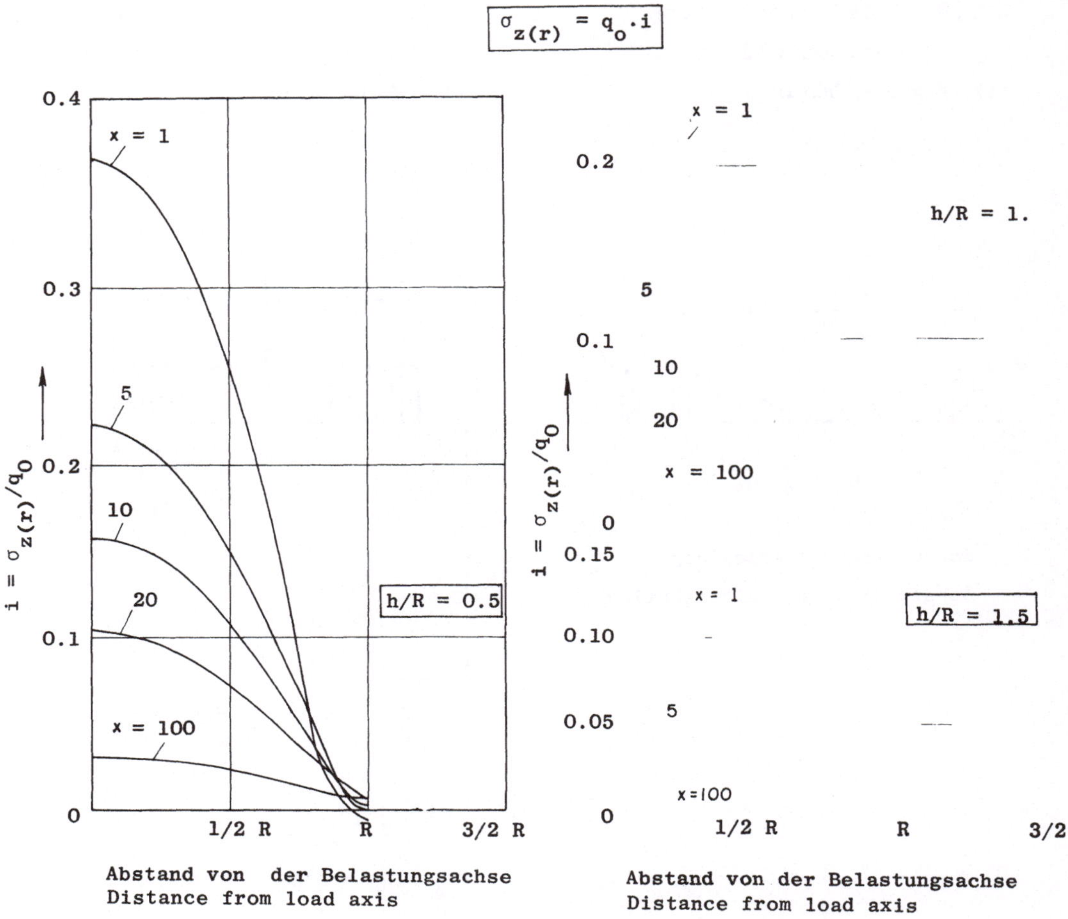

Fig. 1.73b Einflussbeiwerte i
Influence values i

(ii) Voraussetzung: wie im Absch. 1.1.2.7.2.

(ii) Assumption: as in Chap. 1.1.2.7.2.

(iii) Quelle : (S. Seite 273)
Rference (Ref. PP. 273)

* 3 *

Multiple layers – Stresses at interfaces – Circular uniform loading
Mehrere Schichten – Grenzflaechenspannungen – Kreisfoermige Gleichlast

(i) Formeln/Tabellen

(i) Formulae/Tables

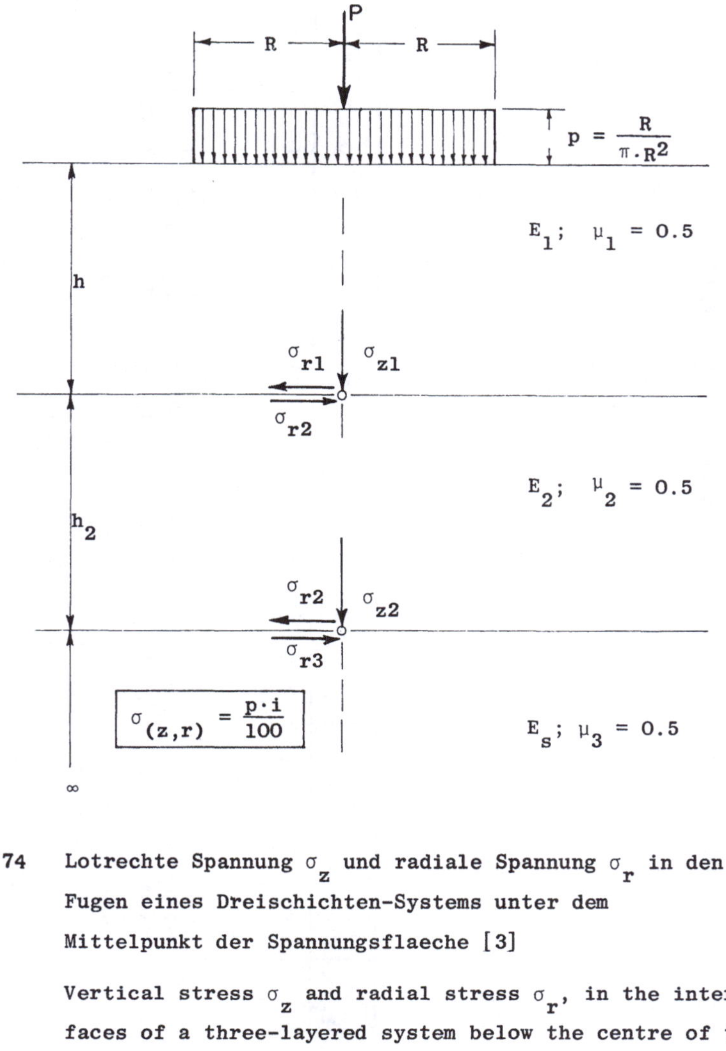

Fig. 1.74 Lotrechte Spannung σ_z und radiale Spannung σ_r in den Fugen eines Dreischichten-Systems unter dem Mittelpunkt der Spannungsflaeche [3]

Vertical stress σ_z and radial stress σ_r, in the interfaces of a three-layered system below the centre of the loaded area [3]

Bezeichnungen — Definitions

$\alpha_1 = R/h_2$ $\qquad \beta = h_1/h_2$

$k_1 = E_1/E_2$ $\qquad k_2 = E_2/E_3$

Anordnung der Tabellenwerte — Arrangement of the tabular values

	k_1	
	σ_{z1}	σ_{z2}
k_2	$\sigma_{z1} - \sigma_{r1}$	$\sigma_{z2} - \sigma_{r2}$
	$\sigma_{z1} - \sigma_{r2}$	$\sigma_{z1} - \sigma_{r3}$

$\beta = 2,\ \alpha_1 = 1$

k_2 \ k_1	5		10		20		50		100		500	
5	9.51 72.70 14.50	3.64 15.30 3.06	6.00 93.90 9.38	2.59 10.20 2.04	3.66 113.00 5.66	1.79 6.44 1.29	1.87 136.00 2.73	1.05 3.30 0.66	1.11 151.00 1.51	0.690 1.90 0.380	0.349 187.00 0.374	0.261 0.513 0.101
10	8.24 79.30 15.80	2.38 18.70 1.86	5.06 103.00 10.30	1.69 12.30 1.23	3.01 124.00 6.21	1.16 7.68 0.77	1.47 148.00 2.96	0.684 3.85 0.385	0.852 164.00 1.64	0.451 2.21 0.221	0.252 203.00 0.405	0.169 0.586 0.059
50	6.74 95.20 19.00	0.860 26.00 0.519	3.97 124.00 12.40	0.612 16.90 0.339	2.23 149.00 7.44	0.424 10.40 0.208	1.01 178.00 3.55	0.252 5.12 0.103	0.554 198.00 1.98	0.166 2.94 0.059		
100	6.43 102.00 20.40	0.552 29.00 0.290	3.75 133.00 13.30	0.397 18.90 0.189	2.08 161.00 8.04	0.277 11.60 0.118	0.921 193.00 3.85	0.163 5.75 0.058				

$\beta = 1, \alpha_1 = 1$

$k_2 \backslash k_1$	5		10		20		50		100		500	
	30.30	9.22	20.80	7.14	13.50	5.33	7.26	4.44	4.43	2.36	1.38	0.921
5	162.00	37.80	229.00	28.10	300.00	19.70	390.00	11.40	457.00	7.10	594.00	2.11
	32.30	7.56	22.90	5.63	15.00	3.96	7.84	2.29	4.57	1.42	1.19	0.420
	27.90	6.11	18.70	4.73	11.80	3.53	6.13	2.26	3.60	1.55	1.04	0.599
10	173.00	47.10	250.00	35.00	330.00	24.40	431.00	13.90	502.00	8.55	645.00	2.46
	34.70	4.70	25.00	3.50	16.50	2.44	8.66	1.39	5.02	0.86	1.29	0.245
	25.00	2.23	16.10	1.74	9.76	1.30	4.68	0.833	2.60	0.572		
50	201.00	68.20	298.00	50.90	399.00	35.30	523.00	19.50	606.00	11.80		
	40.40	1.37	29.80	1.02	20.00	0.70	10.50	0.392	6.05	0.236		
	24.30	1.42	15.60	1.11	9.35	0.839	4.40	0.540				
100	213.00	75.10	319.00	57.80	428.00	39.80	563.00	22.00				
	42.60	0.76	31.80	0.57	21.40	0.400	11.30	0.22				

$\beta = 1/2 \quad \alpha_1 = 1$

k_2 \ k_1	5		10		20		50		100		500	
	66.40	18.10	53.40	15.60	39.90	12.90	24.70	9.42	16.20	7.06	5.48	3.13
5	215.00	72.80	366.00	62.00	561.00	49.80	868.00	34.00	1116.00	23.80	1687.00	8.60
	43.00	14.60	36.70	12.40	28.10	10.00	17.40	6.81	11.20	4.77	3.37	1.72
	63.40	12.10	50.40	10.40	37.00	8.59	22.20	6.29	14.20	4.72	4.43	2.08
10	223.00	91.70	396.00	78.00	615.00	62.80	963.00	42.90	1244.00	29.80	1866.00	10.50
	44.60	9.20	39.40	7.81	30.70	6.28	19.30	4.29	12.40	2.98	3.73	1.05
	59.80	4.36	46.70	3.74	33.50	3.14	10.10	2.34	11.60	17.70		
50	239.00	134.00	449.00	115.00	730.00	93.30	1180.00	63.90	1541.00	44.20		
	47.80	2.69	44.90	2.30	36.50	1.87	23.60	1.28	15.40	0.88		
	59.10	1.77	46.00	2.37	32.80	2.00	18.50	1.51				
100	244.00	152.00	469.00	130.00	776.00	106.00	1270.00	73.10				
	48.80	1.52	46.90	1.30	38.70	1.06	25.40	0.730				

$\beta = 1/4, \quad \alpha_1 = 1$

k_2 \ k_1	5		10		20		50		100		500	
	94.70	26.50	90.70	24.90	82.10	22.80	64.20	19.40	49.00	16.30	20.80	9.19
5	127.00	104.00	301.00	99.00	614.00	90.70	1273.00	75.20	1966.00	61.20	3991.00	30.10
	25.40	20.90	30.20	19.80	30.70	18.20	25.50	15.50	19.70	12.20	8.00	6.03
	92.80	18.00	88.30	16.60	79.60	15.20	61.20	12.90	45.90	10.90	18.30	6.19
10	112.00	134.00	308.00	125.00	657.00	115.00	1404.00	95.60	2195.00	77.90	4500.00	38.00
	22.60	13.40	30.60	12.50	33.00	11.50	28.10	9.56	21.90	7.77	9.00	3.82
	90.20	6.52	85.50	5.91	76.40	5.38	57.60	4.63	42.20	3.97		
50	70.80	197.00	303.00	185.00	725.00	169.00	1657.00	142.00	2668.00	117.00		
	14.20	3.94	30.20	3.69	36.40	3.40	33.10	2.85	26.60	2.33		
	89.70	4.13	84.90	3.71	75.80	3.37	57.00	2.92				
100	50.80	223.00	295.00	208.00	745.00	191.00	1745.00	161.00				
	10.20	2.23	29.40	2.08	37.40	1.91	34.90	1.61				

$\beta = 2, \quad \alpha_1 = 1/2$

k_2 \ k_1	5		10		20		50		100		500	
	2.85	0.950	1.76	0.670	1.06	0.459	0.522	0.268	0.307	0.175	0.093	0.066
5	21.50	4.09	27.20	2.70	32.40	1.69	38.20	0.853	42.10	0.482	50.80	0.131
	4.30	0.817	2.72	0.540	1.62	0.338	0.765	0.170	0.421	0.088	0.101	0.026
	2.51	0.621	1.52	0.435	0.884	0.297	0.422	0.174	0.241	0.114	0.068	0.042
10	23.10	4.96	29.40	3.24	35.00	2.01	41.20	0.999	45.20	0.570	54.90	0.149
	4.62	0.495	2.94	0.323	1.75	0.201	0.825	0.100	0.454	0.057	0.110	0.015
	2.12	0.220	1.23	0.155	0.682	0.108	0.305	0.064	0.165	0.042		
50	27.10	6.80	34.60	4.40	41.20	2.69	48.50	1.32	53.70	0.753		
	5.42	0.136	3.47	0.088	2.06	0.054	0.971	0.026	0.537	0.015		
	2.02	0.140	1.17	0.101	0.644	0.070	0.281	0.041				
100	28.70	7.56	37.00	4.90	44.10	2.99	52.30	1.47				
	5.74	0.076	3.70	0.049	2.21	0.030	1.05	0.015				

$\beta = 1, \quad \alpha_1 = 1/2$

k_2 \ k_1	5		10		20		50		100		500	
	12.00	2.55	7.72	1.93	4.74	1.42	2.39	0.899	1.40	0.610	0.404	0.235
5	60.10	10.90	87.10	7.89	107.00	5.43	132.00	3.07	149.00	1.89	184.00	0.550
	13.20	2.18	8.69	1.58	5.36	1.09	26.64	0.616	1.49	0.376	0.368	0.111
	11.30	1.66	7.13	1.26	4.29	0.929	2.08	0.586	1.18	0.400	0.318	0.152
10	69.10	13.30	92.30	9.68	115.00	6.64	142.00	3.71	160.00	2.26	197.00	0.637
	13.80	1.33	9.22	0.969	5.74	0.664	2.84	0.371	1.60	0.227	0.394	0.064
	10.50	0.586	6.46	0.453	3.74	0.336	1.71	0.213	0.927	0.145		
50	76.10	18.70	104.00	13.70	132.00	9.39	165.00	5.14	186.00	3.08		
	15.20	0.374	10.40	0.275	6.61	0.189	3.30	0.103	1.86	0.062		
	10.30	0.370	6.32	0.287	3.64	0.215	1.64	0.137				
100	79.00	21.10	109.00	15.40	140.00	10.60	175.00	5.75				
	15.80	0.210	10.90	0.154	6.99	0.106	3.50	0.057				

$\beta = 1/2, \quad \alpha_1 = 1/2$

k_2 \ k_1	5		10		20		50		100		500	
	36.20	5.41	25.60	4.52	17.10	3.64	9.41	2.58	5.76	1.89	1.73	0.815
5	144.00	23.20	204.00	19.00	270.00	14.70	363.00	96.69	433.00	6.63	583.00	2.31
	28.80	4.64	20.40	3.81	13.50	2.94	7.26	1.94	4.32	1.32	1.16	0.462
	35.40	3.52	24.70	2.94	16.30	2.39	8.71	1.69	5.20	1.25	1.46	0.538
10	147.00	28.40	211.00	23.30	284.00	18.20	388.00	12.00	465.00	8.18	627.00	2.79
	29.40	2.85	21.10	2.34	14.20	1.83	7.76	1.20	4.65	0.818	1.26	0.279
	34.20	1.20	23.60	1.01	15.30	0.835	7.88	0.612	4.51	0.457		
50	151.00	39.60	226.00	33.00	314.00	26.10	443.00	17.40	540.00	11.80		
	30.10	0.794	22.50	0.659	15.70	0.522	8.87	0.348	5.40	0.237		
	33.90	0.748	23.40	0.633	15.10	0.526	7.72	0.390				
100	152.00	44.10	230.00	36.80	325.00	29.40	466.00	19.70				
	30.30	00.441	23.00	0.369	16.30	0.294	9.32	0.179				

$\beta = 1/4, \quad \alpha_1 = 1/2$

k_2 \ k_1	5		10		20		50		100		500	
	74.10	8.79	61.40	8.03	47.50	7.15	30.80	5.80	20.90	4.72	7.30	2.50
5	189.00	38.00	312.00	34.70	476.00	30.50	747.00	23.80	984.00	18.60	1582.00	8.52
	37.90	7.61	31.30	6.95	23.80	6.10	15.00	4.77	9.87	3.72	3.17	1.71
	73.40	5.70	60.60	5.17	46.60	4.60	29.80	3.76	10.90	3.08	6.61	1.65
10	186.00	46.50	315.00	42.50	489.00	37.40	783.00	29.50	1045.00	23.10	1713.00	10.60
	37.10	4.65	31.50	4.25	24.50	3.74	15.60	2.95	10.50	2.31	3.43	1.06
	72.40	1.90	59.60	1.71	45.60	1.52	28.70	1.28	18.80	1.07		
50	175.00	63.90	314.00	58.40	508.00	51.90	850.00	41.70	1167.00	33.30		
	34.80	1.28	31.30	1.17	25.40	1.04	17.00	0.836	11.67	0.665		
	72.40	1.17	59.40	1.04	45.40	0.934	28.50	0.791				
100	170.00	70.70	311.00	64.50	513.00	57.50	872.00	46.60				
	34.00	0.707	31.20	0.645	25.70	0.575	17.40	0.466				

Tabelle 1.32 Einflussbeiwerte fuer Dreischichtensystem [3]

Table 1.32 Influence value for three-layered system [3]

(ii) Voraussetzung: $\mu = 0.5$

(ii) Assumption: $\mu = 0.5$

(iii) Anwendungsbeispiel:

Gegeben

(iii) Example:

Given

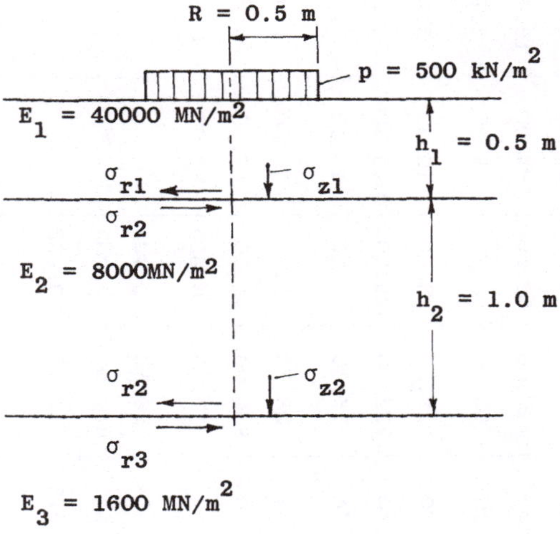

Gesucht

Required

$$\sigma_{z1},\ \sigma_{z2},\ \sigma_{r1},\ \sigma_{r2},\ \sigma_{r3}$$

Loesung

Solution

$$\alpha_1 = R/h_2 = \frac{0.50}{1.00} = 0.50$$

$$\beta = h_1/h_2 = \frac{0.50}{1.00} = 0.50$$

$$k_1 = E_1/E_2 = \frac{40000}{8000} = 5.00$$

$$k_2 = E_2/E_3 = \frac{8000}{1600} = 5.00$$

Aus Tabelle 1.32 From Table 1.32

$i(\sigma_{z1}) = 36.20$ $i(\sigma_{z2}) = 5.41$

$i(\sigma_{z1}) - i(\sigma_{r1}) = 144.00$ $i(\sigma_{z2}) - i(\sigma_{r2}) = 23.20$

$i(\sigma_{z1}) - i(\sigma_{r2}) = 28.80$ $i(\sigma_{z1}) - i(\sigma_{r3}) = 4.64$

$$\sigma_{z1} = \frac{p \cdot i}{100} = \frac{36.20}{100}\ (500) = (0.3620\ 500) = 181.0\ kN/m^2$$

$$\sigma_{z2} = \frac{5.41}{100}\ (500) = (0.0541)(500) = 27.1\ kN/m^2$$

$$\sigma_{r1} = \frac{1}{100}\ (36.20 - 144.00)(500) = -539\ kN/m^2$$

$$\sigma_{r2} = \frac{1}{100}\ (36.20 - 28.80)(500) = 37.0\ kN/m^2$$

$$\sigma_{r2} = \frac{1}{100}\ (5.41 - 23.20)(500) = -\ 88.9\ N/m^2$$

$$\sigma_{r3} = \frac{1}{100}\ (36.20 - 4.64)(500) = 157.8\ kN/m^2$$

(iv) Quellen

References

[1] Fox (1948): Computation of traffic stresses in a simple road structure.
D.S.I.R., Road Res. Tech. Paper No.9
Proc. 2. ICSMFE Rotterdam 2, PP. 236.

[2] Jones (1962): Tables of stresses in three-layer elastic systems.
Highway Research Board Bull. 342, PP. 176.

[3] Kezdi (1970): Handbuch der Bodenmechanik 2, S. 203.